Carboranes

ORGANOMETALLIC CHEMISTRY
A Series of Monographs

EDITORS

P. M. MAITLIS
MCMASTER UNIVERSITY
HAMILTON, ONTARIO
CANADA

F. G. A. STONE
UNIVERSITY OF BRISTOL
BRISTOL, ENGLAND

ROBERT WEST
UNIVERSITY OF WISCONSIN
MADISON, WISCONSIN

BRIAN G. RAMSEY: Electronic Transitions in Organometalloids, 1969.
R. C. POLLER: The Chemistry of Organotin Compounds, 1970.
RUSSELL N. GRIMES: Carboranes, 1970.

Other volumes in preparation.

PETER MAITLIS: The Organic Chemistry of Palladium.

CARBORANES

RUSSELL N. GRIMES

DEPARTMENT OF CHEMISTRY
UNIVERSITY OF VIRGINIA
CHARLOTTESVILLE, VIRGINIA

1970

A C A D E M I C   P R E S S   New York and London

CHEMISTRY

ACADEMIC PRESS, INC.
111 Fifth Avenue, New York, New York, 10003

United Kingdom Edition published by
ACADEMIC PRESS, INC. (LONDON) LTD.
Berkeley Square House, London, W1X 6BA

LIBRARY OF CONGRESS CATALOG CARD NUMBER: 75-127684

PRINTED IN THE UNITED STATES OF AMERICA

Contents

Preface   xiii

1. **Introduction**   1

2. **Structures and Bonding**

    2-1. Structures of the Carboranes   5

        The Polyhedral Carboranes   5
        The Open-Cage Carboranes   9
        Isomerism and Rearrangement   11

    2-2. Nomenclature and Numbering Systems   12

    2-3. Chemical Bonding in the Boranes and Carboranes   13

        The Three-Center Bond   14
        Molecular Orbitals in Boranes and Carboranes   18
        Simplified Molecular Orbital Descriptions   20
        Summary   22

3. **The Small *Nido*-Carboranes: CB_5H_9, $C_2B_4H_8$, $C_3B_3H_7$, $C_4B_2H_6$, and $C_2B_3H_7$**

    3-1. Structures   23

3-2. Synthesis   24

   Borane–Alkyne Reactions   24
   Other Routes to *Nido*-Carboranes   25

3-3. Reactions and Properties   28

   $2\text{-}CB_5H_9$ 28
   $2,3\text{-}C_2B_4H_8$ 28
   $2,3,4\text{-}C_3B_3H_7$ 29
   $2,3,4,5\text{-}C_4B_2H_6$ 29
   $1,2\text{-}C_2B_3H_7$ 31

4. The Small *Closo*-Carboranes: $C_2B_3H_5$, $C_2B_4H_6$, $C_2B_5H_7$, and CB_5H_7

4-1. Synthesis   32

   Borane–Alkyne Reactions   33
   Conversions of *Nido*-Carboranes to *Closo*-Carboranes   36
   Reactions of Alkylboranes with Alkynes   37
   Hydroboration of Alkynylboranes   38
   Dehalogenation of Alkylhaloboranes with Alkali Metals   38
   Pyrolysis and Electric Discharge of Alkylboranes   39
   Reaction of Boranes with Carbon Vapor   39

4-2. Structures and Properties   40

   $C_2B_3H_5$ 40
   $C_2B_4H_6$ 41
   $C_2B_5H_7$ 41
   CB_5H_7 44

5. The Intermediate *Closo*-Carboranes: $C_2B_6H_8$, $C_2B_7H_9$, $C_2B_8H_{10}$, and $C_2B_9H_{11}$

5-1. Synthesis   45

   Reactions of Boranes with Alkynes   45
   Degradation of Icosahedral Carboranes   46

5-2. Structures and Properties   48

   $C_2B_6H_8$ 48
   $C_2B_7H_9$ 48
   $C_2B_8H_{10}$ 48
   $C_2B_9H_{11}$ 50

6. The 1,2-$C_2B_{10}H_{12}$ (*o*-Carborane) System

6-1. Synthesis, Structure, and Properties — 54

Synthesis — 54
Structure — 55
Thermal Stability and Rearrangement — 55
Chemical Properties — 55

6-2. Alkali Metal and Magnesium Derivatives — 66

Alkali Metal Derivatives — 66
Magnesium Derivatives — 67
Zinc Derivatives — 69
Formation of *o*-Carborane Dianions in Liquid Ammonia — 69

6-3. Alkyl, Haloalkyl, and Aryl Derivatives — 70

Synthesis — 70
Substitution at Boron — 72
Properties of Alkyl and Haloalkyl Derivatives — 72
Properties of Aryl Derivatives — 73

6-4. Alkenyl and Alkynyl Derivatives — 75

Synthesis of C-Alkenyl Derivatives — 75
Synthesis of B-Alkenyl Derivatives — 76
Synthesis of Alkynyl Derivatives — 76
Reactions of Alkenyl Derivatives — 77
Reactions of Alkynyl Derivatives — 81

6-5. Carboxylic Acids and Esters — 82

Synthesis of C-Substituted Acids and Esters — 82
Synthesis of B-Substituted Acids — 89
Reactions of Acids and Esters — 89

6-6. Alcohols and Ethers — 92

Synthesis of Alcohols — 92
Synthesis of Ethers — 95
Reactions of Alcohols and Ethers — 97

6-7. Aldehydes and Ketones — 99

Synthesis of Aldehydes — 99
Synthesis of Ketones — 100
Reactions of Aldehydes and Ketones — 103

6-8.   Nitrogen Derivatives   107

    Nitrates and Related Compounds   107
    Amines, Azides, and Diazonium Salts   109
    Amides   113
    Cyano Derivatives   113
    Other Nitrogen-Containing Derivatives   114

6-9.   Phosphorus, Arsenic, and Antimony Derivatives   115

6-10. Silicon Derivatives   120

    o-Carboranyl Silanes   120
    o-Carboranyl Alkoxysilanes   124

6-11. Germanium and Tin Derivatives   127

6-12. Sulfur Derivatives   129

6-13. Halogen Derivatives   133

    Electrophilic Halogenation   134
    Photochemical Halogenation   136
    Fluorination   137
    Synthesis of Halo-*o*-Carboranes from Decaborane(14) Halogen
      Derivatives   137
    Synthesis of 3-Halo-*o*-Carboranes   138
    Synthesis of C-Halo-*o*-Carboranes   139
    Reactions of Halogen Derivatives   141

6-14. Sigma-Bonded Transition Metal Derivatives   146

6-15. Mercury Derivatives   147

7.   The 1,7-$C_2B_{10}H_{12}$ and 1, 12-$C_2B_{10}H_{12}$ (*m*- and *p*-Carborane) Systems

    7-1.   Comparison of *o*- and *m*-Carborane   151

    7-2.   Synthesis and Structure of *m*-Carborane   156

7-3. Mechanisms of Icosahedral Carborane Rearrangements   158

    ortho-meta and *meta-para* Isomerizations   158
    The "Reverse Isomerization" of *m*-Carborane: Conversion of *m*- to
    o-Carborane   159

7-4. Metallation of *m*-Carborane   161

7-5. Alkyl, Aryl, and Alkenyl *m*-Carborane Derivatives   161

    Synthesis   161
    Reactions   162

7-6. *m*-Carboranyl Carboxylic Acids and Esters   163

    Synthesis   163
    Reactions   163

7-7. *m*-Carboranyl Alcohols and Ethers   164

    Synthesis   164
    Reactions   165

7-8. *m*-Carboranyl Aldehydes and Ketones   165

    Reactions   166

7-9. *m*-Carboranyl Nitrogen and Phosphorus Derivatives   167

    Nitrates, Amines, and Diazonium Salts   167
    Amides   168
    Phosphorus Derivatives   168

7-10. *m*-Carboranyl Silicon Derivatives   169

7-11. *m*-Carboranyl Germanium, Tin, and Lead Derivatives   169

7-12. *m*-Carboranyl Sulfur Derivatives   171

7-13. *m*-Carboranyl Halogen Derivatives   172

    Synthesis from Halo-*o*-Carboranes   172
    Electrophilic Halogenation   172
    Fluorination   173
    Photochemical Halogenation   174
    C-Halo-*m*-Carboranes   174
    Reactions and Properties of Halo-*m*-Carboranes   175

7-14. *m*-Carboranyl Mercury Derivatives   175

7-15. *p*-Carborane   177

    Synthesis and Structure   177
    Organic and Organometallic Derivatives   178
    Halogen Derivatives   179

8. **Carborane Polymers**

8-1. Introduction   181

8-2. General Considerations   182

8-3. Class I Polymers   182

    Polyesters   182
    Polyformals   183
    Siloxanes   183
    Polymers with Single Atom Links   187
    Other Class I Polymers   189

8-4. Class II Polymers   190

9. **Degradation of the Icosahedral Cage. Heteroatom Carboranes and Transition Metal π-Complexes**

9-1. Degradation of *o*- and *m*-Carborane   193

    $C_2B_9H_{12}^-$ Ions   193
    $C_2B_9H_{13}$ 196
    $C_2B_9H_{11}^{2-}$ (Dicarbollide) Ions   197

9-2. Dicarbollide–Boron Insertion Reactions   198

9-3. The Monocarbon Carborane Anions: $CB_{10}H_{13}^-$, $CB_{10}H_{11}^-$, $CB_9H_{10}^-$, $CB_{11}H_{12}^-$ 199

    $CB_{10}H_{13}^-$ 199
    $CB_{10}H_{11}^-$ 200
    $CB_9H_{10}^-$ and $CB_{11}H_{12}^-$ 201

9-4.   Carboranes Containing Main-Group Cage Heteroatoms   202

    Group II Heteroatoms   202
    Group III Heteroatoms   202
    Group IV Heteroatoms   204
    Group V Heteroatoms   205

9-5.   Carborane–Transition Metal π-Complexes   207

    General Considerations   207
    Dicarbollyl Complexes of Cr, Mo, and W   212
    Dicarbollyl Complexes of Mn and Re   214
    Dicarbollyl Complexes of Fe   214
    Dicarbollyl Complexes of Co   217
    Dicarbollyl Complexes of Ni and Pd   220
    Dicarbollyl Complexes of Cu and Au   222
    Monocarbollyl–Transition Metal π-Complexes   224
    Carbaphosphollyl–Transition Metal π-Complexes   225
    Nonicosahedral–Transition Metal π-Complexes   226

Supplementary Sources of Information   233

References   236

Author Index   251
Subject Index   265

Preface

The development of borane cage chemistry within the last two decades is surely one of the more exciting and remarkable stories to be found in the chemical literature. Until 1948, not a single boron hydride larger than diborane had been structurally characterized, and the published literature on these compounds could be summarized annually in a few lines in *Chemical Abstracts*. Since then, the combined efforts of structural, synthetic, and theoretical investigators have revealed an area of chemistry of such considerable scope and variety that its synthetic possibilities seem almost unlimited. The diversity of this field is such that detailed treatment of individual topics within boron chemistry is now desirable and necessary. One area of interest concerns the carboranes, or boron–carbon cage molecules, upon which a large part of the recent research in the boron field has centered. Aside from the sheer volume of published and unpublished work on these compounds, a significant aspect of carborane chemistry is its considerable overlap with organic, organometallic, and transition metal coordination chemistry. In both an experimental and a theoretical sense, the carboranes serve to bridge the gap between these older disciplines and boron hydride chemistry, which in truth has been a highly esoteric field with little interaction with other areas.

Although parts of the carborane literature have been reviewed in a number of articles and chapters since the mid-1960's (listed in the section on supplementary sources at the end of this volume), no comprehensive treatment of the carboranes as a class has appeared. This book represents an attempt to satisfy the need for a detailed review of carborane chemistry at this point in its development, and at the same time to introduce the subject to organic and inorganic chemists with no previous knowledge of the borane field. Accordingly, much background material has been included which will be familiar to workers in the area, but which hopefully will be useful to other readers.

Some explanation of the selection and organization of the material in the book seems desirable. Following the generally accepted definition of a carborane as a borane cage molecule containing carbon in the skeletal frame-work, all such compounds are included even when the precise nature of the carbon–boron bonding is unclear (organoboron species in which carbon is present only as a ligand to the boron cage are, of course, excluded).

An attempt has been made to organize the book for optimum usefulness as a reference work, at the cost of some redundancy. Thus, o-carborane and m-carborane derivatives are treated in separate chapters with cross references, and certain reactions are mentioned in more than one context. Although a highly critical approach to the literature seems inappropriate in a new field in which the dust has not yet settled, apparent inconsistencies and doubtful claims are pointed out on occasion.

Certain related topics which have recently been reviewed in detail elsewhere are not extensively treated here. These include the interpretation of nuclear magnetic resonance data, infrared spectra, and mass spectra of boron com-pounds, for which appropriate literature references are given in the section on supplementary information. Tables of carborane derivatives are intended to be comprehensive up to late 1969, except for the o- and m-carboranyl derivatives for which a complete listing would have been inordinately long.

I wish to acknowledge the many direct and indirect contributions of my graduate students, who kept the research going while the writing was in pro-gress, and the assistance of Mrs. Kathi Howard and my wife Nancy in the preparation of the manuscript. I am indebted to Dr. J. F. Sieckhaus and Dr. H. A. Schroeder of the Olin Corporation, who read the manuscript and offered valuable suggestions, to the American Chemical Society and to several individuals for permission to reproduce drawings, and to the Chemistry Department of the University of Virginia for encouragement in the prepar-ation of this book.

1

Introduction

The study of electron-deficient boron cage compounds has developed in recent years into a major area of inorganic–organometallic chemistry, with considerable overlap into organic and polymer chemistry as well. Although the field embraces the binary boron hydrides, the polyhedral borane anions, and the boron halides, a large part of the research interest has centered on boron cage systems in which one or more carbon atoms is present as an integral part of an electron-delocalized borane framework. Compounds of this type have been given the general name carboranes, a term which includes both closed polyhedra and open-cage structures. As this definition suggests, the carboranes are a new class of compounds which are distinct from other organoboron species such as the alkylboranes, in which the carbon is present as a ligand rather than as part of the cage itself.

From the viewpoint of theory, it is convenient to regard the carboranes as derivatives of the boron hydrides in which B^- or BH groups have been replaced by isoelectronic carbon atoms. The relationship is essentially formal, since the carboranes as a class exhibit a chemistry very different from that of the binary boranes. Nevertheless, the theory of structure and bonding in the boranes, as developed by Lipscomb and others (195, 197) is fundamental to any discussion of the carboranes. The following is intended as a brief historical introduction to the field, with detailed treatment of the bonding deferred to later chapters.

The characterization of the small boron hydrides by Alfred Stock and his co-workers (325) was accomplished before the instrumentation was available for detailed structural study, and these chemists assumed hydrocarbon-like chain structures for the boranes. Even so, the fact that boron possesses only three valence electrons forced recognition that the bonding in these compounds could not be precisely analogous to that in the hydocarbons. It was evident, for example, that the 22 valence electrons in B_4H_{10} are insufficient for a butane-type structure, whose 13 bonds would require 26 electrons. This so-called "electron-deficiency" raised formidable difficulties in dealing with the bonding in the boranes, and in the absence of unequivocal structural

1

data the problem remained unsolved. In recent times the structures of a dozen or more boron hydrides have been established, with results quite unlike anything apparently visualized by the earlier workers. The majority of the known boranes resemble fragments of a regular icosahedron, a polyhedron with 20 sides and 12 vertices (Fig. 1-1a); thus $B_{10}H_{14}$, disregarding the hydrogens,

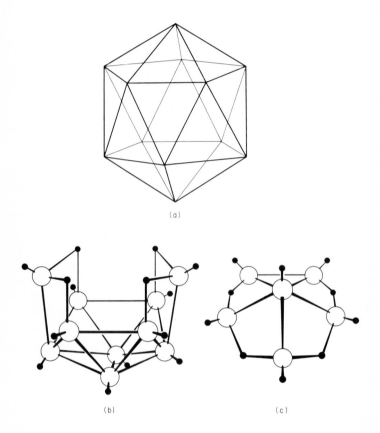

(a)

(b)   (c)

FIG. 1-1. (a) Regular icosahedron. (b) Structure of $B_{10}H_{14}$. (c) Structure of B_6H_{10}.

may be viewed as a slightly distorted icosahedron with two apices missing (Fig. 1-1b), and the boron atoms in B_6H_{10} (Fig. 1-1c) form the bottom of an icosahedron (*195*). Other "icosahedral fragments" among the neutral boranes (*195*) include B_4H_{10}, B_5H_{11}, B_6H_{12}, B_8H_{12}, and B_9H_{15} (B_5H_9, with square pyramidal geometry, is an exception). The icosahedron also occurs as a structural unit in elementary boron, in boron carbide ($B_{12}C_3$), and in some metal borides (*149*). This pattern raised the question of whether a complete icosahedral $B_{12}H_{12}$ hydride might have a stable existence. Molecular orbital calculations by Longuet-Higgins and Roberts (*200*) in 1955 indicated that such a molecule would probably be nonexistent as a neutral species but that an icosahedral $B_{12}H_{12}^{2-}$ ion might well be stable.

The remarkable accuracy of these predictions was revealed, in part, five years later when Pitochelli and Hawthorne (*238*) prepared a salt of $B_{12}H_{12}^{2-}$. A year earlier, in 1959, the same workers had isolated the $B_{10}H_{10}^{2-}$ ion (*131*), which, like $B_{12}H_{12}^{2-}$, had incredible properties for a boron hydride. In contrast to the neutral boranes, both new ions were completely unreactive with air and were unaffected by boiling water or strong acids (*180*). These observations, together with NMR spectra, suggested closed polyhedral geometries. Accordingly, $B_{10}H_{10}^{2-}$ was postulated as a bicapped square antiprism (*198*) and $B_{12}H_{12}^{2-}$ as a regular icosahedron (*238*), both structures eventually being confirmed by single crystal X-ray studies (*49, 355*).

The probability of other stable polyhedral borane cage systems was indicated from molecular orbital calculations carried out by Lipscomb and several associates (*151, 152, 154, 196, 212*), who predicted the existence of a series of borane anions having the general formula $B_nH_n^{2-}$. Thus, $B_5H_5^{2-}$ was postulated as a trigonal bipyramid, $B_6H_6^{2-}$ as an octahedron, and $B_7H_7^{2-}$ as a pentagonal bipyramid. In addition, a series of neutral $C_2B_{n-2}H_n$ carborane cage molecules was suggested (*152, 196*), each one formally derived from its isoelectronic $B_nH_n^{2-}$ analog by replacement of two B^- with two C atoms. Indeed, the first few members of this series had already been prepared by Keilin (*196*, footnote 9) though not publicly reported. The first published accounts of carborane synthesis contained descriptions of a trigonal bipyramidal $C_2B_3H_5$ (*273*), two isomers of an octahedral $C_2B_4H_6$ (*274*), and a $C_2B_5H_7$ (*273*) species which eventually proved to have the predicted pentagonal bipyramidal structure [surprisingly, the carbons were found to be in non-adjacent equatorial positions (*12*)]. In late 1963 the anticipated icosahedral carborane system, $C_2B_{10}H_{12}$, was reported by two different industrial groups (*66, 143*), who at the same time described the preparation of a large number of organic carborane derivatives.

More recently, the polyhedral carboranes $C_2B_6H_8$, $C_2B_7H_9$, $C_2B_8H_{10}$, and $C_2B_9H_{11}$ have each been isolated in at least one isomeric form, thus completing the $C_2B_{n-2}H_n$ series from $n = 5$ to $n = 12$. Of the $B_nH_n^{2-}$ borane dianion series, all members from $n = 6$ to $n = 12$ have been characterized ($B_5H_5^{2-}$ remains unknown) (*215*). All of these cage molecules are relatively stable thermodynamically, some of them extremely so, and available chemical and physical evidence strongly suggests that substantial electron delocalization is responsible for their stability. In the carboranes, it is significant that *the carbon atoms participate in the delocalized bonding*. As a consequence, the usual empirical rules of valency and coordination in organic compounds are irrelevant to these structures; in $C_2B_{10}H_{12}$, for example, each carbon atom is hexacoordinate. As will be seen in Chapter 2, the only generally applicable approach to chemical bonding in these molecules is via molecular orbital theory.

In addition to the closed-cage or polyhedral carborane series, other boron–carbon hydrides are known in which the carbon atoms form part of an *open-cage* boron framework. These carboranes may be viewed in a formal sense as derivatives of neutral boranes obtained by substitution of carbon atoms for BH groups; thus B_6H_{10}, CB_5H_9, $C_2B_4H_8$, $C_3B_3H_7$, and $C_4B_2H_6$ form an isoelectronic series in which all members apparently have the pyramidal B_6H_{10} structure. In the same way, the carborane $C_2B_9H_{13}$ is structurally related to its analog $B_{11}H_{15}$ (*63*), both being known compounds.

The carborane field has been further complicated, but at the same time enriched, by the synthesis of many new species in which atoms other than boron and carbon form a part of the cage framework. To cite just a few, heteroatoms which have been incorporated into carborane systems include beryllium, germanium, phosphorus, arsenic, antimony, tin, lead, gallium, and many transition metals; in all such systems the heteroatoms presumably participate significantly in the delocalized bonding.

The carboranes, of course, constitute only one of many classes of organoboron compounds. The organic chemistry of boron is a vast and well-established field; some of the boron alkyls, for example, were characterized long before the discovery of any of the boron hydrides. For the most part, however, the older types of organoboron species do not involve electron-deficient structures and hence are not closely related to the carboranes. Such systems as the borate esters, boronic acids, boron alkyls, and boron heterocycles are more or less readily accommodated by the classical structural theory of organic chemistry, and the bonding may usually be described in terms of localized two-center bonds or resonance combinations. On the other hand, alkylboron hydrides such as ethyldecaborane are clearly electron-deficient molecules, but since the carbon atoms are external to the cage, their involvement in bonding within the cage is comparatively minor. It is not surprising that most organo-substituted boranes bear a much closer chemical resemblance to the parent boranes than to the carboranes.

In summary, the carboranes form a distinct class of organoboron cage structures which are stabilized by delocalization of valence electrons, and which in fact exhibit many properties typical of aromatic systems. At the same time, the presence of carbon reaction sites in the cage makes possible an extremely versatile and extensive derivative chemistry not paralleled by the boron hydrides.

Structures and Bonding

2-1. STRUCTURES OF THE CARBORANES

The Polyhedral Carboranes

The characterized polyhedral cage systems are listed in Table 2-1 together with species that are unknown at present but whose existence is suggested by simple analogy or predicted from theory (195). The polyhedral borane dianions are included to indicate their isoelectronic and isostructural relationship to the carboranes; for each species in the first vertical column, successive replacement of C with B^- generates an isoelectronic carborane monoanion (column 2) and a borane dianion (column 3). Addition of a proton to any of the carborane monoanions should, in principle, yield a neutral monocarbon carborane of type $CB_{n-1}H_{n+1}$, but only one such molecule, CB_5H_7, has been characterized at this writing.

Other polyhedra isoelectronic with those listed may conceivably exist; for example, formal substitution of C for BH in the $C_2B_{n-2}H_n$ series produces a family of tricarbon carboranes having the formula $C_3B_{n-3}H_{n-1}$. No members of this series have been prepared, but analogs containing germanium, tin, and lead (e.g., icosahedral $GeC_2B_9H_{11}$) are known. In fact, carborane polyhedra with cage heteroatoms in addition to boron and carbon are now fairly numerous (Chapter 9), and include species containing elements from groups II, III, IV, and V of the periodic table as well as many transition elements.

The known or postulated molecular structures of the polyhedral carboranes are presented in Fig. 2-1 (related cage systems containing heteroatoms are discussed in Chapter 9). Although not all of the structures indicated have been confirmed by X-ray or microwave studies, all are strongly supported by boron-11 and proton NMR, mass spectra, and infrared spectra as well as by chemical evidence.

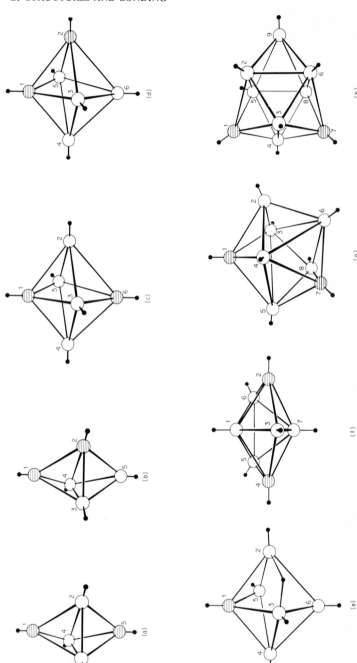

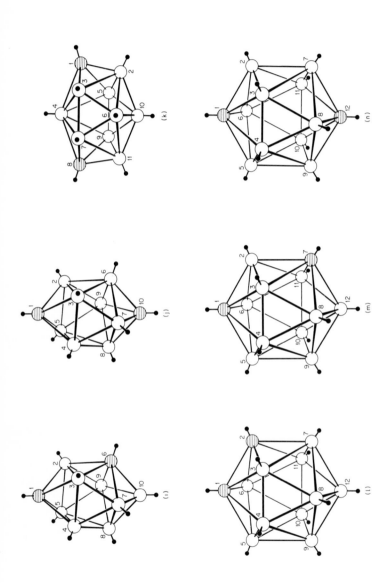

FIG. 2-1. Polyhedral (*closo-*) carborane structures. (a) $1,5-C_2B_3H_5$; (b) $1,2-C_2B_3H_5$; (c) $1,6-C_2B_4H_6$; (d) $1,2-C_2B_4H_6$; (e) CB_5H_7; (f) $2,4-C_2B_5H_7$; (g) $1,7-C_2B_6H_8$; (h) $1,7-C_2B_7H_9$; (i) $1,6-C_2B_7H_9$; (j) $1,10-C_2B_8H_{10}$; (k) $1,8-C_2B_9H_{11}$; (l) $1,2-C_2B_9H_{11}$; (m) $1,7-C_2B_{10}H_{12}$; (n) $1,12-C_2B_{10}H_{12}$.

TABLE 2-1

THE POLYHEDRAL (*Closo*-) CARBORANE SYSTEMS[a]

Valence electrons	The $C_2B_{n-2}H_n$ series[b]		The $CB_{n-1}H_n^-$ series		Borane dianions	
	Carborane	Geometry	Carborane	Geometry	Borane	Geometry
22	$C_2B_3H_5$	(2)[c] Trigonal bipyramid	$[CB_4H_5^-]$	—	$[B_5H_5^{2-}]$	—
26	$C_2B_4H_6$	(2) Octahedron	$CB_5H_6^-$ CB_5H_7	Octahedron Octahedron	$B_6H_6^{2-}$	Octahedron
30	$C_2B_5H_7$	(1) Pentagonal bipyramid	$[CB_6H_7^-]$	—	$B_7H_7^{2-}$	Pentagonal bipyramid
34	$C_2B_6H_8$	(1) Dodecahedron	$[CB_7H_8^-]$	—	$B_8H_8^{2-}$	Dodecahedron
38	$C_2B_7H_9$	(1) Tricapped trigonal prism	$[CB_8H_9^-]$	—	$B_9H_9^{2-}$	Tricapped trigonal prism
42	$C_2B_8H_{10}$	(2) Bicapped square antiprism	$CB_9H_{10}^-$	Bicapped square antiprism	$B_{10}H_{10}^{2-}$	Bicapped square antiprism
46	$C_2B_9H_{11}$	(1) Octadecahedron	$CB_{10}H_{11}^-$	?	$B_{11}H_{11}^{2-}$	Octadecahedron
50	$C_2B_{10}H_{12}$	(3) Icosahedron	$CB_{11}H_{12}^-$	Icosahedron	$B_{12}H_{12}^{2-}$	Icosahedron

[a] Species in brackets are unknown.
[b] Number of known isomers is given in parentheses.
[c] Only one isomer known for parent compound.

The Open-Cage Carboranes

Table 2-2 summarizes the open-cage systems that have been characterized up to the present time, with the structurally related boranes included for comparison.

The probable structures of the open-cage carboranes, a few of which have been subjected to X-ray studies, are shown in Fig. 2-2.

TABLE 2-2

THE OPEN-CAGE (*Nido-*) CARBORANE SYSTEMS

Isoelectronic borane	Carborane[a]	Number of B—H—B bridges	Geometry
B_5H_9		4	Square pyramid
	$C_2B_3H_7$	2	Square pyramid
B_6H_{10}		4	Pentagonal pyramid
	CB_5H_9	3	Pentagonal pyramid
	$C_2B_4H_8$	2	Pentagonal pyramid
	$C_3B_3H_7$	1	Pentagonal pyramid
	$C_4B_2H_6$	0	Pentagonal pyramid
B_9H_{15}		4	9-Atom icosahedral fragment
	$C_2B_7H_{13}$	2	9-Atom icosahedral fragment
$B_{10}H_{14}$		4	10-Atom icosahedral fragment
	$CB_9H_{12}^{-}$ [b]	2	10-Atom icosahedral fragment
$B_{11}H_{15}$		4	11-Atom icosahedral fragment
	$CB_{10}H_{13}^{-}$	2	11-Atom icosahedral fragment
	$C_2B_9H_{13}$ [c]	2	11-Atom icosahedral fragment

[a] One isomer known for each molecule except where otherwise indicated. Not shown are species formed by gain or loss of protons from molecules listed.

[b] Reported as $H_3NCB_9H_{11}$.

[c] Two known isomers.

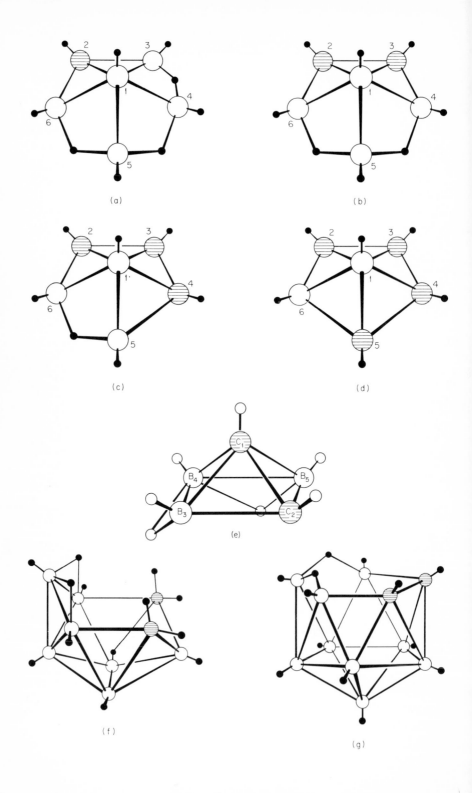

(a)

(b)

(c)

(d)

(e)

(f)

(g)

Isomerism and Rearrangement

The possibility of stereoisomerism based on carbon locations in the cage exists for all members of the $C_2B_{n-2}H_n$ polyhedral series, and it has been observed in several systems. Two isomers have been identified for $C_2B_3H_5$ (one isomer is known only in alkyl derivative form), $C_2B_4H_6$, and $C_2B_8H_{10}$, whereas all of the three possible $C_2B_{10}H_{12}$ icosahedra have been characterized.

A problem of fundamental importance in carborane chemistry concerns the relative stabilities of isomers and the mechanisms of their interconversion. Atomic framework rearrangements have been definitively established in the binary boranes (195), including open frameworks such as B_5H_9, $B_{10}H_{14}$, and $B_{10}H_{16}$ and the closed polyhedron $B_{10}H_{10}^{2-}$. Among the *closo*-carboranes, thermally induced rearrangements have been observed in $C_2B_4H_6$ (Section

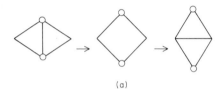

(a)

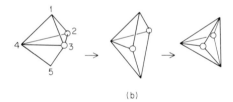

(b)

FIG. 2-3. (a) Schematic drawing of the diamond–square–diamond (dsd) rearrangement mechanism. (b) dsd rearrangement of a trigonal bipyramid (197).

4-2), $C_2B_8H_{10}$ (Section 5-2), $C_2B_{10}H_{12}$ (Sections 7-3 and 7-13), and the $(\pi\text{-}C_2B_7H_9)_2Co^-$ complex ion (Section 9-5). In each case the isomerization process results in greater separation of the relatively positive carbon atoms, with the driving force generally assumed to be carbon–carbon electrostatic repulsion. The mechanisms of these polyhedral rearrangements have not been established, and in fact have been studied experimentally only in the icosahedral species. The general subject has been reviewed by Muetterties and Knoth (215) and in an earlier article by Lipscomb (197), who has suggested some plausible pathways for rearrangements of polyhedra having 5 to 12 atoms in idealized geometries. The common feature of these proposed mechanisms, based on the triangular faces present in all boron polyhedra, involves the

FIG. 2-2. Open-cage (*nido-*) carborane structures. (a) 2-CB_5H_9; (b) 2,3-$C_2B_4H_8$; (c) 2,3,4-$C_3B_3H_7$; (d) 2,3,4,5-$C_4B_2H_6$; (e) 1,2-$C_2B_3H_7$; (f) 1,3-$C_2B_7H_{13}$; (g) 1,2-$C_2B_9H_{13}$.

cooperative stretching of a diamond-shaped group of atoms into a square and back into a diamond (Fig. 2-3a). This diamond–square–diamond, or dsd, process requires relatively little atomic motion and appears reasonable, at least for the smaller polyhedra (Fig. 2-3b). Even in the icosahedral *ortho–meta* and *meta–para* rearrangements, which have been shown to be fairly complex (Chapter 7) the dsd mechanism provides a highly useful working hypothesis (*119, 167*) which, with some modifications, may account for the product distributions observed in the isomerization of halocarboranes (Section 7-3).

2-2. NOMENCLATURE AND NUMBERING SYSTEMS

The systematic naming of polyhedral compounds presents unusual problems, and this is reflected by the lack of uniformity in the carborane literature. No single universally accepted nomenclature exists for such compounds. In this book formulas or trivial names are often used in preference to the bulky systematic nomenclature, since the structures are usually clear from the context; in those instances where systematic names are employed, the system followed is that proposed (*40*) by the Nomenclature Committee of the Division of Inorganic Chemistry of the American Chemical Society. In this scheme, a carborane is named as a borane in which one or more boron atoms has been replaced by carbon, and the familiar oxa-aza convention of organic chemistry is followed. Closed-cage and open-cage systems are designated by the prefixes *closo-* and *nido-*, respectively, and the number of hydrogens is given in parentheses (*1*). Ions are given the ending -borate and the charge is enclosed in parentheses. Examples are given below.

$C_2B_3H_5$	dicarba-*closo*-pentaborane(5)
$C_2B_4H_6$	dicarba-*closo*-hexaborane(6)
$C_2B_5H_7$	dicarba-*closo*-heptaborane(7)
$C_2B_4H_8$	dicarba-*nido*-hexaborane(8)
CB_5H_9	monocarba-*nido*-hexaborane(9)
$C_2B_4H_7^-$	heptahydrodicarba-*nido*-hexaborate(1-)

In numbering either polyhedral or open structures, the procedure (*1*) is to begin with an apex atom and number successive rings or belts in a clockwise direction, as illustrated in Fig. 2-1. When no obvious apex exists, the number 1 atom is that with the smallest coordination number. Carbon atoms in a cage system are given the lowest numbers consistent with these rules. For simplicity, ten- or eleven-particle icosahedral fragments are numbered herein exactly as the closed icosahedron from which they are derived. Thus, the 1,2-dicarbollide ion (Section 9-1), which is formally generated by removal of a boron atom in position 3 from icosahedral $1,2-C_2B_{10}H_{12}$, is designated $(3)-1,2-C_2B_9H_{11}^{2-}$, indicating in parentheses the icosahedral "hole" (137).

Carborane isomers are identified by specifying the cage carbon locations in front of the formula or name, as in $1,6-C_2B_4H_6$ and 1,6-dicarba-*closo*-hexaborane (6). The three isomers of icosahedral $C_2B_{10}H_{12}$ are commonly designated *o*-carborane, *m*-carborane, and *p*-carborane, indicating the 1,2-, 1,7-, and 1,12- isomers, respectively.*

Specific nomenclature problems have arisen with the discovery of unprecedented structures (the carborane–transition metal complexes, for example), and the current usages for such compounds are indicated in the chapters dealing with their synthesis and chemistry.

2-3. CHEMICAL BONDING IN THE BORANES AND CARBORANES

The nature of the bonding in the boron hydrides was formerly one of the classic problems of inorganic chemistry. Fortunately, boron chemistry today benefits from intensive theoretical efforts which have helped to dispel much of the mystery that formerly surrounded these structures. Since a detailed review (195) of the theoretical work on the boranes has recently been provided by Lipscomb, the discussion here will be brief and limited to fundamentals.

Electron deficiency is a term applied to structures in which the total number of valence electrons is less than the number of atomic orbitals available for bonding. It is evident that no such structure can be represented entirely in terms of the usual "two-center" bonds in which two atoms share a pair of electrons. To account for the existence of stable electron-deficient molecules such as the boranes and the carboranes, clearly one must postulate some degree of delocalization of electrons beyond the two-center bond.

This problem has been approached from several points of view, but in boron chemistry three methods (195) have been particularly important: (1) a localized three-center bond approximation, (2) construction of molecular orbitals extending over the entire cage system, and (3) an intermediate approach in

---

* The $C_2B_{10}H_{12}$ system is occasionally referred to by the trivial names "carborane" and "barene," the latter being common in the Russian literature.

which the molecular orbital and valence–bond methods are combined. In the neutral boron hydrides, many of which are of low symmetry and not easily described in molecular orbital terms, Lipscomb has used the three-center bond concept to great advantage and in fact has made it the basis of a detailed topological theory. Molecular orbital theory, on the other hand, has been most usefully applied to the polyhedral systems, whose high symmetry and de-localized bonding are particularly amenable to this method. The third, or intermediate, type of bonding description has proved useful for highly sym-metrical polyhedral and near-polyhedral structures. Each of these approaches is briefly outlined below.

The Three-Center Bond

In open frameworks of low symmetry such as the neutral boranes (some of which contain only a mirror plane), extensive electron delocalization is not expected, and it is reasonable to approximate the bonding in terms of localized electron pairs. Although the classical two-center bond concept is inadequate for electron-deficient situations, localized-bond structures can be written for such molecules by employing the concept of three-center bonding, in which three atoms are linked by a single pair of electrons. In the boron hydrides several types of three-center bonds have been postulated (*195*), as shown in Fig. 2-4. Any such bond requires a contribution of one orbital from each of

FIG. 2-4. Types of three-center bonds: closed BBB, open BBB, and BHB.

the three atoms to form three molecular orbitals, one of which is bonding, a second is antibonding, and the third is either antibonding or nonbonding. A pair of electrons may then, of course, occupy the bonding orbital. Figure 2-5 illustrates in idealized form the atomic orbital contributions.

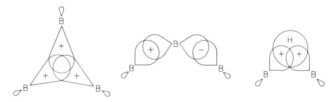

FIG. 2-5. Orbital contributions to three-center bonds.

Each boron atom in a molecule supplies four orbitals and three electrons, and is therefore responsible for a "deficiency" of one electron. Since each three-center bond just compensates for a deficiency of one electron, the number of three-center bonds in a neutral borane or carborane must equal the number of boron atoms. Thus B_2H_6 has two three-center bonds and B_6H_{10} is assigned six (Fig. 2-6). Highly symmetrical structures such as B_5H_9 (C_{4v} point group) may require the use of resonance forms (195) which, when combined, have the molecular symmetry (Fig. 2-6).

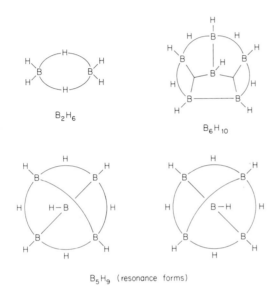

FIG. 2-6. Bonding descriptions of B_2H_6, B_6H_{10}, and B_5H_9.

The number of possible combinations of two- and three-center bonds that may be used to represent a given borane is determined by the numbers of available orbitals, electrons, and boron and hydrogen atoms, and by whatever assumptions are made concerning allowed structural features. The relations between these have been expressed by Lipscomb (195) in a set of "equations of balance" as follows. For a neutral borane B_pH_{p+q} containing s H bridges, x BH_2 groups (boron atoms having two terminal H atoms), t three-center B—B—B bonds, and y two-center B—B bonds, three conditions are set forth:

$$s + t = p \qquad (1)$$
$$s + x = q \qquad (2)$$
$$p = t + y + (q/2) \qquad (3)$$

Equation (1) simply restates the earlier observation that the number of three-center bonds of all types must equal the number of boron atoms in the molecule. Equation (2) is the hydrogen balance, and equation (3) arises from the assumption that each boron is bonded to at least one terminal hydrogen, in which case the molecule may be considered as a collection of BH units. Every such unit has one pair of electrons available for cage bonding, and these may be used in the formation of three-center B—B—B bonds, two-center B—B bonds, and bonds to hydrogen (in this case half of the electron pair is supplied by the H atom). Since none of the algebraic quantities may be negative, one obtains the limitation $q/2 \leqslant s \leqslant$ the smaller of p or q. To illustrate, we apply these considerations to $B_{10}H_{14}$. Since $p = 10$ and $q = 4$, it follows that $2 \leqslant s \leqslant 4$, and the only possibilities may be tabulated as

s	t	y	x
4	6	2	0
3	7	1	1
2	8	0	2

There are no other solutions consistent with the equations of balance. In this case the actual structure (Fig. 1-1b) is known to contain four hydrogen bridges, so that the correct formulation is 4620. A reasonable, but not necessarily unique, representation is shown in Fig. 2-7 (terminal H atoms are

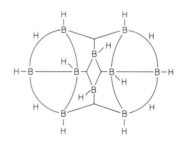

FIG. 2-7. Bonding description of $B_{10}H_{14}$.

omitted). This structure is consistent with the known C_{2v} symmetry of the molecule (195).

   For any given styx combination, the number of viable bonding arrangements is limited, in the complete topological theory, by considerations of symmetry, valence angles, and charge distribution. For example, adjacent boron atoms may not be joined by both a closed and an open three-center bond since an unfavorably small valence angle would result. This and other restrictions have been suggested by Lipscomb (195), largely based upon the structural features present (or absent) in known boron hydrides. From time to time the theory has had to be modified or extended (*196*) to accommodate novel structures. Several of the recently discovered boranes, for example, contain boron atoms having no terminal hydrogens ($B_{10}H_{16}$, $B_{20}H_{16}$, $B_{18}H_{22}$, and iso-$B_{18}H_{22}$), but it is a simple matter to extend the equations of balance to allow for such situations.

Borane ions may be handled in a similar manner: if the charge is m and the general formula $B_pH_{p+q+m}^m$, the equations (*195*) are

$$s + t = p + m$$
$$s + x = q + m$$
$$t + y = p - m - (q/2)$$

Applying these conditions to $B_{10}H_{10}^{2-}$ gives a unique styx combination, 0830, corresponding to eight B—B—B three-center bonds and two B—B two-center bonds. But at this point it must be recognized that the known $B_{10}H_{10}^{2-}$ is a closed polyhedron with high (D_{4d}) symmetry and apparently considerable electron delocalization. Any description of such a system in terms of localized two- and three-center bonds will require an inordinate number of resonance structures, which can be dealt with simultaneously only with the aid of a computer (*152*). In general, the advantages of the three-center bond approximation tend to disappear in polyhedral borane structures, and such molecules are now treated almost exclusively in terms of molecular orbital descriptions.

It might be expected that three-center bond formulations would be appropriate for open-cage carborane structures, since these are structural analogs of the neutral boranes. In principle this is true, but the substitution of carbon for boron in a borane cage system presents problems not encountered with the binary boron hydrides. The principal difficulty is that very little is known about charge distribution and the nature of boron–carbon bonding in the *nido*-carboranes. The situation is illustrated by 2,3-$C_2B_4H_8$ (Fig. 2-2b), for which at least three different bonding arrangements have been proposed (*150, 230, 326*). One suggestion (*326*), depicted in Fig. 2-8, contains a carbon–carbon double bond which donates electrons to an available *sp* orbital on the apex boron atom. Although this proposal is reasonable, it is not readily applied to such isoelectronic structures as $C_3B_3H_7$ and CB_5H_9 (Fig. 2-2).

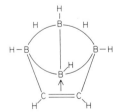

Fig. 2-8. Proposed bonding description of 2,3-$C_2B_4H_8$ (*326*).

Any attempt to formulate a general topological theory for the open-cage carboranes must deal with some unresolved questions, such as the conceivable, but as yet unknown, participation of carbon in C—H—B or C—H—C bridge bonds. The discovery of new cage systems may supply some interesting answers. At the present time, however, the available information on *nido*-carborane structures is probably too limited to permit useful predictions of new compounds on the basis of semiempirical topological rules, in a fashion analogous to Lipscomb's treatment of the boranes. In a more positive vein, it may be noted that the approximate molecular orbital treatment described in the last section of this chapter has been quite usefully applied to the small *nido*-carboranes and has led to the prediction of some new structures in this class.

Molecular Orbitals in Boranes and Carboranes

The characterization of the bonding in polyhedral carboranes and borane anions as highly delocalized is well supported by chemical evidence. The thermal and hydrolytic stability of these molecules is striking when compared to the neutral boranes, and it is hard to avoid the implication of resonance stabilization similar to that in aromatic hydrocarbons. Additional evidence of delocalized bonding is found in the observed inductive effects in cage substitution reactions (Chapters 6 and 7) and in the remarkable discovery by Hawthorne, Young, and Wegner (*139*) of carborane analogs of ferrocene, in which a transition metal atom is π-bonded to a five-membered ring in a carborane cage system (Chapter 9). Unfortunately, there is as yet little quantitative data bearing directly on this problem. The diamagnetic susceptibility of the potassium salt of $B_{10}H_{10}^{2-}$ has a large residual value of 37×10^{-6} cm^3/mole after subtraction of the individual atom susceptibilities, and this has been taken by Lipscomb (*195*, p. 93) as a further indication of electron delocalization. Other measurements relating to this question, such as the dissociation constants of carborane carboxylic acids, are discussed in Chapters 6 and 7.

High symmetry and evidence of aromatic character make the polyhedral boranes reasonable candidates for treatment by molecular orbital methods. One of the earliest calculations led to the prediction of a stable $B_{12}H_{12}^{2-}$ icosahedral anion by Longuet-Higgins and Roberts (*200*). These authors assumed that each B—H bond in an icosahedral $B_{12}H_{12}$ molecule involves one localized electron pair, leaving three orbitals and two electrons on each boron to be utilized in cage framework bonding. The 36 molecular orbitals were found to include 13 bonding orbitals, of which the top four are degenerate.

The hypothetical neutral molecule has but 24 electrons for these 13 orbitals and thus is unlikely to be stable in icosahedral symmetry, but the $B_{12}H_{12}^{2-}$ ion contains two additional electrons and hence a closed-shell configuration.

Subsequent extended Hückel calculations by Lipscomb and Hoffmann (*151, 154*) on a number of polyhedral systems indicated filled bonding orbitals for $B_5H_5^{2-}$ (D_{3h} symmetry), $B_6H_6^{2-}$ (O_h), $B_7H_7^{2-}$ (D_{5h}), $B_8H_8^{2-}$ (D_{3d}), $B_{10}H_{10}^{2-}$ (D_{4d}), $B_{12}H_{12}^{2-}$ (I_h), and several carborane analogs of these ions. By necessity, all orbital interactions were included in this work, since inconsistencies developed when only the interactions between nearest neighbors were considered (*151*). As an illustration of this treatment, the energy levels calculated for $B_5H_5^{2-}$ are shown in Fig. 2-9. The hydrogen orbitals and one orbital on each

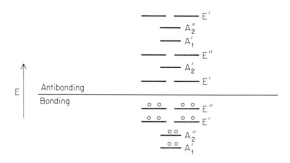

FIG. 2-9. Molecular orbital energy levels calculated for the $B_5H_5^{2-}$ ion (arbitrary energy scale).

boron are assumed to form localized σ B—H bonds, so that 15 atomic orbitals are available for cage bonding. These are combined to form six bonding and nine antibonding molecular orbitals; thus 12 valence electrons, corresponding to the dinegative ion, are required to fill the bonding orbitals.

The best evidence of the essential validity of this approach is the actual synthesis and stability of many of the species in Table 2-1. In addition to orbital energies, LCAO–MO methods have also been used to compute charge distributions (*152*) and sequences of substitution by electrophilic reagents on the polyhedral systems (*153*). Among the most important general results (*152*), so far as the *icosahedral* carborane isomers are concerned, is that the carbon atoms are found to be the most positive locations on the cage, whereas the boron atoms tend to be more negative the greater their distance from carbon. The experimental evidence presently available (discussed in detail in Chapter 6) is in general agreement with these calculations, as may be illustrated by noting that Friedel-Crafts halogenation of *o*-carborane ($1,2\text{-}C_2B_{10}H_{12}$) occurs first at the 9,12 and then at the 8,10 borons. Halogenation of all borons in

o-carborane is not possible under Friedel-Crafts conditions but B-perchlorination has been achieved photochemically; significantly, the 3,6 borons (those closest to carbon) are the last to be substituted (Section 6-13).

Simplified Molecular Orbital Descriptions

Detailed calculations of the type briefly discussed above have given a solid theoretical interpretation to the bonding in polyhedral boranes and carboranes, but it is desirable to have a general picture of the electronic structure of carboranes that is reasonably simple and allows predictions of new species without involved computations. An intermediate approach which lies conceptually between the localized-bond and complete molecular orbital treatments has been employed rather successfully (*151, 195*) to describe both complete polyhedra and polyhedral fragments. As in the full molecular orbital descriptions, it is convenient to first consider the boron hydride anions and then extend the reasoning to the carboranes by the isoelectronic analogy. Several simplifying assumptions are made, however. First, one imagines a polyhedral anion such as $B_{10}H_{10}^{2-}$ to be separated into "equatorial" and "polar" regions. Second, all bonds to hydrogen, whether terminal or bridge, are considered to be localized and not involved in the framework molecular orbitals.

This method has been applied by Lipscomb and Hoffmann (*151, 195, 196*) to the bipyramidal ions $B_5H_5^{2-}$, $B_6H_6^{2-}$, and $B_7H_7^{2-}$. Consider pentagonal bipyramidal $B_7H_7^{2-}$. Applying the artificial equatorial–apical separation, the

Fig. 2-10. Simplified description of bonding in the $B_7H_7^{2-}$ ion.

two apex B—H groups are removed to leave a $B_5H_5^{2-}$ planar ring which may be given the same molecular orbital treatment commonly employed for aromatic hydrocarbon systems. If each ring boron atom is assumed to be sp^2 hybridized, a σ-bonded framework may be set up as shown in Fig. 2-10. The five remaining p_z orbitals are perpendicular to the ring, and these may combine to form a set of five delocalized molecular orbitals which are, in turn, to be used in bonding to the apical B—H groups. However, of the five ring MO's only the σ_u and π_g orbitals are of the correct symmetry to bond with the available pure p orbitals on the apex borons. The consequence of bonding the two apical B—H groups to the ring is, then, the formation of six molecular orbitals of which three are bonding. Three electron pairs are required for a closed-shell structure, corresponding to the dinegative ion $B_7H_7^{2-}$. The application of this argument to trigonal bipyramidal $B_5H_5^{2-}$ and square bipyramidal (octahedral) $B_6H_6^{2-}$ gives similar results (195, 196). Although this method leads to unfavorable electron/orbital ratios in some cases, particularly $B_7H_7^{2-}$, the qualitative indication of a stable existence for these dinegative ions is in agreement with the LCAO calculations (195). The actual isolation of $B_6H_6^{2-}$, $B_7H_7^{2-}$, and of carborane analogs of all three ions has been noted earlier.

Larger polyhedra have been treated in this manner, notably $B_8H_8^{2-}$ (D_{3d}), $B_{10}H_{10}^{2-}$ (D_{4d}), and $B_{12}H_{12}^{2-}$ (I_h), and in each case closed-shell electronic structures are predicted (195), in agreement with the LCAO results. However, the method has perhaps been most useful in its application to polyhedral fragments, since boron frameworks of this type have not been readily described by any other generalized treatment. When the approach described above for bipyramidal species is applied to pyramidal molecules with four, five, and six boron atoms the result is a prediction (195) of closed-shell configurations for both B_nH_n and $B_nH_n^{4-}$. In the B_nH_n case unreasonable charge distributions are obtained, but for $B_nH_n^{4-}$ the distributions, expressed in terms of electron/orbital ratios, are quite good. The ions $B_4H_4^{4-}$, $B_5H_5^{4-}$, and $B_6H_6^{4-}$ are unlikely to be stable because of the large negative charge (195), but the charge may be reduced by the formal addition of H^+ to form B—H—B bridge bonds. In this way the known B_5H_9 and B_6H_{10} structures and the partly characterized $B_4H_7^-$ may be obtained. Alternatively, as Lipscomb (195) has pointed out, one may formally replace B^- with C to give pyramidal carborane analogs such as C_4BH_5, $C_4B_2H_6$, $C_3B_2H_5^-$, $C_5BH_6^-$, and others. The possibilities are further increased if substitution of C and addition of H^+ are combined. Thus, from the hypothetical $B_6H_6^{4-}$ one may account for the known pentagonal pyramidal series $C_4B_2H_6$, $C_3B_3H_7$, $C_2B_4H_8$, CB_5H_9, and B_6H_{10}, in which the number of hydrogen bridges is 0, 1, 2, 3, and 4, respectively (Table 2-2). The only known carborane analog of $B_5H_5^{4-}$ is the very recently synthesized $C_2B_3H_7$, described in the following chapter. No carboranes analogous to $B_4H_4^{4-}$ have been reported.

Summary

In this chapter the discussion has been limited to concepts of bonding that have had reasonably general application to borane and carborane structures. Bonding problems of a more specific nature, such as those associated with the metallocarboranes, are dealt with individually at appropriate points in the book.

3

The Small *Nido*-Carboranes:
CB_5H_9, $C_2B_4H_8$, $C_3B_3H_7$, $C_4B_2H_6$, and $C_2B_3H_7$

3-1. STRUCTURES

In contrast to the $C_2B_{n-2}H_n$ polyhedral carboranes, which fall into a well-defined series, the known open-cage or *nido*-carborane species do not form a single structural class. In addition, predictions of new cage systems of this type are difficult even on a qualitative basis; the open structures and low symmetry seem to preclude the kind of general molecular orbital approach that has worked well with the closed polyhedral systems. Further, the number of known *nido*-carboranes is probably insufficient to allow empirical judgments as to which structural features might or might not be expected in new species. An example is the unique methylene-bridged $C_2B_7H_{13}$ (Section 5-1) which was not predicted prior to its discovery. Larger *nido*-carboranes, such as $C_2B_9H_{13}$ and its derivative species, are closely related chemically to the polyhedral carboranes and are discussed in that context (Chapters 5 and 9).

The known *nido*-carboranes having five or fewer boron atoms do have a close structural relationship and are logically treated as a group. Hexaborane-(10), B_6H_{10}, is a pentagonal pyramid with four bridge hydrogen atoms laced about the base; successive formal replacement of basal B—H_{bridge} units with carbon atoms generates the isoelectronic series 2-CB_5H_9, 2,3-$C_2B_4H_8$, 2,3,4-$C_3B_3H_7$, and 2,3,4,5-$C_4B_2H_6$ (Fig. 2-2). All of these are known (the last only in the form of alkyl derivatives), but only one isomer of each system has been characterized at this writing. Of the structures shown, only that of $C_2B_4H_8$ has been confirmed by an X-ray study (*22, 326*), the others having been deduced from NMR, mass spectra, and infrared spectra.

In recent work, 1,2-$C_2B_3H_7$, a *nido*-carborane isoelectronic with B_5H_9, has been prepared and characterized (*78a*). This molecule apparently has a square pyramidal structure with one apical carbon and one equatorial carbon (Fig. 2-2e) as shown by NMR studies and other data.

3-2. SYNTHESIS

Borane–Alkyne Reactions

Table 3-1 summarizes the known preparative routes to the small *nido*-carboranes and their derivatives. The majority of these reactions involve the interaction of boranes with alkynes under milder conditions than those employed in *closo*-carborane synthesis (Chapter 4). Thus, mixtures of B_5H_9 and C_2H_2, which at 450° or in an electric discharge yield small polyhedral carboranes, interact at 215° to produce $2,3\text{-}C_2B_4H_8$ and several methyl derivatives of $2\text{-}CB_5H_9$ (Table 3-1). Alkynes and B_5H_9 in pyridine solution at room temperature produce alkyl derivatives of $2,3\text{-}C_2B_4H_8$. Similarly, although the less stable boranes B_4H_{10} and B_5H_{11} react explosively with alkynes in the gas phase at 100°, yielding *closo*-carboranes (Chapter 4), at 25–50° the same reactants slowly produce *nido*-carborane species.

Detailed studies (*106, 107*) of the reactions of alkynes with B_4H_{10} and B_5H_{11} have disclosed several clear patterns, including the fact that $C_3B_3H_7$ and its methyl derivatives are formed in much higher yield from B_4H_{10} than from B_5H_{11} (perhaps significantly, the $C_3B_3H_7$ system has not been obtained in any other reaction). Comparison of the reactions of B_4H_{10} with acetylene, methyl-acetylene, and dimethylacetylene shows that the yield of $C_3B_3H_7$ derivatives is highest with acetylene as the reactant and lowest when dimethylacetylene is used; conversely, the $C_2B_4H_8$ system is obtained in highest yield from the dimethylacetylene reaction and in poorest yield from acetylene (*107*). The alkyne–B_5H_{11} reactions show essentially the same trends.

The mechanisms of *nido*-carborane formation in borane–alkyne systems are of interest in several respects. Such reactions differ markedly from borane–*alkene* gas-phase reactions, which produce alkylboranes but not carboranes (*18, 116, 159, 194, 201, 203*). In addition, it is noteworthy that B_2H_6 and alkynes in the gas phase at 70° yield several organoboron products but no carborane species whatsoever (*194*). The formation of $C_2B_4H_8$ from C_2H_2 and B_5H_9 evidently occurs without breakage of either C—C or C—H bonds, as shown in deuterium tracer experiments with C_2D_2 (*291*). On the other hand, the synthesis of $C_3B_3H_7$ and its dimethyl derivatives from C_2H_2 necessarily involves C—C cleavage despite the mild (25–50°) reaction conditions. Here again, tracer studies (*79*) indicate that C—H cleavage, if it occurs, is not extensive.

The synthesis of the 2-methyl, 2,3-dimethyl, and 2,4-dimethyl derivatives of $2,3,4\text{-}C_3B_3H_7$ in the relatively clean reaction of C_2H_2 with B_4H_{10} in the vapor phase at 50° has been examined kinetically (78) by measuring the rates of carborane formation as a function of C_2H_2 and B_4H_{10} concentrations. The essential results are that the reaction is first order in B_4H_{10} but *negative* order in acetylene, for all three carborane products. Since the only major carbon-containing product other than the $C_3B_3H_7$ derivatives is a homogeneous white polymer, it appears that the reaction proceeds via a common intermediate which is a precursor both to the $C_3B_3H_7$ derivatives and to the polymer. Increasing the acetylene concentration favors polymer formation at the expense of carborane production.

When the gas-phase reaction of B_4H_{10} and C_2H_2 at 50° is conducted with a tenfold excess of C_2H_2, the same $C_3B_3H_7$ derivatives are obtained as in the 1:1 reaction; in addition, however, the square pyramidal *nido*-carborane 1,2-$C_2B_3H_7$ is produced in 3–4% yield (this compound is not observed to form in the equimolar B_4H_{10}–C_2H_2 reaction) (78a).

Other Routes to *Nido*-Carboranes

The *C*-methyl-*B*-pentaethyl derivative of 2-CB_5H_9 has been obtained in reactions of ethyldiborane with acetylene and of ethyldiborane with sodium in tetrahydrofuran, and by dehalogenation of alkylhaloboranes (as in the action of lithium metal on ethyldifluoroborane) (94, 184). Treatment of the ionic products of the reaction between 1,7-$C_2B_6H_8$ and BH_4^- salts with dry HCl produces unsubstituted 2-CB_5H_9 and two of its methyl derivatives (54a, 55) (Section 5-2).

The addition of a catalytic amount of diethylchloroborane at 40° to *cis*-1,2-bis(diethylboryl)-1,2-dialkylethylene yields a peralkyl derivative of 2,3,4,5-$C_4B_2H_6$ (17).

$$\underset{(C_2H_5)_2B}{\overset{R}{>}}C=C\underset{B(C_2H_5)_2}{\overset{R}{<}} \xrightarrow[40°]{(C_2H_5)_2BCl} (C_2H_5)_4R_2C_4B_2$$

$$R = CH_3, C_2H_5$$

This is the only reported synthesis of the $C_4B_2H_6$ system which warrants confidence [the reaction of diphenylacetylene with phenyldibromoborane in the presence of potassium metal yields a product for which the composition $(C_6H_5)_6C_4B_2$ has been inferred (165) from rather indirect evidence].

TABLE 3-1

SYNTHESIS OF SMALL *Nido*-CARBORANES

Reaction[a]	Products[b]				References
	2-CB$_5$H$_9$	2,3-C$_2$B$_4$H$_8$	2,3,4-C$_3$B$_3$H$_7$	Other	
B$_5$H$_9$ + RC≡CR′ + pyridine or lutidine (L)		2-CH$_3$ 2,3-(CH$_3$)$_2$ 2-n-C$_3$H$_7$ 2-Isopropenyl 2-Phenyl			222, 230
B$_5$H$_9$ + LiC≡CCH$_3$ (L)	2-C$_2$H$_5$	2-CH$_3$			228
B$_5$H$_9$ + HC≡CH (G, 215°)	2-CH$_3$ 3-CH$_3$ 4-CH$_3$	Parent			222, 226, 275
B$_5$H$_9$ + RC≡CR′ (G, 165–175°)		2-CH$_3$ 2,3-(CH$_3$)$_2$			222
CH$_3$B$_5$H$_8$ + RC≡CR′ (G, 205–250°)		1-CH$_3$ 4-CH$_3$ 5-CH$_3$ 1,2,3-(CH$_3$)$_3$ 2,3,4-(CH$_3$)$_3$ 2,3,5-(CH$_3$)$_3$			227
B$_5$H$_{11}$ + HC≡CH (G, 25–50°)			2-CH$_3$ 2,4-(CH$_3$)$_2$		107
B$_5$H$_{11}$ + RC≡CR′			2-CH$_3$		107

Reaction					Ref.
$B_4H_{10} + HC{\equiv}CH$ (G, 25–50°)	4-CH_3	Parent	2-CH_3 2,3-$(CH_3)_2$ 2,4-$(CH_3)_2$ Parent	1,2-$C_2B_3H_7{}^c$	24, 78a, 106, 107, 157
$B_4H_{10} + RC{\equiv}CR'$ (G, 25–50°)	2-CH_3 2,B-$(CH_3)_2$ 2,B'-$(CH_3)_2$	2-CH_3 2,3-$(CH_3)_2$	2-C_2H_5-3,4-$(CH_3)_2$		107
$C_2H_5B_2H_5$ + $HC{\equiv}CH$ (L)	2-CH_3-B-$(C_2H_5)_5$				94, 184
$C_2H_5BF_2$ + Li (L)	2-CH_3-B-$(C_2H_5)_5$				184
$[(C_2H_5)_2BR]_2C_2$ + $(C_2H_5)_2BCl$ (L)				$(C_2H_5)_6C_4B_2$ $(C_2H_5)_4C_4B_2(CH_3)_2$	17
HCl (g) + solids from reaction of $C_2B_6H_8$ + $BH_4{}^-$ (L, 100°)	Parent 1-CH_3 3-CH_3				54a, 55
$C_2B_4H_8$ + Br_2 or Cl_2 (L)	4-Br 4-Cl				289
B_5H_9 + $SiH_3C{\equiv}CR$ (G, 165–185°)	2-SiH_3-3-R R—SiH_3, H				191
B_5H_9 + $(CH_3)_3SiC{\equiv}CH$ (G, 200–220°)	2-$Si(CH_3)_3$				191

[a] L = liquid phase or solution, G = gas phase.

[b] Only characterized carborane products are included.

[c] Obtained from 10:1 mole ratio of C_2H_2 to B_4H_{10}.

3-3. REACTIONS AND PROPERTIES

2-CB$_5$H$_9$

No chemistry has been reported for the parent compound or partially alkylated derivatives, but the *C*-methyl-*B*-pentaethyl derivative is stable to 180° and is oxidized in air only over a period of months. The same compound is slowly decomposed by 30% H$_2$O$_2$ in acid at 25° (*94*, *184*). Pyrolysis of this material at 200° forms several highly alkylated derivatives of *closo*-1,5-C$_2$B$_3$H$_5$ and *closo*-2,4-C$_2$B$_5$H$_7$ (Chapter 4). No deuterium exchange is observed with tetraethyldideuteriodiborane at 120° (*94*).

2,3-C$_2$B$_4$H$_8$

The synthesis of small *closo*-carboranes from C$_2$B$_4$H$_8$ via pyrolysis, electric discharge, and ultraviolet irradiation is discussed in Chapter 4. As with the larger *nido*-carboranes such as C$_2$B$_9$H$_{13}$ (Chapter 9), C$_2$B$_4$H$_8$ is attacked by hydride ion to remove one of the two hydrogen bridges and form the 2,3-C$_2$B$_4$H$_7^-$ anion; excess hydride fails to remove the remaining bridge proton. Subsequent treatment of the anion with HCl or DCl yields the original

$$C_2B_4H_8 + NaH \xrightarrow{\text{diglyme}} Na^+C_2B_4H_7^- + H_2 \xrightarrow{\text{DCl}} C_2B_4H_7D$$

carborane or its bridge-deuterated derivative (*224*). Deuterium exchange on 2,3-C$_2$B$_4$H$_8$ is effected at B(4), B(5), and B(6) by B$_2$D$_6$ at 100°, both bridge protons being unaffected (*224*); in diglyme at 25° the exchange with B$_2$D$_6$ occurs only at B(4) and B(6) (*291*). In the presence of D$_2$, however, all terminal and bridge hydrogens in C$_2$B$_4$H$_8$ undergo exchange (*291*).

The aluminum halide-catalyzed reactions of C$_2$B$_4$H$_8$ with Cl$_2$ below $-30°$ and with Br$_2$ below $-10°$ form the 4-chloro- and 4-bromo- derivatives, respectively; no evidence for other halogenated products has been found (*289*).

The reaction of 2,3-C$_2$B$_4$H$_8$ with B(CH$_3$)$_3$ at 300° proceeds in a manner similar to the boron-insertion reactions of C$_2$B$_9$H$_{13}$ and the dicarbollide ions (Chapter 9), producing in this case methylated derivatives of *closo*-2,4-C$_2$B$_5$H$_7$ (*267*). A related gas-phase interaction is that of 2,3-C$_2$B$_4$H$_8$ with

$Ga(CH_3)_3$ at 215° (*109*), which forms the novel *closo*-gallacarborane $CH_3GaC_2B_4H_6$ (Section 9-4).

2,3,4-$C_3B_3H_7$

Since the parent compound has only recently been identified (as a very minor product of the C_2H_2–B_4H_{10} gas phase reaction at 50°) (*157*), most chemical studies of this carborane system have involved the more easily accessible *C*-methyl derivatives. The reaction of 2,3- or 2,4-$(CH_3)_2C_3B_3H_5$ with sodium hydride removes the lone bridge hydrogen, forming the respective $(CH_3)_2C_3B_3H_4^-$ ion and generating 1 mole of H_2 per mole of carborane

$$(CH_3)_2C_3B_3H_5 + NaH \rightarrow Na^+(CH_3)_2C_3B_3H_4^- + H_2$$

reactant. Treatment of the ion with anhydrous DCl yields the bridge-deuterated $(CH_3)_2C_3B_3H_4D$ species (*79*). All evidence suggests that the 2,3- and 2,4-$(CH_3)_2C_3B_3H_4^-$ ions contain a C_3B_2 basal ring which, like cyclopentadienide ion, $C_5H_5^-$, contains three electron pairs* and is capable of forming π-bonded "sandwich" complexes with transition metals. Such complexes have been prepared (*156*) both by reaction of $C_3B_3H_7$ methyl derivatives with $Mn_2(CO)_{10}$, and directly from the tricarbahexaborate anions (*157*) (Section 9-5).

2,3,4,5-$C_4B_2H_6$

Little chemistry has been reported for this system, which is known only in the form of peralkyl derivatives. These compounds (Table 3-2) are stable toward acids and hydrolysis; neither boiling 6 N H_2SO_4 nor methanol or ethanol at 150° produces any reaction. Oxidation with O_2 or H_2O_2 occurs very slowly at 20° (*17*). Since this carborane cage system apparently incorporates a planar C_4B ring with no hydrogen bridges, it seems likely to form π-bonded metallocene analogs by direct reaction with appropriate transition metal reagents. No such species have been reported at this writing.

---

* Formal removal of an apical BH^{2+} unit (in the manner of Lipscomb's ring-polar separation (Section 2-3)) from $C_3B_3H_6^-$ generates the hypothetical planar species $C_3B_2H_5^{3-}$, which is isoelectronic with $C_5H_5^-$.

TABLE 3-2

DERIVATIVES OF THE SMALL *Nido*-CARBORANES

Compound	Data[a]	References
2-CB$_5$H$_9$ derivatives[b]		
Parent	B, H, MS, IR	*54a, 55*
1-CH$_3$	B, H, MS, IR	*54a, 55*
2-CH$_3$	B, H, MS, IR	*54a, 226*
3-CH$_3$	B, H, MS, IR	*54a, 226*
4-CH$_3$	B, H, MS	*226*
2,B-(CH$_3$)$_2$	IR, MS	*107*
2,B′-(CH$_3$)$_2$	IR, MS	*107*
2-CH$_3$-B-(C$_2$H$_5$)$_5$	IR, B, H, MS	*94*
2-C$_2$H$_5$	IR, MS, B, H	*228*
2,3-C$_2$B$_4$H$_8$ derivatives		
Parent	IR, MS	*224*
	B, H	*221, 222, 230, 351*
1-CH$_3$	IR, B, H	*227*
2-CH$_3$	IR, H	*107, 222*
4-CH$_3$	B, H, MS	*227*
5-CH$_3$	B, H, MS	*227*
2,3-(CH$_3$)$_2$	IR, B, H, MS	*107, 224, 229*
1,2,3-(CH$_3$)$_3$	IR, B, H, MS	*227*
2,3,4-(CH$_3$)$_3$	B, H, MS	*227*
2,3,5-(CH$_3$)$_3$	B, H, MS	*227*
2-*n*-C$_3$H$_7$	H	*229*
2-C$_6$H$_5$	B, H	*222*
2-Isopropenyl	B, H	*222*
4-Br (mp −69°)	IR, B, H, MS, VP	*289*
4-Cl (mp −74°)	IR, B, H, MS, VP	*289*
2,3,4-C$_3$B$_3$H$_7$ derivatives		
Parent	IR, B, H, MS	*157*
2-CH$_3$	IR, MS, B, H	*24, 106, 107*
2,3-(CH$_3$)$_2$	IR, MS, B, H	*24, 106, 107*
2,4-(CH$_3$)$_2$	IR, MS, B, H	*24, 106, 107*
2-C$_2$H$_5$-3,4-(CH$_3$)$_2$	IR, MS, H	*107*
2,3,4,5-C$_4$B$_2$H$_6$ derivatives		
(C$_2$H$_5$)$_6$	IR, B, H, Raman, MS	*17*
1,2,3,6-(C$_2$H$_5$)$_4$-4,5-(CH$_3$)$_2$	IR, B, H, Raman, MS	*17*
1,2-C$_2$B$_3$H$_7$	IR, B, H, MS	*78a*

[a] IR = infrared spectrum or band positions; MS = mass spectrum (partial or complete) or cutoff m/e value; B = ^{11}B NMR data; H = proton NMR data; VP = vapor pressure data.

[b] Additional data on di- and trialkyl derivatives are given in reference *54a*.

1,2-C$_2$B$_3$H$_7$

The chemical behavior of this compound is unique, in at least one respect, among known carborane species. Although the molecule undergoes no detectable change in the vapor phase at 50°, liquid C$_2$B$_3$H$_7$ rapidly polymerizes at room temperature to a white solid. Even solutions of this carborane in CS$_2$ or CDCl$_3$, at dilutions of 4:1 or 5:1, become gel-like within an hour at room temperature (*78a*). No H$_2$ is evolved during the polymerization.

The characterized derivatives of the small *nido*-carboranes are listed in Table 3-2.

4

The Small *Closo*-Carboranes: $C_2B_3H_5$, $C_2B_4H_6$, $C_2B_5H_7$, and CB_5H_7

The exploration of carborane chemistry has largely focused on the icosahedral $C_2B_{10}H_{12}$ isomers and their derivatives, as a result of which far more is known of these larger systems than any of the smaller members of the *closo*-carborane family. This state of affairs reflects primarily an industrial interest in the development of carborane-based polymers (Chapter 8) from decaborane(14), large stocks of which have been accessible to several industrial laboratories. In contrast to *o*-carborane, $1,2\text{-}C_2B_{10}H_{12}$, which is easily obtained in high yield from $B_{10}H_{14}$, the small parent carboranes have until recently been prepared only in relatively inefficient processes and in minute quantities. On the other hand, certain highly alkylated derivatives of the small carboranes have been synthesized in bench-scale amounts and in reasonable yields. The parent compounds $1,5\text{-}C_2B_3H_5$, $1,6\text{-}C_2B_4H_6$, and $2,4\text{-}C_2B_5H_7$ are now obtainable in good yield from a recently discovered borane–acetylene flow process described below. Despite their limited availability compared to the $C_2B_{10}H_{12}$ isomers, the small carboranes are interesting species and their structural simplicity makes them attractive for many research purposes. The unsubstituted compounds are highly volatile, thermally stable materials which are unreactive with air, water, and nonoxidizing acids at room temperature. Their preparation and handling frequently involves high vacuum techniques and gas chromatography.

4-1. SYNTHESIS

Table 4-1 summarizes the reactions from which small *closo*-carboranes and their alkyl derivatives have been obtained.* Only structurally characterized

---

* Some published reports have claimed the synthesis of "carboranes" but omit characterization data other than elemental analysis or mass spectra of mixtures. These papers have been reviewed elsewhere (*185*) but are not included in this discussion.

carborane products are included; in most of the reactions listed additional products were reported. Yields are not indicated because of the notorious variation of such data with reaction parameters (particularly in gas-phase systems) and on separation and purification techniques. The table indicates the surprising variety of conditions under which carborane cage systems have been prepared, and demonstrates that the number of observed cage isomers having three, four, or five boron atoms is sharply limited.

Small polyhedral carboranes have been isolated from several types of reactions:

1. High-energy borane–alkyne interactions
2. Conversion of *nido*-carboranes to *closo*-carboranes
3. Reactions of alkylboranes with alkynes
4. Hydroboration of alkynylboranes
5. Dehalogenation of alkylhaloboranes with alkali metals
6. Pyrolysis and electric discharge of alkylboranes
7. Reactions of boranes with carbon vapor

Only reactions 1, 2, 6, and 7 have given parent *closo*-carboranes, the other methods producing alkylated species only. In contrast to the synthesis of the icosahedral $C_2B_{10}H_{12}$ isomers, quantitative yields have not been approached for any individual small polyhedral system.

Borane–Alkyne Reactions

The first carboranes of any type to be reported in the literature were the parent compounds $1,5\text{-}C_2B_3H_5$, $1,2\text{-}C_2B_4H_6$, $1,6\text{-}C_2B_4H_6$, and $2,4\text{-}C_2B_5H_7$, prepared in very small yields (1–2% or less) by the circulation of a penta-borane(9)–acetylene mixture through an electric glow discharge (*273, 274*). A new process (*47*) involving direct thermal reaction of B_5H_9 and C_2H_2 at 490° in a rapid stream of H_2 has given high conversions (~70%) to small carboranes, the major products being $1,6\text{-}C_2B_4H_6$ and $2,4\text{-}C_2B_5H_7$ with a smaller yield of $1,5\text{-}C_2B_3H_5$. This is an important technological breakthrough which has opened the way to large-scale production of the small carboranes, and should considerably spur research on the largely neglected chemistry of these compounds.

Other work has revealed that high-energy (explosion or electric discharge) reactions of acetylene and higher alkynes with B_2H_6, B_4H_{10}, and B_5H_{11} produce parent carboranes as well as alkyl derivatives (Table 4-1). Most of these reactions are extremely complex (in contrast to the low-energy borane–alkyne interactions discussed in Chapter 3), and yield considerable quantities of dark nonhomogeneous solids as well as the volatile products, so that the

TABLE 4-1

SYNTHESIS OF SMALL *Closo*-CARBORANES

Reaction[a]	Carborane Products[b]				Refs.
	$1,5\text{-}C_2B_3H_5$	$1,6\text{-}C_2B_4H_6$	$2,4\text{-}C_2B_5H_7$	Other	
$B_2H_6 + C_2H_2$ discharge or flash (G)	Parent 2-CH₃	Parent 1-CH₃ 2-CH₃	Parent 1-CH₃ 3-CH₃ 5-CH₃	$C,3\text{-}(CH_3)_2\text{-}1,2\text{-}C_2B_3H_3$	103–105
$B_4H_{10} + C_2H_2$ flash (G)	2-CH₃	Parent	Parent 1-CH₃ 2-CH₃ 3-CH₃ 5-CH₃		106
$B_4H_{10} + HC_2CH_3$ flash (G)	Parent 2-CH₃	1-CH₃	Parent, 2-CH₃, 1,7-(CH₃)₂	Methyl derivs. of $1,2\text{-}C_2B_3H_5$	108
$B_4H_{10} + CH_3C_2CH_3$ flash (G)	Parent 2-CH₃	1-CH₃ 2-C₂H₅ (C₂H₅)₂ C₂H₅—CH₃	Parent 1,7-(CH₃)₂ 1-C₂H₅ + higher alkyl derivs.	Methyl derivs. of $1,2\text{-}C_2B_3H_5$	108
$B_5H_9 + C_2H_2$ discharge (G)	Parent	Parent	Parent	$1,2\text{-}C_2B_4H_6$	273, 274
$B_5H_9 + C_2H_2$ 490° flow system (G)	Parent	Parent	Parent		47
$B_5H_9 + (CH_3)_2SiH_2$ (G, 170°)	Parent	Parent			191
$B_5H_{11} + C_2H_2$ flash (G)	Parent 2-CH₃	Parent	Parent 1-CH₃ 3-CH₃,5-CH₃		108
$B_5H_{11} + HC_2CH_3$ flash (G)	Parent 2-CH₃	1-CH₃	Parent, 2-CH₃, 1,7-(CH₃)₂	Methyl derivs. of $1,2\text{-}C_2B_3H_5$	108

$2,3$-$C_2B_4H_8$ pyrolysis or discharge (G)	Parent	Parent	Parent	$1,2$-$C_2B_4H_6$	45, 222, 229
$2,3$-$C_2B_4H_8$ ultraviolet irradiation (G)	Parent	Parent		$1,2$-$C_2B_4H_6$	266, 290
$2,3$-$(CH_3)_2$-$C_3B_3H_5$ pyrolysis (G)			$1,7$-$(CH_3)_2$		106
2-CH_3-B-$(C_2H_5)_5$-2-CB_5H_3 pyrolysis (L)	$1,5$-$(CH_3)_2$-$2,3,4$-$(C_2H_5)_3$		$2,4$-$(CH_3)_2$-B-$(C_2H_5)_4$ (3 isomers)		94, 184
2-$Si(CH_3)_3$-$2,3$-$C_2B_4H_7$ pyrolysis (L, 220°)			2-$Si(CH_3)_3$		191
$2,3$-$C_2B_4H_8$ + $B(CH_3)_3$ (G)			B-methyl derivs.		267
$1,3$-$C_2B_7H_{13}$ pyrolysis (G, L)			Parent (trace)	$C_2B_6H_8$ $C_2B_7H_9$ $C_2B_8H_{10}$	54, 329
$C_2H_5B_2H_5$ + C_2H_2 (L)			$2,4$-$(CH_3)_2$-B-$(C_2H_5)_5$		188, 189
$(C_2H_5)_2BC\equiv CCH_3$ + $(C_2H_5)_2BH$ (L)	$1,5$-$(CH_3)_2$-$2,3,4$-$(C_2H_5)_3$ $(C_2H_5)_5$				185, 187
$(C_3H_7)_2BC\equiv CCH_3$ + $(C_3H_7)_4B_2H_4$ (L)	$1,5$-$(C_2H_5)_2$-$2,3,4$-$(C_3H_7)_3$				185, 187
$C_2H_5BF_2$ + Li, in tetrahydrofuran			$2,4$-$(CH_3)_2$-B-$(C_2H_5)_5$		184, 185
$(C_2H_5)_2BCl$ + Li, in tetrahydrofuran			$2,4$-$(CH_3)_2$-B-$(C_2H_5)_5$		185, 186
$C_2H_5B_2H_5$ pyrolysis (G)			2-CH_3-B-$(C_2H_5)_5$ $2,4$-$(CH_3)_2$-B-$(C_2H_5)_5$		185, 186
1-CH_3-B_5H_8 discharge or 590° pyrolysis (G)				CB_5H_7	223, 228
1-C_2H_5-B_5H_8 (G, 520°)				1-CH_3-CB_5H_6 $2,3$-$C_2B_4H_8$	228
B_5H_9 + carbon vapor (G)				CB_5H_7	248

[a] L = liquid phase or solution, G = gas phase.
[b] Only characterized carborane products are included.

only available data on mechanisms of carborane formation is of an indirect nature. A few consistencies have been noted, however (*106, 108*): (1) the identified organoboron products are exclusively carboranes; (2) alkene–borane reactions, with rare exceptions, do not produce carboranes (*18, 116, 159, 194, 201, 203*); (3) the same carborane cage isomers tend to form regardless of the borane or alkyne reactants (specific alkyl derivatives and yields vary considerably, however); (4) carborane yields are higher in reactions of methyl- and dimethylacetylene than in those of acetylene; (5) methyl- and dimethylacetylene tend to favor the formation of the smaller (C_2B_3 and C_2B_4) carborane systems at the expense of the $C_2B_5H_7$ system.

These observations, together with studies of the B_2H_6–C_2H_2 reaction in which the products were frozen out immediately following their formation in an electric discharge (*104*), suggest that the $C_2B_3H_5$ species are produced in early stages of the borane–alkyne reactions, but that $1,5$-$C_2B_3H_5$ reacts further under high-energy conditions unless it is immediately trapped out; alkyl derivatives of $1,2$-$C_2B_3H_5$, on the other hand are relatively stable once formed (*108*). The $C_2B_5H_7$ parent and alkylated species are evidently produced in later stages of the discharge or flash reactions from precursor species formed earlier. In summary, the formation of larger carborane cage systems in high-energy processes is favored by long reaction times, while shorter reaction times favor the smaller species.

A surprising aspect of borane–acetylene high-energy reactions is the abundance of methyl carboranes produced (Table 4-1). The distribution of methyl groups on the carborane cages is nearly random (some reactions yield all possible monomethyl derivatives of $2,4$-$C_2B_5H_7$) but with indications of a preference for attachment at boron rather than carbon. This suggests that during the reaction some acetylenic carbon–carbon bonds may be severed to give boron–carbon bonded fragments which lead eventually to B-alkylated carboranes, whereas other acetylene units are incorporated intact into carborane cage systems. The cages may then rearrange to the thermodynamically favored isomers, usually involving separation of the cage carbons. An apparent exception, however, is the $1,2$-$C_2B_3H_5$ system (see below), alkyl derivatives of which are stable and show no tendency to cage-isomerize under thermal excitation (*105*).

Conversion of *Nido*-Carboranes to *Closo*-Carboranes

When $2,3$-$C_2B_4H_8$ or its alkyl derivatives are subjected to pyrolysis, electric discharge, or ultraviolet radiation, the principal volatile products are polyhedral carboranes (Table 4-1). Since the yields are reasonable and $2,3$-$C_2B_4H_8$

is easily prepared from commercially available B_5H_9 and C_2H_2 (Chapter 3), this is a convenient laboratory route to small parent *closo*-carboranes. It is interesting that $2,4\text{-}C_2B_5H_7$ (a major product of the pyrolysis of $C_2B_4H_8$) is not obtained from the action of ultraviolet light on $2,3\text{-}C_2B_4H_8$, while the comparatively unstable $1,2\text{-}C_2B_4H_6$ is found (*266, 290*).

Polyhedral carboranes have also been obtained from other *nido*-carboranes. As indicated in Table 4-1, certain alkyl derivatives of $2,3,4\text{-}C_3B_3H_7$ and of $2\text{-}CB_5H_9$ yield alkylated *closo*-carboranes on pyrolysis. This type of *nido*-to *closo*-carborane conversion may not be completely general, however, since it has not been observed (*106*) in the pyrolysis of 2-methyl- and 2,4-dimethyl-2,3,4-tricarbahexaborane(7).

A different and potentially important approach to small *closo*-carborane synthesis is the closure of *nido*-carborane cages by insertion reactions. Thus, $2,3\text{-}C_2B_4H_8$ reacts with $B(CH_3)_3$ at $300°$ to yield several methyl derivatives of *closo*-$2,4\text{-}C_2B_5H_7$ (*267*). Similar insertions involving icosahedral fragments such as $C_2B_9H_{13}$ have been employed to reconstruct the icosahedral $C_2B_{10}H_{12}$ cage or to insert heteroatoms into the framework (Chapter 9).

Reactions of Alkylboranes with Alkynes

In this method, mole quantities of alkylboranes and alkynes are allowed to react in the liquid phase or in solution. Addition of acetylene at room temperature to liquid tetraethyldiborane, $(C_2H_5)_4B_2H_2$, gives a mixture of organoboron products; upon heating these in an autoclave at $200°$, the colorless liquid $1,5\text{-}(CH_3)_2\text{-}2,3,4\text{-}(C_2H_5)_3\text{-}1,5\text{-}C_2B_3$ is obtained in 10 to 15% yield, together with several noncarborane compounds (*188*). This process is assumed (*185*) to involve formation of diethylethynylborane initially, followed by hydroboration to 1,1,1-tris(diethylboryl)ethane. The latter compound presumably disproportionates, forming triethylborane and 1,5-dimethyl-2,3,4-triethyl-1,5-dicarba-*closo*-pentaborane(5):

$$H_5C_2 \diagdown \overset{H}{\underset{H}{\diagup B \diagup B \diagdown}} \diagup \overset{C_2H_5}{C_2H_5} \quad \xrightarrow{2HC\equiv CH} \quad 2(C_2H_5)_2BC\equiv CH \quad \xrightarrow[\text{hydroboration}]{(C_2H_5)_4B_2H_2}$$
$$+ 2H_2$$

$$[(C_2H_5)_2B]_3C\text{—}CH_3 \quad \longrightarrow \quad (CH_3)_2C_2B_3(C_2H_5)_3 + 3(C_2H_5)_3B$$

An analogous series of reactions starting with ethyldiborane and acetylene is reported (*189*) to yield 1,3,5,6,7-pentaethyl-2,4-dimethyl-dicarba-*closo*-heptaborane(7) $[(CH_3)_2C_2B_5(C_2H_5)_5]$ and other higher carboranes, but very little data have been published on these experiments.

Hydroboration of Alkynylboranes

Since dialkylalkynylboranes are assumed to be intermediates in the synthesis of carboranes from alkylboranes and alkynes (see above) the hydroboration of alkynylboranes might be expected to give carboranes directly. Accordingly, pentaethyl-1,5-dicarba-*closo*-pentaborane(5) is obtained in better than 50% yield in the reaction of diethyl-1-propynylborane with diethylborane (*185, 187*):

$$(C_2H_5)_2BC\equiv CCH_3 + 2(C_2H_5)_2BH \rightarrow C_2B_3(C_2H_5)_5 + \text{other organoboranes}$$

In an analogous manner, dipropyl-1-propynylborane and tetrapropyldiborane react to form the 1,5-diethyl-2,3,4-tripropyl derivative of $1,5\text{-}C_2B_3H_5$. The alkynylboranes used in these reactions are prepared from an alkynyl lithium and the trimethylamine adduct of a dialkylfluoroborane (*185*):

$$Li-C\equiv C-CH_3 + (C_2H_5)_2\overset{F}{B}:N(CH_3)_3 \longrightarrow$$

$$LiF + H_3C-C\equiv C-\overset{N(CH_3)_3}{\underset{\cdot\cdot}{B}}(C_2H_5)_2 \xrightarrow{(C_2H_5)_2O:BF_3}$$

$$(CH_3)_3N:BF_3 + (C_2H_5)_2O + (C_2H_5)_2B-C\equiv C-CH_3$$

Removal of the amine with boron trifluoride etherate is necessary before the carborane synthesis is attempted; in the presence of Lewis bases, the hydroboration of alkynylboranes does not yield *closo*-carboranes, instead giving derivatives of 2,3,4,5-tetracarba-*nido*-hexaborane(6), $C_4B_2H_6$.

Dehalogenation of Alkylhaloboranes with Alkali Metals

Peralkyl derivatives of $2,4\text{-}C_2B_5H_7$ may be prepared by the reaction of lithium metal with certain alkylhaloboranes under carefully controlled conditions. Dialkylchloroboranes, dialkylfluoroboranes, and alkyldifluoroboranes are dehalogenated by lithium in tetrahydrofuran at temperatures near 0°, forming a mixture of organoboron products from which *closo*-carboranes may be isolated by vapor phase chromatography (*184–186*). For example, 1,3,5,6,7-pentaethyl-2,4-dimethyl-dicarba-*closo*-pentaborane(7) $[(CH_3)_2C_2B_5(C_2H_5)_5]$ has been obtained from both $C_2H_5BF_2$ and $(C_2H_5)_2BCl$.

The total yield of volatile carboranes may be increased by heating triethylborane together with the solid residues from the above reactions (*185*). This treatment generates additional distillable alkyl derivatives of $1,5\text{-}C_2B_3H_5$ and $2,4\text{-}C_2B_5H_7$.

In addition to the polyhedral carboranes mentioned, the dehalogenation reactions yield less stable carborane compounds, some of which have been identified as alkyl derivatives of 2-carba-*nido*-hexaborane(9), CB_5H_9. Since many of these products are converted by heat into *closo*-carboranes, larger quantities of the latter may be obtained by simply heating the dehalogenation reaction mixture to 150° or 200° (*184, 185*). Mass spectroscopic analysis of the products is indicative of numerous unidentified carborane species containing up to seven boron atoms.

As is the case with most reactions in which small carboranes are formed, the mechanism of the dehalogenation process has not been established. Some evidence exists for the presence of free radicals; the addition of pyridine to the reaction mixture, for example, produces stable, identifiable radicals containing B—N bonds (*185*). It has been suggested (*185*) that the free radicals involved are of two types, with the unpaired electron at B and C locations, respectively, and that combination of these leads to B—C bond formation. Fragments formed in this manner then presumably rearrange to carboranes and other organoboron molecules.

Pyrolysis and Electric Discharge of Alkylboranes

Ethyldiborane at 170° yields a complex mixture of products from which at least one carborane derivative, $2\text{-}CH_3\text{-}1,3,5,6,7\text{-}(C_2H_5)_5\text{-}2,4\text{-}C_2B_5H$, has been isolated and identified (*186*). Other *closo*-carborane products are believed from mass spectroscopic analysis to be formed in the pyrolysis of several alkylboranes and alkyldiboranes (*185*), but these compounds have not been characterized.

The pyrolysis of 1-methylpentaborane(9), $CH_3B_5H_8$, at 590° in a fast flow system (*228*), and the reaction of the same compound in an electric discharge (*223*), yield the novel monocarbahexaborane CB_5H_7 (see below). Pyrolysis of 1-ethylpentaborane(9) at 520° produces a $CH_3CB_5H_6$ and other small carboranes (*228*).

Reaction of Boranes with Carbon Vapor

Pentaborane(9) undergoes an apparent insertion reaction with carbon vapor (*248*) to yield several products including CB_5H_7 (identical with the species obtained from 1-methypentaborane).

4-2. STRUCTURES AND PROPERTIES

$C_2B_3H_5$

Three isomers based on a trigonal bipyramid are possible, having carbon atoms in the 1,5, 1,2, and 2,3 positions, respectively. The only known *unsubstituted* isomer is $1,5$-$C_2B_3H_5$, in which the carbons occupy nonadjacent apex positions, despite the fact that most of the known reactions leading to this carborane involve acetylene or higher alkynes. The failure, thus far, to find any evidence for the parent 1,2- or $2,3$-$C_2B_3H_5$ may be an indication that these adjacent-carbon isomers are highly unstable with respect to rapid rearrangement to the 1,5- species. Molecular orbital calculations (*152*) indicate an order of decreasing thermodynamic stability $1,5 > 1,2 > 2,3$ for the three isomers.

$1,5$-$C_2B_3H_5$

The trigonal bipyramidal structure with apical carbons (Fig. 2-1a) is clearly indicated from ^{11}B and ^1H NMR studies supported by infrared and mass spectra (*273*). Very little is known of the chemistry of the highly volatile parent compound, but it is thermally stable below 150° (*222*) and fails to react with air, water, acetone, trimethylamine, or carbon dioxide at room temperature (*273*). In the presence of B_2D_6 the boron-bonded H atoms in $C_2B_3H_5$ are replaced by deuterium (*273*).

Several alkyl derivatives are known (Table 4-2), which in general have significantly greater thermal stability than the parent compound. For example, $1,5$-$(CH_3)_2$-$2,3,4$-$(C_2H_5)_3$-$1,5$-C_2B_3 is unreactive with air at 171° and is not oxidized by H_2O_2 at 100° (*188*) [in contrast, $1,5$-$C_2B_3H_5$ slowly decomposes to tan solids and H_2 at 150° *in vacuo* (*222*)].

$1,2$-$C_2B_3H_5$

The parent compound (Fig. 2-1b) is unknown, but a C,3-dimethyl derivative has been isolated in 10-mg quantity and characterized with reasonable certainty from ^{11}B and ^1H NMR, infrared spectra, mass spectra, and chemical studies (*103–105, 108*). In addition, tentative evidence exists for several other alkyl derivatives of this system (Table 4-2) (*105, 108*). The pyrolysis of C,3-$(CH_3)_2$-$1,2$-$C_2B_3H_3$ in both the liquid and vapor phase gave no evidence for a cage rearrangement to the 1,5 isomer (*105*), although higher alkyl derivatives of the original compound were formed (significantly, two different B-trimethyl-C-monomethyl derivatives of $C_2B_3H_5$ were obtained—a result accountable only in terms of nonequivalent cage carbon atoms). The available evidence suggests that the unknown and presumably unstable parent $1,2\text{-}C_2B_3H_5$ is stabilized with respect to cage rearrangement by attached alkyl groups. The nature of the stabilization is, however, open to question, and suggestions of both inductive (104) and steric effects (197, footnote 48) have been put forward.

$C_2B_4H_6$

The two possible octahedral isomers (Figs. 2-1c and 2-1d) are known in both parent and alkylated form, and octahedral structures have been assigned to both isomers on the basis of ^1H and ^{11}B NMR, infrared and mass spectra (274). In addition, it is relevant to note that an X-ray diffraction study has established the octahedral structure of the isoelectronic $B_6H_6^{2-}$ ion (255). At room temperature both 1,6 and $1,2\text{-}C_2B_4H_6$ are stable to air, water, ammonia, acetone, and trimethylamine (274). The boron-bonded H atoms in $1,6\text{-}C_2B_4H_6$ undergo H—D exchange with B_2D_6 at room temperature, although less rapidly than does $1,5\text{-}C_2B_3H_5$. The 1,6- isomer is thermally more stable than the 1,2- isomer, and at 250° the latter compound quantitatively rearranges to $1,6\text{-}C_2B_4H_6$ (222).

$C_2B_5H_7$

Of the four geometrically possible isomers based on a pentagonal bipyramid, only $2,4\text{-}C_2B_5H_7$ (Fig. 2-1f) is known. This molecule is remarkably resistant to thermal degradation, significantly more so than the $C_2B_3H_5$ and $C_2B_4H_6$ carboranes, and parent $C_2B_5H_7$ and its alkyl derivatives have frequently been obtained as major products of borane–alkyne discharge and flash reactions and pyrolyses of nido-carboranes (Table 4-1). The bipyramidal structure with the carbons in nonadjacent equatorial positions has been established by microwave spectroscopy (12, 13), ^1H and ^{11}B NMR, (104, 222, 225, 229), and infrared spectra (104, 225) (the ^{11}B NMR spectrum of unsubstituted $2,4\text{-}C_2B_5H_7$ contains three B—H doublets in a 1:2:2 ratio; unequivocal assignment of the two larger resonances was not possible until the spectra of several B-methyl (104) and B-deuterio (225) derivatives became available).

The exploration of the chemistry of this carborane has only begun, but preliminary indications are that it has a derivative chemistry resembling that

TABLE 4-2

DERIVATIVES OF THE SMALL *Closo*-CARBORANES

Compound	mp (°C)	bp (°C)	Other data[a]	References
1,5-$C_2B_3H_5$ derivatives				
Parent	−126.4	−3.7	IR, MS, B, H, VP	104, 222, 273
2-CH_3			IR, MS, B, H	104
$(C_2H_5)_5$	−61.5	84–86 (9 mm)	—	187
1,5-$(CH_3)_2$-2,3,4-$(C_2H_5)_3$	−84	58 (11 mm)	B, H, Raman	188
1,2-$C_2B_3H_5$ derivatives				
C,3-$(CH_3)_2$			IR, MS, B, H	103, 104
C,5-$(CH_3)_2$			IR, MS	105
3,4,5-$(CH_3)_3$			IR, MS	105
C,3,4,5-$(CH_3)_4$			IR, MS	105
C′,3,4,5-$(CH_3)_4$			IR, MS	105
1,6-$C_2B_4H_6$ derivatives				
Parent			IR, MS, B, H, VP	104, 222, 229, 274
2-CH_3			IR, MS, B, H	104
1-CH_3			IR, MS	108
1,6-$(CH_3)_2$	−2		MS, B, H	229
1-n-C_3H_7	−88		MS, B, H	229

	m.p.	b.p.	Spectroscopic data	References
1,2-$C_2B_4H_6$ derivatives				
Parent			IR, MS, B, H	222, 274
2,4-$C_2B_5H_7$ derivatives				
Parent			IR, MS, B, H	104, 222, 225, 229
			Microwave, dipole moment	12, 13, 225
1-CH_3			IR, MS, B, H	104
2-CH_3			IR, MS, B, VP	106, 219
3-CH_3			IR, MS, B, H	104
5-CH_3			IR, MS, B, H	104
1,7-$(CH_3)_2$	−30 to −28		IR, MS	106
2,4-$(CH_3)_2$			B, H, IR, VP, MS	219, 229
1-C_2H_5			IR, MS, H	108
1-n-C_3H_7	−99 to −97		B, H	229
2-Br			IR, MS, B	219
5-Br			IR, MS, B	219
2-$(CH_3)_3Si$	87		IR, MS	219
2,4-$(CH_3)_2$-B-$(C_2H_5)_5$		65 (10^{-3} mm)	IR, MS, B, H	186
2-CH_3-B-$(C_2H_5)_5$			IR, MS, B, H	186
B-$(CH_3)_5$			IR, MS, B, H	267
CB_5H_7 derivatives				
Parent		26 (503 mm)	IR, MS, B, H	223, 228

ᵃ IR = infrared spectrum or band positions; MS = mass spectrum (partial or complete) or cutoff m/e value; B = boron-11 NMR data; H = proton NMR data; VP = vapor pressure data.

of the $C_2B_{10}H_{12}$ icosahedral carboranes. Thus, reaction with *n*-butyllithium yields the dilithio derivative, which in turn may be used as a precursor to other derivatives in which the ligands are bonded to the cage carbon atoms (*219*).

$$C_2B_5H_7 + 2C_4H_9Li \rightarrow Li_2C_2B_5H_5 + 2C_4H_{10}$$
$$Li_2C_2B_5H_5 + 2CH_3I \rightarrow (CH_3)_2C_2B_5H_5 + 2LiI$$

The reaction with *n*-butyllithium is sluggish in ether or ether–pentane (extremely so in pentane itself), indicating a marked decrease in the electron-acceptor properties of this system as compared with 1,2- and 1,7-$C_2B_{10}H_{12}$ (Chapters 6 and 7). The C,C'-dilithio derivative reacts as expected to give the lithium salt of the dicarboxylic acid, but subsequent acidification with aqueous HCl results in complete destruction of the carborane cage (parent $C_2B_5H_7$ is unaffected by this reagent).

The parent carborane reacts with Br_2 in CS_2 over $AlBr_3$ to yield a single monobromo derivative, shown by NMR and other evidence to be 5-Br-2,4-$C_2B_5H_6$ (*219*).

$$Li_2C_2B_5H_5 \xrightarrow[-LiX]{RX} 2\text{-R-4-}LiC_2B_5H_5 \xrightarrow[-LiCl]{HCl} 2\text{-}RC_2B_5H_6$$

$$R = CH_3, X = I$$
$$R = Si(CH_3)_3, X = Cl$$

$$Li_2C_2B_5H_5 \xrightarrow{2CO_2} (LiOOC)_2C_2B_5H_5 \xrightarrow[H_2O]{5HCl} 5H_2 + \text{boric acid}$$

$$C_2B_5H_7 \xrightarrow[AlBr_3]{Br_2} C_2B_5H_6Br$$

CB_5H_7

The structure of this molecule, the only known polyhedral borane containing a bridge hydrogen atom, is unique at this writing. From mass spectra, infrared spectra, and NMR studies, a distorted octahedral geometry (Fig. 2-1e) has been proposed (*223*). The molecule is isoelectronic with the octahedral $B_6H_6^{2-}$ and $C_2B_4H_6$ species, and is formally derived from the latter by the replacement of one carbon atom with a BH unit. Reaction with sodium hydride produces one mole of H_2, forming the $CB_5H_6^-$ ion. Protonation of the latter species with HCl regenerates the original carborane (*248*).

$$CB_5H_7 + NaH \xrightarrow{-H_2} Na^+CB_5H_6^- \xrightarrow{HCl} CB_5H_7 + NaCl$$

The Intermediate *Closo*-Carboranes: $C_2B_6H_8$, $C_2B_7H_9$, $C_2B_8H_{10}$, and $C_2B_9H_{11}$

The reactions that have been utilized in the preparation of small *closo*-carboranes, described in the preceding chapter, have not in general produced carboranes larger than $C_2B_5H_7$ in appreciable yield. Small amounts of $C_2B_6H_8$ and $C_2B_8H_{10}$ have been obtained in reactions of boranes with alkynes, but an entirely different approach based upon the readily available $1,2\text{-}C_2B_{10}H_{12}$ (*o*-carborane) has given bench-scale quantities of $C_2B_6H_8$, $C_2B_7H_9$, $C_2B_8H_{10}$, and $C_2B_9H_{11}$. Opening the way to this synthetic route was the key discovery that the $1,2\text{-}C_2B_{10}H_{12}$ icosahedron, despite its general chemical stability, may be partially degraded by strong bases to give the open-faced $C_2B_9H_{12}^-$ ion; subsequent reactions of this species ultimately lead to all of the intermediate *closo*-carboranes. In this manner the advantages of the relatively high availability of $1,2\text{-}C_2B_{10}H_{12}$ have been extended to all of the known carborane systems having six or more boron atoms.

5-1. SYNTHESIS

Reactions of Boranes with Alkynes

Just two instances of the preparation of intermediate *closo*-carboranes directly from boron hydrides have been reported. The C,C'-dimethyl derivative of $1,7\text{-}C_2B_6H_8$ has been obtained in 12% yield from the reaction of hexaborane(10), B_6H_{10}, with dimethylacetylene (parent $C_2B_6H_8$ is formed from B_6H_{10} and C_2H_2 in extremely small yield) (*350*). Second, small amounts of $1,6\text{-}C_2B_8H_{10}$ and a $B\text{-}CH_3\text{-}1,6\text{-}C_2B_8H_9$ have been isolated from the products of flash reactions of B_4H_{10} and B_5H_{11} with acetylene (*106, 108*).

45

The use of boranes as precursors to intermediate-size carboranes may increase as studies of the relatively new octaboranes and nonaboranes progress. Until very recently, B_9H_{15} was the only identified boron hydride having seven to nine boron atoms and little was known of its chemistry. However, recent investigations of this hydride (*38, 48, 52*) and the isolation of iso-B_9H_{15} (*51, 52*), B_8H_{12} (*48, 53, 64, 65*), B_8H_{14} (*53*), and B_8H_{18} (*50, 324*) suggest new routes to the carboranes based on these compounds.

Degradation of Icosahedral Carboranes

As is described in detail in Chapter 9, both 1,2- and 1,7-$C_2B_{10}H_{12}$ (and their C-alkyl derivatives) are attacked by strong Lewis bases to yield, respectively, the (3)-1,2-$C_2B_9H_{12}^-$ and (3)-1,7-$C_2B_9H_{12}^-$ ions*:

$$C_2B_{10}H_{12} + CH_3O^- + 2CH_3OH \rightarrow C_2B_9H_{12}^- + H_2 + B(OCH_3)_3.$$

The $C_2B_9H_{12}^-$ isomers are icosahedral fragments with the "extra" hydrogen atom presumably occupying the open face (Fig. 9-1). Protonation of either isomeric ion gives the corresponding dicarba-*nido*-undecaborane(13) species (*348*), either of which on pyrolysis yields the *closo*-carborane 1,8-$C_2B_9H_{11}$ (*84, 327, 329*).

$$\text{(3)-1,2-}C_2B_9H_{12}^- \xrightarrow{H^+} \text{(3)-1,2-}C_2B_9H_{13} \searrow^{\Delta}$$
$$\qquad\qquad\qquad\qquad\qquad\qquad\qquad\qquad 1,8\text{-}C_2B_9H_{11}$$
$$\text{(3)-1,7-}C_2B_9H_{12}^- \xrightarrow{H^+} \text{(3)-1,7-}C_2B_9H_{13} \nearrow_{\Delta}$$

Degradation of 1,8-$C_2B_9H_{11}$ with oxidizing agents produces the open-cage structure 1,3-$C_2B_7H_{13}$, a novel organoborane containing two bridging methylene groups (Fig. 2-2f) (*81, 328, 329, 342*). An improved synthesis of this compound utilizes the chromic acid oxidation of the 1,7-$C_2B_9H_{12}^-$ ion (Fig. 5-1), which probably proceeds with a loss of six electrons as shown (*81*).

$$\text{(3)-1,7-}C_2B_9H_{12}^- + 6H_2O \rightarrow C_2B_7H_{13} + 2B(OH)_3 + 5H^+ + 6e^-$$

---

* We follow here the suggestion of Hawthorne (*137*) that these icosahedral fragment species be numbered in the same manner as the complete icosahedron, with the location of the "missing" atom in parentheses.

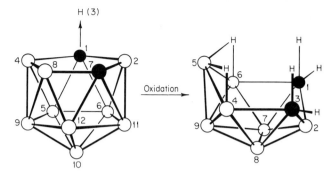

FIG. 5-1. Structures of the (3)-$1,7$-$C_2B_9H_{12}^-$ ion (left) and $1,3$-$C_2B_7H_{13}$ (right). Solid circles indicate carbon and open circles, boron (81).

The pyrolysis of $1,3$-$C_2B_7H_{13}$ *in vacuo* or in diphenyl ether yields $1,7$-$C_2B_6H_8$, $1,7$-$C_2B_7H_9$, $1,6$-$C_2B_8H_{10}$, and traces of $2,4$-$C_2B_5H_7$ (54, 82). Similar treatment of monosubstituted and disubstituted C-methyl and C-phenyl derivatives of $C_2B_7H_{13}$ generally gives the corresponding *closo*-carborane derivatives (82, 329, 330). Since the $C_2B_6H_8$ and $C_2B_8H_{10}$ products are formed in nearly equal amounts, whereas $C_2B_7H_9$ is produced in relatively low yield, it has been suggested (82) that a disproportionation of the $C_2B_7H_{13}$ starting material is the predominant reaction. Addition of B_2H_6 during the pyrolysis increases the yield of $C_2B_8H_{10}$ considerably, at the same time reducing that of $C_2B_6H_8$ and $C_2B_7H_9$ to nearly zero (54, 82). In contrast, good yields of $C_2B_6H_8$ are obtained in the pyrolysis of pure $C_2B_7H_{13}$ *in vacuo* (54, 82). Not surprisingly, the slow reaction of $1,7$-$C_2B_6H_8$ with B_2H_6 at room temperature yields $1,7$-$C_2B_7H_9$ (30%) and $1,6$-$C_2B_8H_{10}$ (10%), suggesting that the initial carborane product of $1,3$-$C_2B_7H_{13}$ pyrolysis is $1,7$-$C_2B_6H_8$ (54). The diagram below summarizes these observations.

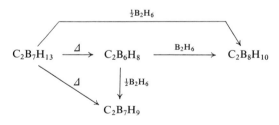

An alternative route to $1,7$-$C_2B_7H_9$ involves the conversion of $1,3$-$C_2B_7H_{13}$ to its mono- or disodium salt (128) followed by pyrolysis in diphenyl ether at $200°$ (82).

$$C_2B_7H_{13} + 2NaH \xrightarrow{-H_2} 2Na^+C_2B_7H_{11}^{2-} \xrightarrow{\Delta} 1,7\text{-}C_2B_7H_9$$

5-2. STRUCTURES AND PROPERTIES

$C_2B_6H_8$

The structure of the only known isomer has been shown by an X-ray diffraction study (*117, 118*) of the C,C′-dimethyl derivative to consist of a dodecahedron slightly distorted from D_{2d} symmetry, with the carbon atoms at positions 1 and 7 (Fig. 2-1g). This geometry is essentially that suggested earlier on the basis of the ^{11}B NMR spectrum (*14, 350*), which, however, requires an assumption of coincidental overlap of two nonequivalent boron resonances. The carbon atoms are located as predicted by molecular orbital theory for maximum stability in a D_{2d} dodecahedron, although the possible existence of less stable isomers is not precluded (*117, 118*). It may be noted that the related boron cage molecules $B_8H_8^{2-}$ (*170*) and B_8Cl_8 (*164*) have similar D_{2d} dodecahedral geometry.

Other than the reaction with B_2H_6 to give 1,7-$C_2B_7H_9$, mentioned above, the only chemistry to be reported for $C_2B_6H_8$ concerns its reaction with BH_4^- ion in diglyme at 100°, which yields ionic products of unknown nature. Treatment of these solids with dry HCl produces *nido*-1-CB_5H_9 (Chapter 3) and its 1-methyl and 3-methyl derivatives (*55*).

$C_2B_7H_9$

The structure originally predicted from ^1H and ^{11}B NMR spectra and later confirmed by the X-ray method (*181*) is that of a tricapped trigonal prism (Fig. 2-1h) with the carbon atoms occupying two of the apices. No other isomers are known. Very recently, electrophilic substitution reactions have been shown to occur preferentially at B(9), and the methyl, ethyl, bromo, and tetrabromo derivatives, have been prepared (*54b*).

$C_2B_8H_{10}$

The expected geometry is that of the isoelectronic $B_{10}H_{10}^{2-}$ ion, which has been established (*49, 198*) as a bicapped square antiprism. The ^{11}B and ^1H NMR spectra (*329*) of the C,C′-$(CH_3)_2C_2B_8H_8$ isomer obtained by pyrolysis of $(CH_3)_2C_2B_7H_{11}$ are in agreement with this model and suggest that the framework carbon atoms occupy one apical and one equatorial position (Fig. 2-1i). Since an adjacent-carbon structure is unlikely, the carbon atoms are presumed to occupy the 1,6 locations.

On heating the C,C'-dimethyl derivative of $1,6\text{-}C_2B_8H_{10}$ at 300–350° in diphenyl ether, a second isomer is obtained in nearly quantitative yield (*82, 329*). This carborane has been identified as C,C'-dimethyl-$1,10\text{-}C_2B_8H_{10}$ from its ^{1}H and ^{11}B NMR spectra, which indicate the equivalence of the two cage carbon atoms and of the eight boron atoms (Fig. 2-1j).

The reaction of $1,6\text{-}(CH_3)_2\text{-}1,6\text{-}C_2B_8H_8$ with B_2H_6 at 225° gives $1,7\text{-}(CH_3)_2\text{-}1,7\text{-}C_2B_{10}H_{10}$ (C,C'-dimethyl-*m*-carborane) in 27% yield (*329*), a result of

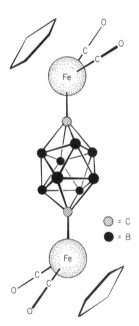

= C
= B

FIG. 5-2. Proposed structure of 1,10-$[(\pi\text{-}C_5H_5)Fe(CO)_2]_2\text{-}1,10\text{-}(\sigma\text{-}C_2B_8H_8)$ (*277*).

some interest as a route to the *m*-carborane system which does not involve thermal rearrangement of *o*-carborane. Treatment of $1,10\text{-}C_2B_8H_{10}$ with excess butyllithium produces $1,10\text{-}Li_2C_2B_8H_8$, which in turn reacts with $(\pi\text{-}C_5B_5)Fe(CO)_2I$ in diethyl ether to yield an interesting σ-bonded complex (*277*) whose proposed structure is indicated in Fig. 5-2.

In general, the chemistry of 1,6- and $1,10\text{-}C_2B_8H_{10}$ resembles that of 1,7- and $1,12\text{-}C_2B_{10}H_{12}$ (*m*- and *p*-carborane), respectively. Thus, $1,6\text{-}C_2B_8H_{10}$ is easily degraded by ethanolic KOH or piperidine, producing a $C_2B_7H_{12}^-$ ion, while $1,10\text{-}C_2B_8H_{10}$ is highly resistant to degradation by basic reagents (*83*). Lithiation of both the 1,6 and 1,10 isomers by *n*-butyllithium proceeds normally to give mono- and dilithio derivatives (monolithiation of $1,6\text{-}C_2B_8H_{10}$

results in a 2:1 mixture of the 1-lithio and 6-lithio species). The lithio derivatives of both 1,6- and $1,10\text{-}C_2B_8H_{10}$ react as expected with methyl iodide, forming C-methylated products, and with carbon dioxide, yielding the mono- or dicarboxylic acid.

$$1,10\text{-}C_2B_8H_{10} \xrightarrow{n\text{-}C_4H_9Li} LiC_2B_8H_9 \xrightarrow{n\text{-}C_4H_9Li} Li_2C_2B_8H_8$$

$$\begin{array}{cc} & \begin{array}{c} 1.\ CO_2 \\ 2.\ H_3O^+ \end{array} & \Bigg| CH_3I \\ (HOOC)_2C_2B_8H_8 & (CH_3)_2C_2B_8H_8 \end{array}$$

The pK_a values of the C-monomethyl- and C-monophenyl carboxylic acids, and of the C,C'-dicarboxylic acid, have been determined in 50% aqueous ethanol as 4.2, 4.1, and 3.8, respectively (the latter figure apparently corresponds to ionization of the second proton).

The reaction of $1\text{-}Li\text{-}10\text{-}C_6H_5\text{-}1,10\text{-}C_2B_8H_8$ with N_2O_4 yields a nitro derivative which may be reduced to the 1-amino-10-phenyl compound. Similarly, $1\text{-}Li\text{-}10\text{-}CH_3\text{-}1,10\text{-}C_2B_8H_8$ is readily converted to $1\text{-}I\text{-}10\text{-}CH_3C_2B_8H_8$ by iodine in ether.

Direct chlorination of $1,10\text{-}C_2B_8H_{10}$ in CCl_4 at $25°$ produces the B-octachloro derivative, but complete chlorination is effected by treatment of $1,10\text{-}(CH_3)_2\text{-}1, 10\text{-}C_2B_8H_8$ with a stream of Cl_2 in refluxing CCl_4 under ultraviolet light (*83*).

$C_2B_9H_{11}$

The single known isomer of this carborane was originally postulated (*14*) from its NMR spectra and degradation products, to have the polyhedral structure indicated in Fig. 2-1k. A later X-ray diffraction study (*336*) confirmed this geometry, in which the carbons are in the 1,8 positions. Since this polyhedral system is obtained by pyrolysis of either (3)-1,2- or $(3)\text{-}1,7\text{-}C_2B_9H_{13}$ (see above), a rearrangement of the cage is clearly required in the case of the (3)-1,2 isomer in order to separate the carbon atoms.

Several aspects of the chemistry of $C_2B_9H_{11}$ have been explored, one of which is the oxidative degradation to the $C_2B_7H_{13}$ system described earlier. Another type of reaction, so far unique among the carboranes, involves the behavior of $C_2B_9H_{11}$ or an alkyl derivative as a Lewis acid in the reversible formation of acid–base adducts (*329*).

$$C_2B_9H_{11} + :L \rightleftharpoons C_2B_9H_{11}L$$

L = triethylamine, triphenylphosphine, hydroxide ion, ethyl isocyanide

These adducts are isoelectronic with the $(3)\text{-}1,7\text{-}C_2B_9H_{12}^-$ ion and are probably of similar structure.* In this case the attacking base must open the polyhedral $C_2B_9H_{11}$ cage, very likely at $B(4)$, which is uniquely bonded to both carbon atoms and is a reasonable site for nucleophilic attack (329).

The reaction of $1,8\text{-}C_2B_9H_{11}$ with methyllithium in ether-hexane generates an ion formulated as $C_2B_9H_{11}CH_3{}^-$ (129, 230a), which is apparently a methyl derivative of the icosahedral-fragment $(3)\text{-}1,7\text{-}C_2B_9H_{12}^-$ ion described earlier. Indeed, the latter ion itself is produced in good yield by the action of borohydride ion on $1,8\text{-}C_2B_9H_{11}$ (129, 230a). NMR studies indicate that the

$$1,8\text{-}C_2B_9H_{11} \xrightarrow{\text{CH}_3\text{Li}} C_2B_9H_{11}CH_3{}^-$$

$$1,8\text{-}C_2B_9H_{11} \xrightarrow{\text{NaBH}_4} (3)\text{-}1,7\text{-}C_2B_9H_{12}{}^-$$

methyl group is attached at $B(4)$; a number of other 4-substituted derivatives of the $(3)\text{-}1,2\text{-}C_2B_9H_{12}^-$ ion have been prepared in analogous reactions (230a).

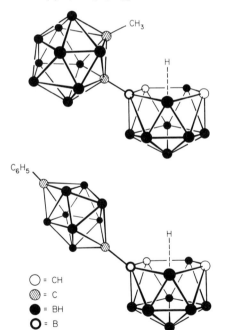

$\bigcirc$ = CH
$\oslash$ = C
$\bullet$ = BH
$\mathbf{O}$ = B

FIG. 5-3. Proposed structures of the $C_2B_9H_{11}C_2B_{10}H_{10}CH_3{}^-$ ion (top) and the $C_2B_9H_{11}C_2B_8H_8C_6H_5{}^-$ ion (bottom) (129).

* Species formulated as $C_2B_9H_{11}L$, where L is a Lewis base, are formed in reactions of $C_2B_9H_{12}^-$ salts with Lewis bases in the presence of $FeCl_3$ (357) (Section 9-1). The possibility that these adducts may be identical with those obtained from $1,8\text{-}C_2B_9H_{11}$ does not seem to have been discussed.

TABLE 5-1

DERIVATIVES OF THE INTERMEDIATE *Closo*-CARBORANES

Compound	mp (°C)	bp (°C)	Other data[a]	References
1,7-$C_2B_6H_8$ derivatives				
Parent			IR, H	*82*
1-CH_3			IR	*82*
1,7-$(CH_3)_2$	−40	63 (134 mm)	IR, MS, B, H	*329, 330*
	−63 to −58		MS, B	*350*
1-C_6H_5			IR, H	*82*
1,7-$C_2B_7H_9$ derivatives[b]				
Parent			IR, H	*82*
1-CH_3			IR, H	*82*
1-C_6H_5			IR, H	*82*
1,7-$(CH_3)_2$	−22		IR, MS, B, H	*329, 330*
1,6-$C_2B_8H_{10}$ derivatives				
Parent			IR, MS, H	*82, 106*
1-CH_3			IR, H	*82*
6-CH_3			H	*82*
B-CH_3			IR, MS	*106*
1-C_6H_5			IR, H	*82*
1,6-$(CH_3)_2$	1.0 to 1.6	73 (32 mm)	IR, MS, B, H	*329, 330*
1,10-$C_2B_8H_{10}$ derivatives				
Parent			IR, H	*82*
1-CH_3			IR, H	*82*
1-C_6H_5			IR, H	*82*
1,10-$(CH_3)_2$	27		IR, MS, B, H	*329, 330*
B-Cl_8			MS, B, H	*83*
1,10-$(CCl_3)_2$-B-Cl_8			B	*83*
1,8-$C_2B_9H_{11}$ derivatives				
Parent	212		IR, MS, B, H	*327, 329*
1-CH_3	84		IR, MS	*327, 329*
1-C_6H_5	37		IR, MS	*327, 329*
1-p-C_6H_4Br	101		MS	*329*
1,8-$(CH_3)_2$	57		IR, MS	*327, 329*

[a] IR = infrared spectrum or band positions; MS = mass spectrum (partial or complete) or cutoff m/e value; B = boron-11 NMR data; H = proton NMR data; VP = vapor pressure data.

[b] For additional data on $C_2B_7H_9$ derivatives, see reference (*54b*)

When $1,8\text{-}C_2B_9H_{11}$ is allowed to react with 1-lithio-2-methyl-o-carborane in ether, a linked-cage ion is obtained; a similar species is produced from $1,8\text{-}C_2B_9H_{11}$ and the 1-lithio-10-phenyl derivative of $1,10\text{-}C_2B_8H_{10}$ (*129, 230a*).

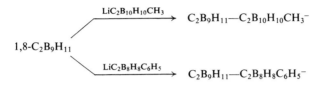

$$1,8\text{-}C_2B_9H_{11} \xrightarrow{\text{LiC}_2B_{10}H_{10}CH_3} C_2B_9H_{11}\text{—}C_2B_{10}H_{10}CH_3^-$$

$$\xrightarrow{\text{LiC}_2B_8H_8C_6H_5} C_2B_9H_{11}\text{—}C_2B_8H_8C_6H_5^-$$

The proposed structures, shown in Fig. 5-3, illustrate the versatility inherent in the chemistry of the polyhedral carboranes and underline the fact that the exploration of this field has barely started.

Characterized derivatives of the $C_2B_6H_8$, $C_2B_7H_9$, $C_2B_8H_{10}$, and $C_2B_9H_{11}$ polyhedral systems are listed in Table 5-1.

6

The 1,2-$C_2B_{10}H_{12}$ (*o*-Carborane) System

6-1. SYNTHESIS, STRUCTURE, AND PROPERTIES

Synthesis

Decaborane (14) (Fig. 1-1b) is attacked by Lewis bases such as alkylamines, alkylsulfides, and acetonitrile, forming bis(ligand) derivatives of general formula $B_{10}H_{12}L_2$, which in turn react with acetylene to give the white crystalline *o*-carborane in high yield* (*66, 143, 456, 457*):

$$B_{10}H_{12}(CH_3CN)_2 + HC{\equiv}CH \longrightarrow \underset{B_{10}H_{10}}{HC{-}CH} + 2CH_3CN + H_2$$

The use of substituted acetylenes frequently leads to the corresponding C-substituted carborane derivative. Thus, isopropenylacetylene produces 1-isopropenyl-*o*-carborane:

$$B_{10}H_{12}(CH_3CN)_2 + \underset{CH_3}{CH_2{=}C{-}}C{\equiv}CH \longrightarrow \underset{CH_3\ \ \ B_{10}H_{10}}{CH_2{=}C{-}C{-}CH} + 2CH_3CN + H_2$$

Analogous reactions have been reported with alkyl, haloalkyl, acetoxymethyl-, alkenyl-, alkynyl-, carbethoxy-, and arylacetylenes, as described in this chapter. Acetylenes containing hydroxy groups, however, destroy the decaborane cage and fail to give carborane products at all, so that *o*-carborane derivatives

---

* The symbol $\underset{B_{10}H_{10}}{HC{-}CH}$ is commonly used in the literature to represent the icosahedral 1,2-$C_2B_{10}H_{12}$ isomer, whereas $HCB_{10}H_{10}CH$ and $\overline{HCB_{10}H_{10}CH}$ designate 1,7-$C_2B_{10}H_{12}$ and 1,12-$C_2B_{10}H_{12}$, respectively.

containing such functional groups as $-CH_2OH$ and $-COOH$ must be prepared via other routes (see below). Since isolation of the bis(ligand) decaborane species is usually unnecessary, the carborane preparation may be carried out between $B_{10}H_{14}$ and the alkyne directly in the presence of the Lewis base.

In analogous fashion, diacetylene reacts with $B_{10}H_{12}L_2$ species to yield 1,1'-bis-o-carboranyl and a lesser amount of 1-ethynyl-o-carborane (56–58).

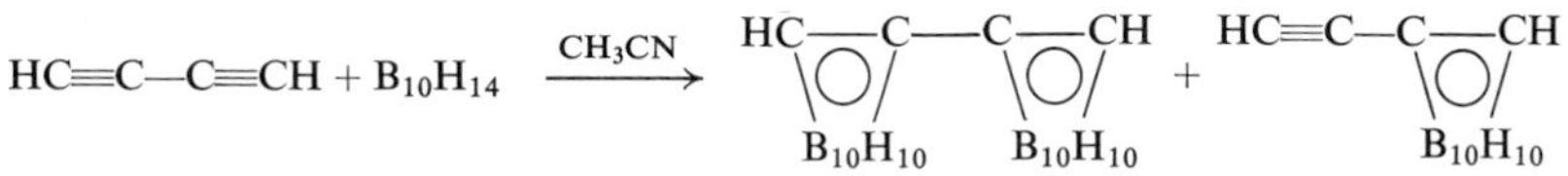

$$HC{\equiv}C{-}C{\equiv}CH + B_{10}H_{14} \xrightarrow{\text{CH}_3\text{CN}} \underset{B_{10}H_{10}}{HC{-}C{-}C{-}CH} \quad \underset{B_{10}H_{10}}{C{-}CH} + \underset{B_{10}H_{10}}{HC{\equiv}C{-}C{-}CH}$$

Structure

Despite an early incorrect X-ray investigation (457) and an inconclusive electron diffraction study (339), the icosahedral geometry of $1,2$-$C_2B_{10}H_{12}$ (Fig. 2-11) has been established beyond question by X-ray diffraction (10, 241–245, 321, 341) and ^{11}B NMR evidence (337). Two additional icosahedral isomers, 1,7- and $1,12$-$C_2B_{10}H_{12}$ (m- and p-carborane) are possible and both are known. The chemistry of these isomers differs significantly in some respects from that of the $1,2$-$C_2B_{10}H_{12}$ system and is accordingly dealt with separately in Chapter 7.

Thermal Stability and Rearrangement

Below 400° $1,2$-$C_2B_{10}H_{12}$ is unaffected by heat, but at 400° to 500° in an inert atmosphere it rearranges quantitatively to the 1,7 isomer (92, 93, 234, 314, 315, 403). The latter compound decomposes near 620° with formation of $1,12$-$C_2B_{10}H_{12}$ (233, 234) (the mechanisms of these isomerizations are discussed in Chapter 7). Substituted derivatives of $1,2$-$C_2B_{10}H_{12}$ undergo analogous rearrangements; thus, 1-CH_3-$1,2$-$C_2B_{10}H_{11}$ is converted to 1-CH_3-$1,7$-$C_2B_{10}H_{11}$ at 400° (92), and $1,2$-$[(CH_3)_3Si]_2$-$1,2$-$C_2B_{10}H_{10}$ isomerizes to $1,7$-$[(CH_3)_3Si]_2$-$1,7$-$C_2B_{10}H_{10}$ at 300° (252).

Chemical Properties

The o-carborane cage exhibits a resistance to degradation that is extraordinary when compared to the binary boron hydrides and conventional

TABLE 6-1

SELECTED o-CARBORANE DERIVATIVES

Compounds are classified according to groups bonded directly to the carborane cage or linked to the cage via methylene units (e.g., 1-chloro-o-carborane is listed as a halo derivative and 1-p-chlorophenyl-o-carborane is classified as an aryl derivative). Within each category, compounds are listed in the order of increasing number of substituents on the cage. Only the more common and structurally simple derivatives are tabulated here; for data on other compounds consult the appropriate references cited in the text.

Substituent groups	mp ($^\circ$C)	Other data[a]	References
Parent	294.5–295.5		*233, 455, 457*
		IR	*7, 92, 193, 260*
		B	*260, 264*
		MS	*92*
		DM	*62, 190, 205*
		Raman	*37*
Alkyl derivatives			
1-CH_3	218–219		*91*
	217–218		*455*
1-C_2H_5	38.5–39.5		*454*
	52–53		*455*
		bp 75–80 (0.5 mm)	*69*
1-n-C_3H_7	68–69		*69, 455*
1-$CH(CH_3)_2$	−4	bp 100 (4 mm), IR, UV	*66*
	−14.5	bp 98–100 (1–1.5 mm)	*455*
	−16.9	bp 98–99 (2 mm)	*320*
1-n-C_4H_9	11.2–11.3	bp 105–106 (1–1.5 mm)	*320, 454, 455*
1-n-C_6H_{13}		bp 101–102 (0.5 mm)	*69, 143*
1-$CH_2CH_2CH(CH_3)_2$	33		*143*
B-n-C_3H_7	41–44	bp 78–81 (1 mm)	*209*
B-n-C_4H_9	22–23	bp 104–105 (1 mm)	*209*
3-C_2H_5	15–17	H, B	*135*
1,2-$(CH_3)_2$	265		*281*
	262–263		*364*
1,2-$[CH(CH_3)_2]_2$	104–105		*69*
1-CH_3-2-n-C_4H_9	100–101		*425*
Haloalkyl derivatives			
1-CH_2Cl	91.5		*364, 455*
		IR	*365*
	83–85		*69*
1-CH_2Br	47–49		*66*
	31.5	IR	*365*
1-CH_2I		bp 100–110 (2 mm)	*364*
1-$(CH_2)_2Cl$	82	IR	*365*

(*continued*)

TABLE 6-1—Continued

Substituent groups	mp (°C)	Other data[a]	References
1-$(CH_2)_2Br$	110	IR	365
1-$(CH_2)_2I$	130.5–131		364, 455
1-CH_3-2-CH_2Br	125–127		69, 265
1,2-$(CH_2Cl)_2$	114–115		69, 457
	119–120		143
	113–114	IR	365
1,2-$(CH_2Br)_2$	68–69.5		66
	67–67.5	IR	365
Aryl derivatives			
1-C_6H_5	69.5–70		69
		IR, Raman	300
1-o-$C_6H_4C_2H_5$	61.5		364
3-C_6H_5	109–110	H, B	135
3-m-$C_6H_4CH_3$	64–65		419
1-$CH_2C_6H_5$	57–58		310
1,2-$(CH_2C_6H_5)_2$	151–152		425
1-C_6H_5-2-n-C_3H_7	96–97		425
1-C_6H_5-2-$CH_2C_6H_5$	84.5–86		364
	98–99		425
1,2-$(C_6H_5)_2$	148–149		66, 320
1,3-$(C_6H_5)_2$	105–106	H, B	135
3,6-$(C_6H_5)_2$	173.5–174.5	H, B	135
1,2-$(CH_3)_2$-3-C_6H_5	158–159	H, B	135
1,2-$(CH_3)_2$-3,6-$(C_6H_5)_2$	214–216	H, B	135
1-m-C_6H_4F	68–68.5		127
1-p-C_6H_4F	138–139		127
1-o-C_6H_4Cl	87–87.5	UV	388
1-p-C_6H_4Cl	142–142.5	UV	382, 388
1-o-C_6H_4Br	104–104.5	UV	388
1-m-C_6H_4Br	107–108		127
1-p-C_6H_4Br	134.5–134.7		455
	132–132.5	UV	382, 388
3-p-C_6H_4Br	123–124		419
1-o-C_6H_4I	105–106	UV	388
1-p-C_6H_4I	112.5–113	UV	382, 388
1-o-C_6H_4COOH	201–202		410
1-o-C_6H_4COCl	72–74		410
1-o-$C_6H_4COOC_2H_5$	64–65		410
1-m-C_6H_4COOH	280–281	pK_a (Table 6-2)	127
1-p-C_6H_4COOH	224–225	pK_a (Table 6-2)	127
3-m-C_6H_4COOH	241	pK_a (Table 6-2)	420
3-p-C_6H_4COOH	232	pK_a (Table 6-2)	420
1-o-$C_6H_4NH_2$	147–148		410
1-m-$C_6H_4NH_2$	81–82		127, 388
1-p-$C_6H_4NH_2$	104–105		127, 388

(continued)

TABLE 6-1—*Continued*

Substituent groups	mp (°C)	Other data[a]	References
$1\text{-}o\text{-}C_6H_4NO_2$	161–161.5		*127, 419*
$1\text{-}m\text{-}C_6H_4NO_2$	140–141		*127*
	144–145		*382*
$1\text{-}p\text{-}C_6H_4NO_2$	167–168	H	*127, 388*
	164	IR	*300*
$3\text{-}o\text{-}C_6H_4NO_2$	146–147		*419, 420*
$3\text{-}m\text{-}C_6H_4NO_2$	127–128		*419*
$3\text{-}p\text{-}C_6H_4NO_2$	193–194		*419, 420*
Alkenyl and alkynyl derivatives			
$1\text{-}CH{=}CH_2$	76–77	bp 75–77 (0.5 mm)	*69, 142*
	78–79		*143*
$1\text{-}C(CH_3){=}CH_2$	46.7–47.7	bp 50 (1.6 mm), UV, IR	*66*
$1\text{-}CH_2CH{=}CH_2$	63–65		*142*
$1\text{-}(CH_2)_2CH{=}CH_2$	45–46	bp 99–101 (0.2 mm)	*91*
$3\text{-}CH{=}CH_2$	43–44		*398*
$1,2\text{-}[C(CH_3){=}CH_2]_2$	82–83	bp 105–107 (1 mm)	*69, 454*
$1\text{-}CH_3\text{-}2\text{-}CH{=}CH_2$	164–165		*425*
$1\text{-}CH_3\text{-}2\text{-}CH_2CH{=}CH_2$	17–18		*91*
$1\text{-}(CH_2)_2CH_3\text{-}2\text{-}CH{=}CH_2$	33–34		*425*
$1\text{-}CH_2C_6H_5\text{-}2\text{-}CH{=}CH_2$	110–111		*425*
$1\text{-}C_6H_5\text{-}2\text{-}CH_2CH{=}CH_2$	70–71		*425*
$1\text{-}CH{=}CH_2\text{-}2\text{-}CH_2CH{=}CH_2$	92–93		*425*
$1\text{-}C{\equiv}CH$	75–78		*56*
$1\text{-}CH_3\text{-}2\text{-}CH_2C{\equiv}CH$	92–94		*425*
$1\text{-}C{\equiv}CI$	144–145		*59*
Carboxylic acids			
$1\text{-}COOH$	150–150.5		*142, 463*
	150.5–151	IR, pK_a (Table 6-2)	*313, 372, 380, 298, 376*
$3\text{-}COOH$	95	pK_a (Table 6-2)	*398*
$1\text{-}CH_2COOH$	193–194	IR, pK_a (Table 6-2)	*298, 373, 380*
$1\text{-}(CH_2)_2COOH$	146–148	pK_a (Table 6-2)	*293, 373*
	147–149		*371*
$1\text{-}(CH_2)_3COOH$	157–158		*293*
$1\text{-}(CH_2)_4COOH$	153–154		*293*
$1,2\text{-}(COOH)_2$	232		*142*
	216–217		*463*
$1\text{-}CH_3\text{-}2\text{-}COOH$	194–195	IR, pK_a (Table 6-2)	*92, 380*
	201–203		*92*
$1\text{-}C_6H_5\text{-}2\text{-}COOH$	142–143	IR, pK_a (Table 6-2)	*372, 373*
$1\text{-}p\text{-}C_6H_4NO_2\text{-}2\text{-}COOH$			*382*
$1\text{-}o\text{-}C_6H_4NO_2\text{-}2\text{-}COOH$			*382*
$1,2\text{-}(CH_2COOH)_2$	203–205		*142*

(*continued*)

TABLE 6-1—*Continued*

Substituent groups	mp (°C)	Other data[a]	References
1-C_6H_5-2-CH_2COOH	196–198		*361*
1-CH_3-2-CH_2COOH	121	IR	*379, 380*
1-CH=CH_2-2-COOH	142.5–143	pK_a (Table 6-2)	*373, 463*
1-CH=CH_2-2-CH_2COOH	188		*379*
Esters and acyl halides			
1-$COOCH_3$	73		*143*
1-$COOC_2H_5$	61.5–62	IR	*372*
1-$CH_2COOC_2H_5$	10–11	bp 110–112 (2–3 mm), IR	*317*
3-OCCH$_3$ || O	76–76.5		*397, 407*
1,2-$(COOCH_3)_2$	66–67		*143*
	62–62.5	IR	*317*
1,2-$(COOC_2H_5)_2$	10–11		*371*
1,2-$(CH_2COOCH_3)_2$	47–48		*142*
1,2-$(CH_2COOC_2H_5)_2$	40		*142*
1-CH_3-2-$COOC_2H_5$		bp 101–103	*312*
	7.5–8.5	IR	*317*
1-C_6H_5-2-$COOC_2H_5$	34–35	IR	*317*
1-CH_3-3-OCCH$_3$ || O	74–75		*397, 407*
1-COCl	39–41	IR	*372*
1-CH_2COCl	39	IR	*372*
1,2-$(COCl)_2$	69–70		*142*
1,2-$(CH_2COCl)_2$			*142*
1-C_6H_5-2-COCl		bp 145–146 (2 mm)	*410*
Alcohols and B-hydroxy derivatives			
1-CH_2OH	229–230	IR	*365*
3-OH	356–358		*397, 407*
B-OH	412–414	IR	*418*
1-$(CH_2)_2OH$	51–52	IR	*365*
1-$(CH_2)_3OH$	53–53.5	IR	*365*
1-$(CH_2)_4OH$	48.6	IR	*365*
1-$CH_2CH(CH_3)OH$	77.5		*364*
1-CH_3-3-OH	259–261		*397, 407*
1-CH_3-B-OH	264–266		*411*
1,2-$(CH_3)_2$-B-OH	292–293	IR	*418*
1,2-$(CH_2OH)_2$	299–300	IR	*365*
	303–304		*142*
1-COOH-2-CH_2OH		IR	*90*
1,2-$(CH_2CH_2OH)_2$	124–125		*142*
1-CH_3-2-CH_2OH	268–269		*426*
1-C_6H_5-2-CH_2OH	89–90		*426*

(continued)

TABLE 6-1—*Continued*

Substituent groups	mp (°C)	Other data[a]	References
$1\text{-}CH_3\text{-}2\text{-}CH(CH_3)OH$	148–150		*426*
$1\text{-}C_6H_5\text{-}2\text{-}CH(C_6H_5)OH$	120–121		*426*
$1\text{-}CH_3\text{-}2\text{-}CH(C_6H_5)OH$	107–108		*430*
$1,2\text{-}(C_6H_5COH)_2$	134–135		*426*

Ethers and epoxides

$1\text{-}C(CH_3)CH_2$ (epoxide)		bp 96–100 (0.3 mm), IR	*66*
$1\text{-}CH_2CHCH_2$ (epoxide)	68		*142*
$1\text{-}(CH_2)_2CHCH_2$ (epoxide)	60–62	bp 132–135 (0.4 mm)	*91*
$1\text{-}CH_2OCH_2\text{-}2\text{-}(1\text{-}o\text{-}C_2B_9H_{11})$	344		*455*
$1\text{-}C_6H_5\text{-}2\text{-}CH_2OC_6H_5$	59–60		*372*
$1,2\text{-}(CH_3)_2\text{-}B\text{-}OCH_3$	98–99		*411*

Aldehydes

$1\text{-}CHO$	212–213		*202, 438*
	208–209	IR	*304*
$B\text{-}CHO$	242–244		*398*
$1\text{-}CH_2CHO$	94–95		*202*
$1\text{-}CH_3\text{-}2\text{-}CHO$	222–223		*438*
$1\text{-}CH_3\text{-}2\text{-}CH_2CHO$	140–142		*202*
$1\text{-}C_6H_5\text{-}2\text{-}CHO$	57–58	IR	*437*

Ketones

$1\text{-}COCH_3$	54–55		*439*
$1\text{-}COC_6H_5$	57.5–58.0	IR	*317*
$1\text{-}CH_3\text{-}2\text{-}COC_6H_5$	67	IR	*435, 436*
	65.5–66.0	IR	*317*
$1\text{-}C_6H_5\text{-}2\text{-}COCH_3$	67–68	IR	*362, 435, 436*
$1\text{-}CH_3\text{-}2\text{-}COCH_3$		IR	*442*
	118–119		*439*
$1\text{-}C_6H_5\text{-}2\text{-}COC_6H_5$	76–77	IR	*362, 435, 436*
	80–81		*430*
	86.5–87	IR	*317*
$1\text{-}C_6H_5\text{-}2\text{-}COCH(CH_3)_2$	46–47		*362*

Nitro and nitroso derivatives and nitrates

$1\text{-}NO_2$	100–101	IR	*462*
$1\text{-}NO_2\text{-}2\text{-}C_6H_5$	54–55		*382*
$1\text{-}NO$	196.5–197.5		*169*
$B\text{-}ONO_2$	103–104	IR	*418*
$1\text{-}CH_3\text{-}B\text{-}ONO_2$	99–100		*411*

(*continued*)

TABLE 6-1—*Continued*

Substituent groups	mp (°C)	Other data[a]	References
$1,2\text{-}(CH_3)_2\text{-}B\text{-}ONO_2$	68–69	IR	*418*
$1,2\text{-}(CH_2ONO_2)_2$	19–20	IR	*365*
$1\text{-}CH_2OH\text{-}2\text{-}CH_2ONO_2$	15	IR	*365*
Amines			
$1\text{-}NH_2$	304–305		*402*
$3\text{-}NH_2$	218–219		*396, 407*
$1\text{-}CH_2N(C_2H_5)_2$	33–35		*143*
$1\text{-}(CH_2)_2N(C_2H_5)_2$	46.5–48		*364, 371*
$1\text{-}B[N(CH_3)_2]_2$	44–48	IR	*23*
$3\text{-}N(C_6H_5)_2$	124–126	IR	*250*
$1\text{-}CH_3\text{-}2\text{-}NH_2$	303		*402*
$1\text{-}C_6H_5\text{-}2\text{-}NH_2$	98–99		*402*
$1\text{-}C_6H_5\text{-}2\text{-}CH_2NH_2$	61–62	IR	*437*
$1\text{-}CH_3\text{-}3\text{-}NH_2$	231–232		*396, 407*
$1\text{-}C_2H_5\text{-}3\text{-}NH_2$	91–93		*407*
$1\text{-}C_6H_5\text{-}3\text{-}NH_2$	81–82		*396, 407*
$1\text{-}C_6H_5\text{-}2\text{-}NHOH$	98–99		*382*
$1,2\text{-}(CH_3)_2\text{-}3\text{-}NH_2$	319–320		*396, 407*
Amides, azides, cyanides, and isocyanates			
$1\text{-}CONH_2$	114–115	IR	*317*
	118.5–119		*312*
$1\text{-}CON(C_2H_5)_2$	109–109.5		*364*
$1\text{-}CONHC_6H_5$	132–133		*142*
$1\text{-}CH_2CONH_2$	149–149.5	IR	*317*
	144–145		*364*
$1\text{-}CH_2CON(C_2H_5)_2$	82–83		*364, 371*
$1\text{-}C_6H_5\text{-}2\text{-}CONH_2$	111–111.5	IR	*317*
$1\text{-}CH_3\text{-}2\text{-}CONH_2$	217–218	IR	*317*
$1\text{-}CON_3$	84–85		*402*
$1\text{-}CH_3\text{-}2\text{-}CON_3$	42–43		*402*
$1\text{-}C_6H_5\text{-}2\text{-}CON_3$	92–93		*402*
$1\text{-}C_6H_5\text{-}2\text{-}CN$	105–106		*437*
$1,2\text{-}(NCO)_2$		IR	*142*
Phosphorus and arsenic derivatives			
$1,2\text{-}[P(C_6H_5)_2]_2$	208–210		*466*
	219		*7*
$1,2\text{-}[P(C_6H_5)Cl]_2$	172–174		*7*
$1,2\text{-}[P(C_6H_5)N_3]_2$	126–128		*7*
$1\text{-}C_6H_5\text{-}2\text{-}P(n\text{-}C_6H_{13})_2$	9–10		*367*
$1,2\text{-}[As(CH_3)_2]_2$	111	H, B, IR	*278, 358*
$1,2\text{-}[As(C_6H_5)_2]_2$	204.5–205.5	IR	*278*

(*continued*)

TABLE 6-1—*Continued*

Substituent groups	mp (°C)	Other data[a]	References
Sulfur derivatives			
1-SH	205–207		*465*
1,2-$(SH)_2$	265–267		*282*
1,2-$(SCH_3)_2$	101–102		*282*
1,2-$(SC_6H_5)_2$	189–191		*282*
1,2-$(SCH_2C_6H_5)_2$	102–104		*282*
1-CH_3-2-SH	244–246		*465*
1-C_6H_5-2-SH	67–69		*465*
1-CH_3-2-SCH_3	124–125		*465*
1-C_6H_5-2-SCH_3	90–92		*465*
1,2-$(SCl)_2$	79–80		*269*
Silicon derivatives			
1-$Si(CH_3)_3$	94–95		*142*
1-$Si(C_6H_5)_3$	165–167		*142*
1-$CH_2Si(CH_3)_3$	52		*142*
1-$CH_2SiCl_2CH_3$		IR	*265*
1,2-$[Si(CH_3)_3]_2$	141–142		*252*
1-C_6H_5-2-$Si(CH_3)_3$	105–106		*367*
1,2-$(SiCl_3)_2$	121–122		*231*
1,2-$[Si(CH_3)_2Cl]_2$	112.5–113.5		*231*
1,2-$[Si(C_6H_5)_2Cl]_2$	244–245		*231*
	249–251		*252*
1,2-$[Si(CH_3)Cl_2]_2$	119–120		*231*
Monohalo derivatives[b]			
3-F	270–271	IR	*250*
	242		*407*
		^{19}F NMR	*408*
1-CH_3-3-F	207–208		*407*
		^{19}F NMR	*408*
1,2-$(CH_3)_2$-3-F	234–235		*407*
		^{19}F NMR	*408*
1-CH_3-2-COOH-3-F	240		*407*
1-Cl	210–212		*296*
	210–212	IR	*299a*
1-CH_3-2-Cl	235–236		*296*
	235–236	IR	*299a*
1-C_6H_5-2-Cl	51–52		*447*
3-Cl	169.5–170		*397, 407*
1-CH_3-3-Cl	184–185		*397, 407*
1-C_6H_5-8-Cl	70–70.5		*391*
9-Cl	237–238		*389*
		IR	*193, 318*
		DM	*62, 307*
1-CH_3-9-Cl	145–146		*318*

(*continued*)

TABLE 6-1—*Continued*

Substituent groups	mp (°C)	Other data[a]	References
1-C_6H_5-9-Cl	63–64		*389*
1-CH_3-12-Cl	160–161	IR	*318*
1-C_6H_5-12-Cl	85–86		*391*
1-CH_3-2-Br	224–225	IR	*299a*
1-C_6H_5-2-Br	76–77		*447*
1-n-C_3H_7-2-Br	44–45	bp 107 (1 mm)	*143*
1-n-C_4H_9-2-Br		bp 85–90 (0.5 mm)	*143*
1-Br	177–179	IR	*299a*
3-Br	118.5–119		*397, 407*
	121–122	IR	*250*
8-Br	186.5–187		*390*
		IR	*193*
1-CH_3-8-Br	120–121		*391*
1-C_6H_5-8-Br	87–88		*391*
9-Br		IR	*193, 318*
		DM	*62, 307*
1,2-$(C_6H_5)_2$-9-Br	128–129		*389*
1-CH_3-12-Br	167–168		*391*
1-I	131–136		*91*
1-C_6H_5-2-I	96–97		*447*
1-C_6H_5-8-I	127.5–128		*391*
9-I	117–118		*389*
		IR	*193, 318*
		DM	*62, 307*
1-CH_3-9-I	129–130		*389*
1-C_6H_5-9-I	93–94		*389*
1-CH_3-12-I	157		*389*
1-C_6H_5-12-I	146–147		*391*
Dihalo derivatives[b]			
1,2-Cl_2	231–233	IR	*299a*
9,12-Cl_2		DM	*62, 307*
1-CH_3-9,12-Cl_2	152–153		*318*
9-Cl-12-Br	238–239		*389*
1,2-Br_2		IR	*193*
		DM	*205*
8,9-Br_2	202–203		*390*
1-CH_3-8,9-Br_2	110–111		*391*
1-C_6H_5-8,9-Br_2	75–76		*391*
1-CH_3-8,12-Br_2	154–155		*391*
1-C_6H_5-8,12-Br_2	108–109		*391*
9,10-Br_2		DM	*205*
9,12-Br_2		B	*247*
		IR	*193*
		DM	*62, 205, 307*

(*continued*)

TABLE 6-1—*Continued*

Substituent groups	mp (°C)	Other data[a]	References
$1,2\text{-}I_2$	165–166	IR	*299a*
$9,12\text{-}I_2$	185–186		*389*
		IR	*193, 318*
		DM	*62, 307*
$1\text{-}CH_3\text{-}9,12\text{-}I_2$	158–159		*389*
Trihalo derivatives[b]			
$8,9,12\text{-}Br_3$			*390*
		IR	*193, 318*
		B	*247*
$1\text{-}CH_3\text{-}8,9,12\text{-}Br_3$	168–170		*391*
Tetrahalo derivatives[b]			
$8,9,10,12\text{-}Cl_4$	351–352		*389*
$1\text{-}C_6H_5\text{-}8,9,10,12\text{-}Cl_4$	209–211		*389*
$8,9,10,12\text{-}Br_4$	352–353	IR	*193, 461*
$8,9,10,12\text{-}I_4$	325–326		*461*
$1,2\text{-}(CH_3)_2\text{-}8,9,10,12\text{-}Br_4$		B	*247*
Decahalo and dodecahalo derivatives[b]			
$B\text{-}F_{10}$	258–261	MS	*182*
$B\text{-}Cl_{10}$	273–275	pK_a (Table 6-3)	*182*
	274.5–276		*444*
		IR	*192, 193, 461*
		DM	*62, 307*
$1\text{-}CH_3\text{-}B\text{-}Cl_{10}$	>400	pK_a (Table 6-3)	*443*
		IR	*192*
$1\text{-}C_2H_5\text{-}B\text{-}Cl_{10}$		IR	*192*
	239–240		*461*
$1\text{-}CH_2\text{-}CH{=}CH_2\text{-}B\text{-}Cl_{10}$	Dec. 330–332		*443*
$1\text{-}C_6H_5CH_2\text{-}B\text{-}Cl_{10}$	182–183		*443*
$1,2\text{-}(CH_3)_2\text{-}B\text{-}Cl_{10}$	>400		*443*
$1\text{-}CH_3\text{-}2\text{-}C_2H_5\text{-}B\text{-}Cl_{10}$	>400		*443*
$1,2\text{-}(CH_2C_6H_5)_2\text{-}B\text{-}Cl_{10}$	216–217		*443*
$1,2\text{-}(CH_2CH{=}CH_2)_2\text{-}B\text{-}Cl_{10}$	Dec. 422–424		*443*
Cl_{12}	426–430	IR	*193, 461*

Derivatives with small expolyhedral rings

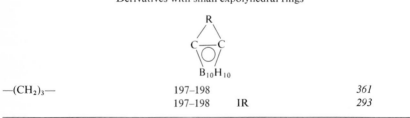

$-(CH_2)_3-$	197–198		*361*
	197–198	IR	*293*

(continued)

TABLE 6-1—*Continued*

Substituent groups	mp (°C)	Other data[a]	References
—(CH$_2$)$_4$—	122–123		361
—CH$_2$—O—CH$_2$—	259.5–260.5	IR	91
—CH$_2$—O—CH$_2$—O—CH$_2$—	90–91		99, 364
$-CH_2-O-\overset{\overset{\displaystyle O}{\|}}{C}-$	253–254.5	IR	90
$-CH_2-\overset{\overset{\displaystyle O}{\|}}{C}-CH_2-$	170		142
$-\overset{\overset{\displaystyle O}{\|}}{C}-O-\overset{\overset{\displaystyle O}{\|}}{C}-$	180		142
—CH$_2$⟨benzene⟩CH$_2$—			425
 ⟨structure⟩—CH$_2$—	133–134		372
$-\overset{\overset{\displaystyle O}{\|}}{C}$⟨structure⟩	165.5–166		372, 455
$-CH_2-\overset{\overset{\displaystyle O}{\|}}{C}$⟨structure⟩	142–144		361
—Si(CH$_3$)$_2$—O—Si(CH$_3$)$_2$—	160–161	IR	231 236
—CH$_2$Si(CH$_3$)$_2$CH$_2$—	149–150		142

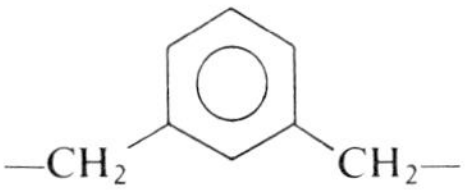

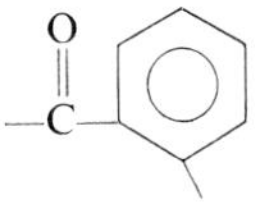

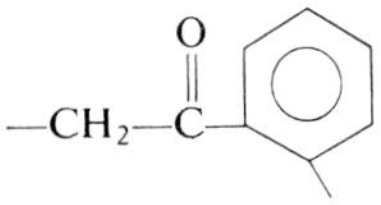

[a] IR = infrared spectra or band positions; B = ^{11}B NMR spectra; H = ^{1}H NMR spectra; MS = mass spectra or cutoff m/e value; DM = dipole moment; bp = boiling point (°C); UV = ultraviolet spectra or band positions.

[b] Numbering of cage positions is as shown in Fig. 2-11 (Chapter 2), regardless of systems used in individual papers in the literature.

organoboron structures. As a consequence, it is possible to carry out a variety of reactions on substituent groups attached to the carborane while leaving the cage system intact. The *o*-carborane unit displays strong electron-withdrawing character with respect to substituents attached at the carbon atoms. This property is reflected, for example, in the high acid strength of carborane carboxylic acids, the ease of metallation of the carborane C—H group, and the relative unreactivity of halomethylcarboranes with respect to nucleophilic reagents. The electron-attracting properties of o-carborane are much weaker, however, toward substituents attached to cage boron atoms; similarly, m- and p-carborane are far less electrophilic than the o-carborane system. These observations are illustrated and discussed in detail in the following sections and in Chapter 7.

Derivatives of o-carborane substituted at carbon are prepared by two general methods, one of which is the reaction of substituted acetylenes with bis(ligand)decaborane compounds as described above. A second approach involves the metallation of o-carborane followed by treatment with reagents such as carbon dioxide, halogens, organic and inorganic halides, epoxides, aldehydes, and others, to yield mono- or di-C-functional o-carboranes. The synthesis and properties of these derivatives are described below. A partial listing of structurally characterized o-carborane derivatives is given in Table 6-1.

6-2. ALKALI METAL AND MAGNESIUM DERIVATIVES

Alkali Metal Derivatives

The mildly acidic C—H bonds in o-carborane react easily with n-butyl-lithium (*91, 142, 463*) or phenyllithium (*91, 142*) to form C-lithio-o-carborane; an excess of the organometallic reagent yields the C,C′-dilithio derivative.

$$\underset{B_{10}H_{10}}{\underset{\diagdown O\diagup}{HC-CH}} + C_4H_9Li \xrightarrow[60-80°]{C_6H_6} \underset{B_{10}H_{10}}{\underset{\diagdown O\diagup}{HC-CLi}} + C_4H_{10} \xrightarrow{C_4H_9Li} \underset{B_{10}H_{10}}{\underset{\diagdown O\diagup}{LiC-CLi}} + C_4H_{10}$$

In ether or ether–benzene solvent (but not in benzene itself) an equilibrium exists between the mono- and dilithio species (*377*). Lithium exchange has also been noted in attempts to prepare o-carboranyl monocarboxylic acid (Section 6-5) by reaction of the monolithio derivative with CO_2 in ether, which gives instead the dicarboxylic acid (*142*). In benzene, however, the monofunctional acid is obtained in good yield (*376, 455*).

$$2\ \underset{\substack{B_{10}H_{10}\\\sim80\%}}{\underset{\diagdown O\diagup}{HC-CLi}} \rightleftharpoons \underset{\substack{B_{10}H_{10}\\\sim10\%}}{\underset{\diagdown O\diagup}{LiC-CLi}} + \underset{\substack{B_{10}H_{10}\\\sim10\%}}{\underset{\diagdown O\diagup}{HC-CH}}$$

In the *m*-carborane system the equilibrium lies much further to the left, with only 2% dilithio-*o*-carborane in equilibrium with the monolithio derivative (Section 7-4).

Metallation of *o*-carborane has also been accomplished with metal amides in liquid ammonia. The amides of Li, Na, K, and Ca have been utilized for this purpose (*360, 363, 425, 458*), but such reactions have not been widely employed in carborane derivative syntheses.

Magnesium Derivatives

Grignard reagents are formed in reactions of *o*-carborane with alkyl-magnesium halides (*91*) and by the action of magnesium on 1-halomethyl-*o*-carboranes (*91, 142, 364, 455*). The chemistry of the halomagnesio-*o*-carboranes is complicated, however, in tetrahydrofuran (THF) solutions by equilibria of the type observed for lithiocarboranes (*377*).

$$\underset{B_{10}H_{10}}{HC\!\!-\!\!CH} + C_2H_5MgBr \xrightarrow{\text{THF}} \underset{B_{10}H_{10}}{HC\!\!-\!\!CMgBr} + C_2H_6$$

$$2\ \underset{\substack{B_{10}H_{10}\\80\%}}{HC\!\!-\!\!CMgBr} \xrightleftharpoons{\text{THF}} \underset{\substack{B_{10}H_{10}\\10\%}}{BrMgC\!\!-\!\!CMgBr} + \underset{\substack{B_{10}H_{10}\\10\%}}{HC\!\!-\!\!CH}$$

A consequence is that reactions of 1-bromomagnesio-*o*-carborane with alkyl halides (Section 6-3) produce mixtures of mono- and disubstituted carborane products. Additional evidence for an equilibrium of this sort is the reaction of 1-methyl-2-bromomagnesio-*o*-carborane with *o*-carborane, in which magnesium is exchanged for hydrogen (*377*).

$$\underset{B_{10}H_{10}}{CH_3C\!\!-\!\!CMgBr} + \underset{B_{10}H_{10}}{HC\!\!-\!\!CH} \rightleftharpoons \underset{B_{10}H_{10}}{CH_3C\!\!-\!\!CH} + \underset{B_{10}H_{10}}{HC\!\!-\!\!CMgBr}$$

Reactions of 1-bromomethyl-*o*-carboranes with magnesium were expected to yield the corresponding 1-halomagnesiomethyl derivatives. However, Grafstein *et al.* (*91*) found that in THF solution rearrangement to 1-methyl-2-bromomagnesio-*o*-carborane is extensive. Thus, carbonation of the Grignard prepared from 1-bromomethyl-*o*-carborane in THF yields 1-methyl-2-carboxylic acid-*o*-carborane (*292*). Carbonation of the same Grignard *in ether* produces mostly o-carboranyl acetic acid, indicating that rearrangement occurs to a much lesser extent in this solvent (*292, 380*).

$$\text{HC}\underset{B_{10}H_{10}}{\overline{\diagdown\!\!\text{O}\!\!\diagup}}\text{CCH}_2\text{Br} \xrightarrow[\text{THF}]{\text{Mg}} \left[\text{HC}\underset{B_{10}H_{10}}{\overline{\diagdown\!\!\text{O}\!\!\diagup}}\text{CCH}_2\text{MgBr}\right] \xrightarrow{\text{THF}} \text{BrMgC}\underset{B_{10}H_{10}}{\overline{\diagdown\!\!\text{O}\!\!\diagup}}\text{CCH}_3$$

| Mg/ether (down) | | 1. CO_2 / 2. H_3O^+ (down) |

$$\text{HC}\underset{B_{10}H_{10}}{\overline{\diagdown\!\!\text{O}\!\!\diagup}}\text{CCH}_2\text{MgBr} \xrightarrow[\text{2. }H_3O^+]{\text{1. }CO_2} \text{HC}\underset{B_{10}H_{10}}{\overline{\diagdown\!\!\text{O}\!\!\diagup}}\text{CCH}_2\text{COOH} \qquad \text{HOOCC}\underset{B_{10}H_{10}}{\overline{\diagdown\!\!\text{O}\!\!\diagup}}\text{CCH}_3$$

In contrast to the 1-bromomethyl derivative, 1-chloromethyl-o-carborane on reaction with magnesium in ether forms primarily the rearranged Grignard, 1-methyl-2-chloromagnesio-o-carborane (*380*).

$$\text{HC}\underset{B_{10}H_{10}}{\overline{\diagdown\!\!\text{O}\!\!\diagup}}\text{CCH}_2\text{Cl} \xrightarrow[\text{ether}]{\text{Mg}} \underset{40\text{–}60\%}{\text{ClMgC}\underset{B_{10}H_{10}}{\overline{\diagdown\!\!\text{O}\!\!\diagup}}\text{CCH}_3} + \underset{12\text{–}28\%}{\text{HC}\underset{B_{10}H_{10}}{\overline{\diagdown\!\!\text{O}\!\!\diagup}}\text{CCH}_2\text{MgCl}} + \text{minor products}$$

The difference has been explained in terms of higher acidity of the cage proton in the chloromethyl derivative, probably as a consequence of the greater electronegativity of chlorine compared with bromine (*380*). Consistent with this is the reaction of 1-bromomagnesiomethyl-o-carborane with 1-chloro-methyl-o-carborane, in which the latter compound is C-metallated (*380*):

$$\text{HC}\underset{B_{10}H_{10}}{\overline{\diagdown\!\!\text{O}\!\!\diagup}}\text{CCH}_2\text{MgBr} + \text{HC}\underset{B_{10}H_{10}}{\overline{\diagdown\!\!\text{O}\!\!\diagup}}\text{CCH}_2\text{Cl} \xrightarrow{\text{ether}} \text{HC}\underset{B_{10}H_{10}}{\overline{\diagdown\!\!\text{O}\!\!\diagup}}\text{CCH}_3 + \text{BrMgC}\underset{B_{10}H_{10}}{\overline{\diagdown\!\!\text{O}\!\!\diagup}}\text{CCH}_2\text{Cl}$$

Similarly, it has been observed that while 1-chloromethyl-o-carborane is easily metallated at the cage carbon atom by n-butyllithium or ethylmagnesium bromide, with 1-bromomethyl-o-carborane the main effect produced by these reagents is not metallation but lithium–bromine exchange followed by iso-merization (*299, 380*).

$$\text{HC}\underset{B_{10}H_{10}}{\overline{\diagdown\!\!\text{O}\!\!\diagup}}\text{CCH}_2\text{Cl} + C_4H_9\text{Li} \xrightarrow{\text{ether}} \text{LiC}\underset{B_{10}H_{10}}{\overline{\diagdown\!\!\text{O}\!\!\diagup}}\text{CCH}_2\text{Cl}$$

$$\text{HC}\underset{B_{10}H_{10}}{\overline{\diagdown\!\!\text{O}\!\!\diagup}}\text{CCH}_2\text{Br} + C_4H_9\text{Li} \xrightarrow{\text{ether}} \text{HC}\underset{B_{10}H_{10}}{\overline{\diagdown\!\!\text{O}\!\!\diagup}}\text{CCH}_2\text{Li} \longrightarrow \text{LiC}\underset{B_{10}H_{10}}{\overline{\diagdown\!\!\text{O}\!\!\diagup}}\text{CCH}_3$$

Available evidence strongly suggests that the isomerization of o-carboranyl Grignards does not involve an intramolecular rearrangement, as was first believed (*91, 142*), but instead occurs via an intermolecular metal atom transfer (*293, 297, 298*). A detailed mechanism has been proposed by Zakharkin and associates (*380*) for the reaction of 1-chloromethyl-o-carborane with magnesium in ether.

The possibility of rearrangement is eliminated in Grignards having a methyl or other alkyl group on the second o-carboranyl carbon atom. Grignard reagents of this type give the expected reactions of organomagnesium compounds (*69, 91, 142, 365, 375*).

$$CH_3C \underset{B_{10}H_{10}}{\overset{CCH_2Br}{\diagdown O \diagup}} \xrightarrow{\text{Mg}} CH_3C \underset{B_{10}H_{10}}{\overset{CCH_2MgBr}{\diagdown O \diagup}} \xrightarrow[\text{2. } H_3O^+]{\text{1. } CO_2} CH_3C \underset{B_{10}H_{10}}{\overset{CCH_2COOH}{\diagdown O \diagup}}$$

$$\downarrow H_2O$$

$$CH_3C \underset{B_{10}H_{10}}{\overset{CCH_3}{\diagdown O \diagup}}$$

o-Carboranyl Grignards have also been prepared from 1-halo-o-carboranes in reactions with magnesium in ether (*142*) or with ethylmagnesium bromide (*423*) (Section 6-13).

Zinc Derivatives

Diethylzinc and 1-phenyl-o-carborane at elevated temperatures yield bis(1-phenyl-o-carboranyl) zinc in hexamethyltriamidophosphate (*217*).

Formation of o-Carborane Dianions in Liquid Ammonia

Sodium, potassium, and lithium metals react with o-carborane and its alkyl or aryl derivatives in liquid ammonia *without evolution of hydrogen*, forming adducts containing two metal atoms per carborane molecule (*66, 92, 453*). The adducts dissolve in cold water, again without loss of hydrogen, yielding dicarbadodecaborane anions in which the icosahedral carborane cage remains intact [an earlier report (*92*) that derivatives of $C_2B_9H_{12}^-$ are formed seems incorrect]. Air oxidation in tetrahydrofuran solution produces the original carboranes, but heating the salts in aqueous solution results in disintegration of the carborane cage (*453*).

$$C_6H_5C\!\!-\!\!CC_6H_5 \quad \underset{}{\overset{NH_3(l)}{\longrightarrow}} \quad \left[C_6H_5C\!\!-\!\!CC_6H_5 \right]^{2-} \quad \overset{H_2O}{\underset{10\text{--}20°}{\longrightarrow}} \quad \left[C_6H_5C\!\!-\!\!CC_6H_5 \right]^{-}$$

$$B_{10}H_{10} \qquad + 2M \qquad\qquad B_{10}H_{10} \qquad\qquad\qquad B_{10}H_{11}$$

M = Li, Na, K

$$\cdot 2M^+ \qquad\qquad M^+$$

$$\Big\downarrow O_2$$

$$\overset{H_2O}{\underset{80\text{--}100°}{\nearrow}}$$

$$C_6H_5C\!\!-\!\!CC_6H_5$$
$$B_{10}H_{10}$$

$$C_6H_5CH_3 + H_2 + B(OH)_3$$

The reactions of m- and p-carborane with alkali metals in liquid ammonia give analogous dianions which undergo cage rearrangements to the *ortho* and *meta* dianions, respectively (Section 7-3). Related reactions of arylcarboranes with alkali metals in organic solvents are discussed in Section 6-3.

6-3. ALKYL, HALOALKYL, AND ARYL DERIVATIVES

Synthesis

C-alkyl- and C-aryl-o-carboranes are easily prepared from a bis-ligand decaborane derivative and the appropriate substituted acetylene, as described above (*66, 69, 143, 320, 361, 365, 419, 454, 455*). Even alkenyl and alkynyl carboranes may be obtained in this way; vinylacetylene, for example, reacts with bisacetonitriledecaborane to give 1-vinyl-o-carborane nearly quantitatively (since carborane derivatives with unsaturated side chains have unusual properties, discussion of these compounds is deferred to the following section). Alternatively, alkyl- and aryl-substituted carboranes may be prepared via reactions of primary alkyl bromides or iodides with C-metallated derivatives of o-carborane, usually in ether (*7, 142, 281, 375, 425*), benzene (*425*), or liquid ammonia (*360, 361, 363, 425*).

$$RC\!\!-\!\!CM \qquad + \; R'X \qquad \overset{}{\underset{-MX}{\longrightarrow}} \qquad RC\!\!-\!\!CR'$$
$$B_{10}H_{10} \qquad\qquad\qquad\qquad\qquad B_{10}H_{10}$$

M = Li, Na, K

Secondary and tertiary alkyl halides apparently do not react with metallo-carboranes, and primary alkyl chlorides react only very slowly (*425*). The method is quite general for primary bromides and iodides, however, and may be used to form exocyclic ring derivatives from dimetallo-*o*-carboranes (*361, 425*):

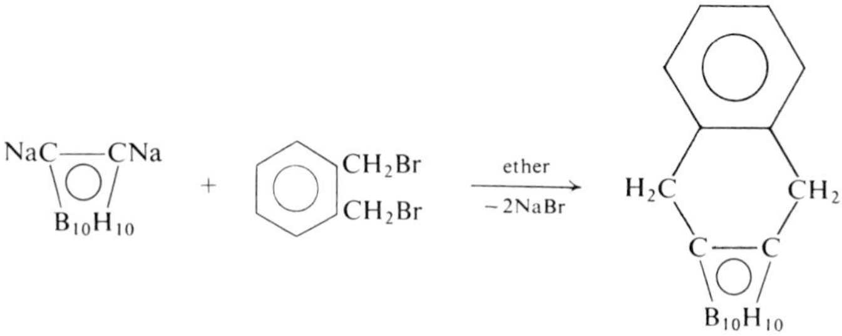

In a similar reaction, *cis*-1,4-dichloro-2-butene combines with dilithio-*o*-carborane to give 4,5-dihydrobenzocarborane. The latter compound is then dehydrogenated with *N*-bromosuccinimide (NBS) to form a product which is apparently the novel species benzocarborane, in which delocalized aryl and *o*-carboranyl systems are fused (*155*).

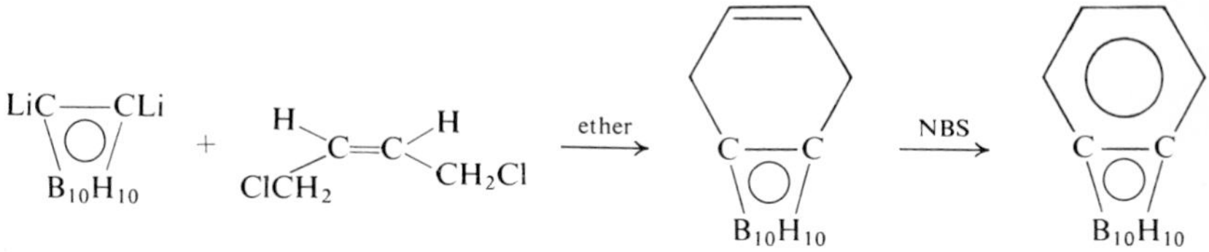

The characterization of benzocarborane is based partly upon proton NMR and ultraviolet spectra, both of which indicate aromatic character for the exopolyhedral ring. Benzocarborane is not attacked by concentrated sulfuric acid at 100° or by bromine in carbon tetrachloride (*155*).

Alkyl and aryl substitution may also be achieved with *o*-carboranyl Grignards, which are prepared as described in Section 6-2. The alkylation proceeds with primary alkyl bromides and iodides, dialkyl sulfates, alkyl *p*-toluene-sulfonates, and benzyl halides (*377*):

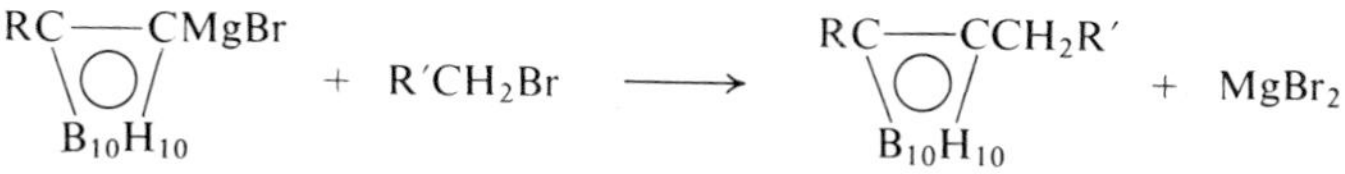

$$R = H, CH_3, C_6H_5, CH=CH_2$$

Again, secondary and tertiary alkyl halides and primary alkyl chlorides are nonreactive. Bis(halomagnesium) carboranes react with α,ω-dihaloalkanes to give exocyclic species such as 1,2-o-carboranylcyclopentane (*377*):

$$\text{BrMgC}\underset{\underset{B_{10}H_{10}}{}}{\overset{}{\diagdown O \diagup}}\text{CMgBr} \xrightarrow{\text{Br(CH}_2)_3\text{Br}} \overset{\overset{\overset{CH_2}{H_2C \diagup \diagdown CH_2}}{}}{C \underset{\underset{B_{10}H_{10}}{}}{\diagdown O \diagup} C} + 2\text{MgBr}_2$$

As a consequence of the equilibrium between mono- and difunctional carborane Grignards (Section 6-2), reactions of o-carboranyl magnesium halides in tetrahydrofuran always produce some disubstituted carborane products together with the expected monosubstituted species (*377*).

The hydrolysis of 1-bromomagnesiomethyl-o-carborane (Section 6-2) forms 1-methyl-o-carborane in good yield (*91, 142*).

Substitution at Boron

The preparation of several B-alkyl and B-aryl o-carboranes from the $C_2B_9H_{11}^{2-}$ (dicarbollide) ion is described in Chapter 9. A mixture of two isomeric B-ethyl-o-carboranes has been obtained from the base-catalyzed reaction of ethyldecaborane(14) with acetylene (*143*), and direct B-alkylation of o-carborane has been reported in the patent literature (*39a*).

Properties of Alkyl and Haloalkyl Derivatives

o-Carboranyl hydrocarbons and haloalkyls are in most cases stable crystalline solids that are unaffected by acids, bases, and ordinary oxidizing agents (*320, 365*). They dissolve readily in many organic solvents, and the lighter species sublime readily *in vacuo* (*320*).

The 1-haloalkyl derivatives of o-carborane illustrate very well the strong inductive ($-$I) effect of the o-carboranyl system. Halomethyl-o-carboranes are essentially unreactive with nucleophilic reagents (*91, 142*), whereas the β-haloethyl and γ-halopropyl carboranes behave normally, for example, reacting easily with sodium iodide (*364*):

$$\text{HC}\underset{\underset{B_{10}H_{10}}{}}{\overset{}{\diagdown O \diagup}}\text{C(CH}_2)_n\text{Cl} + \text{NaI} \xrightarrow[\text{acetone}]{58°,\ 4\ \text{hr}} \text{HC}\underset{\underset{B_{10}H_{10}}{}}{\overset{}{\diagdown O \diagup}}\text{C(CH}_2)_n\text{I} + \text{NaCl}$$

$$n = 2,\ 3,\ 4$$

Similarly, while neither 1-chloromethyl- nor 1-bromomethyl-o-carborane reacts with benzene over aluminum chloride, benzene is alkylated by both β-haloethyl- and γ-halopropyl-o-carborane (364):

$$\underset{B_{10}H_{10}}{HC\!-\!C(CH_2)_nX} \quad + \quad C_6H_6 \quad \xrightarrow{AlX_3} \quad \underset{B_{10}H_{10}}{HC\!-\!C(CH_2)_nC_6H_5}$$

$$n = 2, 3; \; X = Cl, Br$$

1-Halomethylcarboranes do react with magnesium (Section 6-2) (142, 364, 455), though more slowly than do the homologs with longer alkyl chains. The Grignards so obtained are nearly inert toward certain reagents with which reaction would ordinarily be expected, such as simple aldehydes and ketones. A complicating factor, however, is the possibility of rearrangements (91, 142, 375, 380) such as that of 1-bromomagnesiomethyl-o-carborane to 1-methyl-2-bromomagnesio-o-carborane, described in Section 6-2. The latter compound is more reactive than the former, and many reactions of the original Grignard, such as that with benzaldehyde, show evidence of rearrangement:

$$\underset{B_{10}H_{10}}{HC\!-\!CCH_2MgBr} \quad + \quad C_6H_5CHO \quad \xrightarrow{THF} \quad \underset{OH \; B_{10}H_{10}}{C_6H_5CHC\!-\!CCH_3}$$

The same Grignard, on the other hand, reportedly reacts with acetaldehyde to give the "normal" product, 1-(β-hydroxypropyl)-o-carborane (142, 364).

Properties of Aryl Derivatives

Studies of aryl o-carboranes have been concerned primarily with the electronic interaction between the ring and the carborane cage, as discussed elsewhere in this chapter (see especially Sections 6-5, 6-8, and 6-13). In addition, the behavior of aryl carboranes with alkali metals has been described. The reaction of potassium with 1-phenyl-o-carborane in dimethoxyethane or tetrahydrofuran takes place readily, forming colored paramagnetic solutions without evolution of hydrogen. The solutions are decolorized by air, water, and alcohol, and are postulated to contain a stable phenylcarborane anion radical:

$$\underset{B_{10}H_{10}}{C_6H_5C\!-\!CH} \quad + \quad e^- \quad \longrightarrow \quad \left[\underset{B_{10}H_{10}}{C_6H_5C\!-\!CH}\right]^{\overset{\cdot}{-}}$$

Metallation of the carboranyl C—H bond does not occur under the conditions used, and the C—H group is not essential in radical formation since 1-*n*-propyl-2-phenyl-*o*-carborane undergoes a similar reaction. A lack of fine structure in the EPR spectra of the solutions is attributed to delocalization of the unpaired electron in the phenylcarborane anion (*15, 16*).

The anion radical reacts with excess potassium to form a diamagnetic dianion, apparently of the same type as those obtained from *o*-carborane derivatives in liquid ammonia (Section 6-2) (*16*). Acid hydrolysis of the dianion yields a monoanion; acid hydrolysis of the radical anion produces the same monoanion, plus the original 1-phenyl-*o*-carborane.

$$\left[\begin{array}{c} \text{C}_6\text{H}_5\text{C}\!\!-\!\!\text{CH} \\ \diagdown\!\!\text{O}\!\!\diagup \\ \text{B}_{10}\text{H}_{10} \end{array} \right]^{\!\bullet-} \xrightarrow{\text{K}(e^-)} \left[\begin{array}{c} \text{C}_6\text{H}_5\text{C}\!\!-\!\!\text{CH} \\ \diagdown\!\!\text{O}\!\!\diagup \\ \text{B}_{10}\text{H}_{10} \end{array} \right]^{2-} \xrightarrow{\text{H}^+} \left[\begin{array}{c} \text{C}_6\text{H}_5\text{C}\!\!-\!\!\text{CH} \\ \diagdown\!\!\text{O}\!\!\diagup \\ \text{B}_{10}\text{H}_{10} \end{array} \right]^{-} + \begin{array}{c} \text{C}_6\text{H}_5\text{C}\!\!-\!\!\text{CH} \\ \diagdown\!\!\text{O}\!\!\diagup \\ \text{B}_{10}\text{H}_{10} \end{array}$$

$$\text{H}^+$$

The radical anion is quantitatively oxidized to 1-phenyl-*o*-carborane by HgCl$_2$ (*16*).

o-Carborane derivatives with seven-membered aryl rings have been reported (*114*): the reaction of 1-methyl-2-lithio-*o*-carborane with tropenyl methyl ether yields 1-methyl-2-(7-cyclohepta-1,3,5-trienyl)-*o*-carborane. This compound rearranges at 165° to the 1-methyl-2-(3-cyclohepta-1,3,5-trienyl) derivative; the rearranged product, upon reaction with triphenylcarbonium ion, gives the 1-methyl-2-tropenyliumyl-*o*-carborane cation*:

$$\begin{array}{c} \text{CH}_3\text{C}\!\!-\!\!\text{CLi} \\ \diagdown\!\!\text{O}\!\!\diagup \\ \text{B}_{10}\text{H}_{10} \end{array} + \text{C}_7\text{H}_7\text{OCH}_3 \longrightarrow \begin{array}{c} \text{CH}_3\text{C}\!\!-\!\!\text{C} \\ \diagdown\!\!\text{O}\!\!\diagup \\ \text{B}_{10}\text{H}_{10} \end{array} \xrightarrow{\Delta}$$

$$\begin{array}{c} \text{CH}_3\text{C}\!\!-\!\!\text{C} \\ \diagdown\!\!\text{O}\!\!\diagup \\ \text{B}_{10}\text{H}_{10} \end{array} \xrightarrow{(\text{C}_6\text{H}_5)_3\text{C}^+\text{SbF}_6^-} \left[\begin{array}{c} \text{CH}_3\text{C}\!\!-\!\!\text{C} \\ \diagdown\!\!\text{O}\!\!\diagup \\ \text{B}_{10}\text{H}_{10} \end{array} \right]^{+} \text{SbF}_6^- + (\text{C}_6\text{H}_5)_3\text{CH}$$

This cation, like the benzoic acid-*o*-carborane derivatives discussed in Section 6-5, has been examined for evidence of electron withdrawal by the *o*-carboranyl group. Ultraviolet and proton NMR spectra indicate a strong −I effect by

---

* Reaction of triphenylcarbonium hexafluoroantimonate with the first cycloheptatrienyl derivative (before rearrangement) yields triphenylmethane but no other identifiable products.

the carborane cage but virtually no +T electron transfer from the cage to the ring. [In contrast, tropenyliumyl-substituted borane anions such as $B_{10}H_9(C_7H_6)^-$ and $B_{12}H_{11}(C_7H_6)^-$ show significant +I, +T cage-to-ring donation in ground as well as excited states (*112*, *113*)]. Since $B_{12}H_{12}^{2-}$ and $C_2B_{10}H_{12}$ are isoelectronic and isostructural, the −I effect induced in the *o*-carboranyl system is attributed (*127*) to the relatively positive carbon atoms in the *o*-carborane cage. Consistent with this is the fact that the −I effect is much weaker in *m*-carborane (1,7-$C_2B_{10}H_{12}$), in which the carbon atoms are less positively charged.

6-4. ALKENYL AND ALKYNYL DERIVATIVES

Synthesis of C-Alkenyl Derivatives

C-alkenyl-*o*-carboranes have been prepared by the reaction of bis(ligand) decaborane species with alkenylacetylenes (*66*, *69*, *72*, *143*, *320*, *454*, *455*). This seems to be the preferred method, although alkenylcarboranes have also been obtained by the action of alkenyl halides on C-metallocarboranes (*142*, *363*, *425*, *443*) or carboranyl Grignards (*91*, *377*). The pyrolysis of certain *o*-carboranyl esters also produces alkenylcarboranes (*142*).

$$(HC\!\!-\!\!CCH_2)_2CHOCCH_3 \quad \xrightarrow{550°} \quad HC\!\!-\!\!CCH_2CH\!\!=\!\!CHC\!\!-\!\!CH \quad + \; CH_3COOH$$

Similarly, 1-vinyl-*o*-carborane is formed by elimination of acetic acid from 1-(α-acetoxyethyl)- or 1-(β-acetoxyethyl)-*o*-carborane.

$$HC\!\!-\!\!CCH_2CH_2OCCH_3 \quad \xrightarrow{500°} \quad HC\!\!-\!\!CCH\!\!=\!\!CH_2 \quad + \; CH_3COOH$$

$$HC\!\!-\!\!CCH(OCCH_3)CH_3$$

o-Carboranyl-β-chlorovinyl ketones have been prepared by the reaction of acetylene with the acyl chlorides of o-carboranyl carboxylic acids (379).

$$RC{-}CCH_2COCl \quad + \quad HC{\equiv}CH \quad \xrightarrow[5°,\ 1,2\text{-}C_2H_4Cl_2]{AlCl_3} \quad RC{-}CCOCH{=}CHCl$$
$$B_{10}H_{10} \qquad\qquad\qquad\qquad\qquad\qquad\qquad\qquad B_{10}H_{10}$$
$$(55\text{--}75\%)$$

Synthesis of B-Alkenyl Derivatives

The reaction of dicarbollide ion (Chapter 9) with vinyldichloroborane forms 3-vinyl-o-carborane in good yield (398). No other B-alkenyl derivatives, and no B-alkynyl species, have been reported.

$$C_2B_9H_{11}{}^{2-} + CH_2{=}CHBCl_2 \quad\longrightarrow\quad \begin{array}{c} HC{-}CH \\ CH_2{=}CHB_{10}H_9 \end{array} + 2Cl^-$$

Synthesis of Alkynyl Derivatives

Alkynylcarboranes are formed in reactions of bis-ligand decaboranes with dialkynes (56, 57, 59, 69) and by the action of alkynyl halides on metallated o-carborane derivatives in liquid ammonia (425). The former type of reaction may produce 1,1'-bis(o-carboranyl) as well as 1-alkynyl-o-carborane (56–58).

$$B_{10}H_{12}(CH_3CN)_2 + HC{\equiv}C{-}C{\equiv}CH \longrightarrow \underset{B_{10}H_{10}}{HC{-}C}\ \underset{B_{10}H_{10}}{C{-}CH} + \underset{B_{10}H_{10}}{HC{-}CC{\equiv}CH}$$
$$4\% \qquad\qquad\qquad 35\%$$

Certain dialkynes, such as 1,6-heptadiyne, are reported to give only the biscarboranyl derivative (66, 69). Commonly, however, the main product

$$HC{\equiv}C(CH_2)_3C{\equiv}CH + B_{10}H_{12}(CH_3CN)_2 \quad\longrightarrow\quad \underset{B_{10}H_{10}}{HC{-}C}(CH_2)_3\underset{B_{10}H_{10}}{C{-}CH}$$

of such reactions is the alkynylmonocarborane.

Reactions of Alkenyl Derivatives

C-alkenyl derivatives of o-carborane were among the first carboranes to be prepared, and their high chemical stability and the relative inertness of alkenyl double bonds adjacent to the carboranyl group provided some of the earliest evidence that a wholly new type of borane structure was involved. The remarkable behavior of 1-isopropenyl-o-carborane, for example, was noted well before the molecular geometry was established. This compound, like most alkenes having an o-carboranyl group adjacent to the double bond, is far less reactive than most alkenes toward cationic species. Indeed, early investigators failed to note any reaction of bromine with 1-vinyl-, 1-isopropenyl-, or 1-allyl-2-methyl-o-carborane (66, 91, 320). It has been shown, however, that 1 mole of bromine slowly adds to 1-vinyl-o-carborane in carbon tetrachloride over a very long period (150 hr) at 20°, and in less time (1.5 to 2 hr) at the boiling point of the solvent (394). The presence of aluminum chloride is not required.

$$HC{-}CCH{=}CH_2 \quad + \quad Br_2 \quad \longrightarrow \quad HC{-}C\overset{\displaystyle Br}{\overset{|}{C}}H{-}CH_2Br$$
$$\underset{B_{10}H_{10}}{\diagdown O\diagup} \qquad\qquad\qquad \underset{B_{10}H_{10}}{\diagdown O\diagup}$$

With aluminum chloride present, further bromination results in addition to the carboranyl group to give 1-(α,β-dibromoethyl)-8,9,10,12-tetrabromo-o-carborane. The latter compound can be readily converted to 1-vinyl-8,9,10,12-tetrabromo-o-carborane (394):

$$HC{-}CCHBr{-}CH_2Br \quad \xrightarrow[\text{AlCl}_3]{\text{Br}_2}$$
$$\underset{B_{10}H_{10}}{\diagdown O\diagup}$$

$$HC{-}CCHBr{-}CH_2Br \quad \xrightarrow[\text{C}_2\text{H}_5\text{OH}]{\text{Zn}} \quad HC{-}CCH{=}CH_2$$
$$\underset{B_{10}H_6Br_4}{\diagdown O\diagup} \qquad\qquad\qquad \underset{B_{10}H_6Br_4}{\diagdown O\diagup}$$

Addition of bromine to the 1-vinyltetrabromo derivative (or to the iodo analog) in carbon tetrachloride is much slower than is addition to 1-vinyl-o-carborane itself, indicating that the presence of halogen atoms on the carborane cage substantially increases the electron-withdrawing (−I) effect (394).

The double bond of 1-isopropenyl-o-carborane is rapidly dibrominated by bromine in carbon tetrachloride under ultraviolet light, but no halogenation of the carboranyl group is observed (148). The same compound, however, fails to react with bromine in ethanol in the dark (148).

Chlorine in carbon tetrachloride attacks both the double bond and the carboranyl group of 1-vinyl-o-carborane, with or without an aluminum chloride catalyst* (394). In contrast, iodine evidently fails to add to the double bond of 1-vinyl-o-carborane, even over aluminum chloride, but electrophilic substitution on the carborane cage does occur:

$$
\underset{B_{10}H_{10}}{HC\!-\!CCH\!=\!CH_2} + I_2 \xrightarrow[80°]{AlCl_3} \underset{B_{10}H_{10-n}I_n}{HC\!-\!CCH\!=\!CH_2}
$$

$$n = 1, 2, 3, 4$$

Alkenyl double bonds that are separated from the o-carboranyl system by methylene or other insulating groups react more or less normally with electrophilic reagents. Thus, 1 mole of bromine in carbon tetrachloride adds readily to 1-(γ-butenyl)-o-carborane to produce the corresponding dibromobutyl derivative (91). Similarly, o-carboranyl β-chlorovinyl ketones react as expected with 2,4-dinitrophenylhydrazine and with diazomethane, the latter reaction giving acyl pyrazoles (379).

$$
\underset{B_{10}H_{10}}{RC\!-\!CCH_2COCH\!=\!CHCl} \xrightarrow{CH_2N_2} \underset{B_{10}H_{10}}{RC\!-\!CCH_2COC\!-\!\!-CH}
$$

$$R = H, CH_3, CH_2\!=\!CH, BrCH_2$$

Only one mention of hydrogen halide addition to alkenylcarboranes appears in the literature: hydrogen chloride and hydrogen bromide reportedly add readily to 1-vinyl-o-carborane in carbon disulfide over aluminum bromide (394). In accordance with Markovnikov's rule, the products are 1-(α-chloroethyl)- and 1-(α-bromoethyl)-o-carborane, respectively.

The $-I$ effect of the o-carboranyl cage in alkenylcarboranes has been noted with a number of reagents. No reaction of 1-vinyl-, 1-isopropenyl-, or 1-allyl-2-methyl-o-carborane is observed with hydrogen peroxide, peracetic acid, hypochlorous acid, or iodine monobromide in glacial acetic acid (91). Trifluoroperacetic acid does attack both 1-allyl- and 1-isopropenyl-o-carborane, forming stable epoxides (66, 142). The epoxide ring of the allyl compound may be opened to give 1-(β-γ-dihydroxypropyl)-carborane (I); an apparent stereoisomer (II) of the latter compound is produced by conversion of 1-allyl-o-carborane to the hydroxyformal derivative followed by hydrolysis.

---

* Vinylcarborane undergoes an exchange reaction in refluxing carbon tetrachloride, forming the 1-vinyl-9,10-dichloro- and 1-vinyl-8,9,12-trichloro derivatives (446).

$$\underset{B_{10}H_{10}}{HC\text{---}CCH_2CH\text{==}CH_2} \xrightarrow{\quad CF_3CO_3H \quad}$$

$$\underset{B_{10}H_{10}}{HC\text{---}CCH_2CH\text{---}CH_2} \underset{O}{\diagdown\diagup} \xrightarrow{\quad 5\% H_2SO_4 \quad} \underset{B_{10}H_{10}}{HC\text{---}CCH_2CH\text{---}CH_2} \overset{OH\ OH}{\underset{}{}}$$

(I)

$$\xrightarrow[HCOOH]{H_2O_2} \underset{B_{10}H_{10}}{HC\text{---}CCH_2CHCH_2OH} \underset{O\text{---}CH}{\overset{}{}} \xrightarrow{\quad OH^- \quad} \underset{B_{10}H_{10}}{HC\text{---}CCH_2CH\text{---}CH_2} \overset{OH\ OH}{\underset{}{}}$$

(II)

Isomers **I** and **II** have identical elemental and infrared analyses but different melting points.

In contrast to the allyl and isopropenyl derivatives, the reaction of 1-(γ-butenyl-*o*-carborane with trifluoroperacetic acid forms a glycol trifluoroacetate adduct (*91*). Acid-catalyzed opening of an epoxy ring is evidently inhibited by an *o*-carboranyl group when it is adjacent to the ring, but not when it is insulated from the ring by methylene groups as in the γ-butenyl derivative. However, a γ-butenyl epoxide can be prepared by carrying out the reaction under basic conditions.

$$\underset{B_{10}H_{10}}{HC\text{---}CCH_2CH_2CH\text{==}CH_2} + CF_3CO_3H \xrightarrow[CH_2Cl_2]{Na_2CO_3} \underset{B_{10}H_{10}}{HC\text{---}CCH_2CH_2CH\text{---}CH_2} \underset{O}{\diagdown\diagup}$$

The carboranyl −I effect has also been noted in reactions of alkenylcarboranes with the electrophilic dichlorocarbene radical (*270*). Although 1-allyl-, 1-vinyl,- and 1-isopropenyl-*o*-carborane all react with phenyl(bromodichloromethyl)mercury, the allylcarborane is most reactive, in accord with both steric and electronic considerations. In each case the product is the corresponding *gem*-dichlorocyclopropane.

$$\underset{B_{10}H_{10}}{HC\text{---}CCH\text{==}CH_2} \xrightarrow{\quad C_6H_5HgCCl_2Br \quad} \underset{B_{10}H_{10}}{HC\text{---}C\text{---}CH\text{---}CCl_2} \underset{CH_2}{\diagdown\diagup}$$

All three alkenylcarboranes, however, are unreactive compared to cyclohexene; competition between the latter compound and an alkenylcarborane for the organomercury reagent gave 72–75% 7,7-dichloronorcarane but no detectable carborane products.

Alkenyl-*o*-carboranes tend to react as normal alkenes with standard oxidizing and hydrogenating agents. Thus, 1-isopropenyl-*o*-carborane is attacked by ozone (Section 6-7), by alkaline permanganate in acetone (*66*) (oxidizing the alkenyl group only), and by hydrogen over Raney nickel (*66, 454*), giving 1-isopropyl-*o*-carborane*:

$$\underset{B_{10}H_{10}}{HC\!\!-\!\!CC\!=\!CH_2} \overset{CH_3}{} \quad \overset{H_2,\ 50\ psi}{\underset{Raney\ Ni}{\longrightarrow}} \quad \underset{B_{10}H_{10}}{HC\!\!-\!\!CCH(CH_3)_2}$$

1-Isopropenyl-*o*-carborane is also quite susceptible to radical addition, reacting readily with dinitrogen tetroxide, tetrafluorohydrazine, and bromine under ultraviolet light (mentioned earlier) (*148*). With dinitrogen tetroxide the product is a mixture of the nitro nitrite and the dinitro adduct, which are converted by silica gel to the nitro alcohol and a nitro olefin, respectively:

$$\underset{B_{10}H_{10}}{HC\!\!-\!\!CC\!=\!CH_2} \overset{CH_3}{} + N_2O_4 \longrightarrow \underset{B_{10}H_{10}}{HC\!\!-\!\!C\!\!-\!\!C\!\!-\!\!CH_2NO_2} \overset{CH_3}{\underset{X}{}} \overset{silica\ gel}{\longrightarrow}$$

$$\underset{B_{10}H_{10}}{HC\!\!-\!\!C\!\!-\!\!C\!\!-\!\!CH_2NO_2}\overset{CH_3}{\underset{OH}{}} \quad + \quad \underset{B_{10}H_{10}}{HC\!\!-\!\!C\!\!-\!\!C\!=\!C\!\!-\!\!NO_2}\overset{CH_3}{\underset{H}{}}$$

$$X = NO_2,\ ONO$$

Tetrafluorohydrazine reacts to give the corresponding bis(difluoramino) derivative, which in turn can be converted to the difluoraminonitrile by treatment with base (*148*):

$$\underset{B_{10}H_{10}}{HC\!\!-\!\!C\!\!-\!\!C\!=\!CH_2}\overset{CH_3}{} \overset{N_2F_4,\ 150°}{\underset{150\ psi}{\longrightarrow}}$$

$$\underset{B_{10}H_{10}}{HC\!\!-\!\!C\!\!-\!\!C\!\!-\!\!CH_2NF_2}\overset{CH_3}{\underset{NF_2}{}} \overset{NaOCH_3}{\underset{CH_2Cl_2}{\longrightarrow}} \underset{B_{10}H_{10}}{HC\!\!-\!\!C\!\!-\!\!C\!\!-\!\!C\!\equiv\!N}\overset{CH_3}{\underset{NF_2}{}}$$

---

* Hydrogenation of 1,2-diisopropenyl-*o*-carborane, however, requires much higher pressure (1800 psi) (*69*).

Tetrafluorohydrazine reportedly gives analogous reactions with other 1-alkenyl-*o*-carboranes as well (*332*).

The reactions of alkylsilanes, alkoxysilanes, and halosilanes with alkenylcarboranes to give the corresponding addition products are described in Section 6-10.

Although most known reactions of alkenylcarboranes involve the alkenyl group itself, metallation of the carboranyl carbon may be conducted with no effect on the double bond. Consequently, the conversion of 1-alkenyl-*o*-carboranes into difunctional derivatives is straightforward, as shown by the synthesis of 1-isopropenyl-2-carboxy-*o*-carborane from 1-isopropenyl-*o*-carborane (*142*).

$$
\begin{array}{c}
\overset{\displaystyle CH_3}{|} \\
HC\!-\!\!-\!C\!-\!C\!=\!CH_2 \quad \xrightarrow{\ C_4H_9Li\ } \\
\diagdown O \diagup \\
B_{10}H_{10}
\end{array}
$$

$$
\begin{array}{c}
\overset{\displaystyle CH_3}{|} \\
LiC\!-\!\!-\!C\!-\!C\!=\!CH_2 \\
\diagdown O \diagup \\
B_{10}H_{10}
\end{array}
\xrightarrow[\ 2.\ HCl\ ]{\ 1.\ CO_2\ }
\begin{array}{c}
\overset{\displaystyle CH_3}{|} \\
HOOCC\!-\!\!-\!C\!-\!C\!=\!CH_2 \\
\diagdown O \diagup \\
B_{10}H_{10}
\end{array}
+ \ LiCl
$$

Reactions of Alkynyl Derivatives

Treatment of 1-ethynyl-*o*-carborane with a Grignard reagent leads to metallation of the acetylenic group rather than the carborane cage. The Grignard complex may be used to prepare a variety of carboranylacetylene derivatives, such as 2-*o*-carboranyl propiolic acid (*59*).

$$
\begin{array}{c}
HC\!-\!\!-\!C\!-\!C\!\equiv\!CH \\
\diagdown O \diagup \\
B_{10}H_{10}
\end{array}
\xrightarrow[\ C_2H_5OH\ ]{\ C_2H_5MgBr\ }
$$

$$
\begin{array}{c}
HC\!-\!\!-\!C\!-\!C\!\equiv\!CMgBr \\
\diagdown O \diagup \\
B_{10}H_{10}
\end{array}
\xrightarrow[\ 2.\ H_3O^+\ ]{\ 1.\ CO_2\ }
\begin{array}{c}
HC\!-\!\!-\!C\!-\!C\!\equiv\!CCOOH \\
\diagdown O \diagup \\
B_{10}H_{10}
\end{array}
$$

Analogous reactions of the same Grignard with iodine, bromine, ethylene oxide, formaldehyde, alkyl halides, and alkyl sulfates yield, respectively, the corresponding monoiodide, monobromide, β-hydroxyethyl, hydroxymethyl, and monoalkyl derivatives.

6-5. CARBOXYLIC ACIDS AND ESTERS

Synthesis of C-Substituted Acids and Esters

Derivatives of o-carborane containing COOH or CH_2OH groups cannot be prepared directly from bis(ligand) decaborane compounds (Section 6-1) and acetylenic acids or alcohols, since these reagents attack and destroy the borane cage. However, o-carboranyl carboxylic acids have been prepared by several methods, including the reaction of CO_2 with metallocarboranes and carborane Grignards, hydrolysis of esters and acyl halides, and oxidation of carborane alcohols.

Dilithio-o-carborane reacts readily with CO_2, giving on subsequent acidification the dicarboxylic acid (*76, 91, 142, 463*).

$$
\underset{B_{10}H_{10}}{HC\text{---}CH} \xrightarrow{2n\text{-}C_4H_9Li} \underset{B_{10}H_{10}}{LiC\text{---}CLi} \xrightarrow[\text{2. } H_3O^+]{\text{1. } 2CO_2} \underset{B_{10}H_{10}}{HOOCC\text{---}CCOOH}
$$

Reaction of CO_2 with the monolithio derivative in diethyl ether has been reported to give only the dicarboxylic acid, and a lithium exchange (Section 6-2) has been suggested as the mechanism responsible (*142*). Recently, however, the monocarboxylic acid has been prepared in 80% yield by this route in benzene solvent; no dicarboxylic acid is obtained in this medium (*376, 455*).

A similar reaction is obtained with monosodio-o-carborane, which is easily prepared from sodamide and o-carborane (the disodio derivative could not be prepared) (*363, 458*).

The synthesis of C-substituted monocarboxylic acids is straightforward, since metal exchange is blocked (*76, 91, 92, 410, 455, 463*).

$$
\underset{B_{10}H_{10}}{RC\text{---}CH} \xrightarrow{n\text{-}C_4H_9Li} \underset{B_{10}H_{10}}{RC\text{---}CLi} \xrightarrow[\text{2. } H_3O^+]{\text{1. } CO_2} \underset{B_{10}H_{10}}{RC\text{---}CCOOH}
$$

R = alkyl, alkenyl, or aryl group (Table 6-2)

The generality of this route is illustrated by the synthesis of bis(2-carboxy-1-o-carboranylmethyl) ether in high yield from bis(1-o-carboranylmethyl) ether [obtained by reaction of $B_{10}H_{12}(CH_3CN)_2$ with propargyl ether] (*90*).

$$
\underset{B_{10}H_{10}}{(HC\text{---}CCH_2)_2O} \xrightarrow{2n\text{-}C_4H_9Li} \underset{B_{10}H_{10}}{(LiC\text{---}CCH_2)_2O} \xrightarrow[\text{2. } H_3O^+]{\text{1. } CO_2} \underset{B_{10}H_{10}}{(HOOCC\text{---}CCH_2)_2O}
$$

The use of only one equivalent of butyllithium leads to the corresponding monobasic acid.

Reaction of ethylmagnesium bromide with o-carborane followed by carbonation yields the monocarboxylic acid (293, 298).

$$\underset{B_{10}H_{10}}{HC\text{---}CH} \xrightarrow[\text{THF}]{C_2H_5MgBr} \underset{B_{10}H_{10}}{HC\text{---}CMgBr} \xrightarrow[\text{2. } H_3O^+]{\text{1. } CO_2} \underset{B_{10}H_{10}}{HC\text{---}CCOOH}$$

(45–60%)

Carboxylic acids containing one or more methylene units between the carboxyl group and the cage system are obtained from 1-haloalkyl-o-carboranes via the Grignard reagent, but the monocarboxylic acid ($n = 0$ in the series shown) is not obtainable by this route (298, 378).

$$\underset{B_{10}H_{10}}{HC\text{---}C(CH_2)_nBr} \xrightarrow[\text{ether}]{Mg} \underset{B_{10}H_{10}}{HC\text{---}C(CH_2)_nMgBr} \xrightarrow[\text{2. } H_3O^+]{\text{1. } CO_2} \underset{B_{10}H_{10}}{HC\text{---}C(CH_2)_nCOOH}$$

$n = 1\text{--}4$

As described in Section 6-2, rearrangements of the Grignards obtained from 1-bromomethyl-o-carborane in tetrahydrofuran, and from 1-chloromethyl-o-carborane in tetrahydrofuran or ether, lead to the formation of 1-methyl-2-carboxylic acid-o-carborane instead of o-carboranyl acetic acid (such rearrangements are not observed in 1-alkyl-2-halomethyl-o-carboranes, which lack carborane C—H bonds) (91, 142, 292, 375). Attempts to prepare a homologous series of acids (containing higher alkyl groups in place of methyl) by utilizing such rearrangements have been unsuccessful. For example, γ-bromopropyl-o-carborane undergoes cyclization (293, 298):

$$\underset{B_{10}H_{10}}{HC\text{---}C(CH_2)_3Br} \xrightarrow[\text{THF}]{Mg} \underset{B_{10}H_{10}}{\overset{\displaystyle H_2C\overset{CH_2}{\diagup}\diagdown CH_2}{\underset{C\text{---}C}{| \quad |}}}$$

o-Carboranyl acetic acids are conveniently prepared by the action of sodium haloacetates on sodium derivatives of o-carborane in liquid ammonia, followed by acidification with HCl (379):

$$\underset{B_{10}H_{10}}{RC\text{---}CNa} + BrCH_2COONa \xrightarrow[\text{2. HCl}]{\text{1. } NH_3(l)} \underset{B_{10}H_{10}}{RC\text{---}CCH_2COOH}$$

$R = H, CH_3, CH_2{=}CH, CH_2{=}CHCH_3, C_6H_5, BrCH_2$

TABLE 6-2

IONIZATION CONSTANTS OF CARBORANE CARBOXYLIC ACIDS

R	R′	pK_a	Solvent[a]	Refs.
		C-substituted o-carboranyl acids		

$$R\text{—}C\text{——}C\text{—}R'$$
$$\overset{}{\underset{B_{10}H_{10}}{\bigcirc}}$$

R	R′	pK_a	Solvent[a]	Refs.
H	COOH	2.48	A	313
H	COOH	2.49	A	373, 463
H	COOH	2.61	B	398
CH_3	COOH	2.74	B	92
CH_3	COOH	2.53	A	92
H	$m\text{-}C_6H_4COOH$	5.84	C	127
H	$m\text{-}C_6H_4COOH$	5.79	C	419
H	$m\text{-}C_6H_4COOH$	6.57	D	419
H	$p\text{-}C_6H_4COOH$	5.86	C	127
H	$p\text{-}C_6H_4COOH$	5.88	C	419
H	$p\text{-}C_6H_4COOH$	6.55	D	419
H	CH_2COOH	4.06	A	373
H	CH_2COOH	3.83	E	373
H	$(CH_2)_2COOH$	4.58	E	373
$(CH_3)_2CH$	COOH	2.56	A	373
C_4H_9	COOH	3.01	E	373
$CH_2{=}CH$	COOH	2.72	A	373
C_6H_5	COOH	3.12	A	373
CH_3OCH_2	COOH	2.33	A	373

B(3)-substituted o-carboranyl acids

$$HC\text{——}CH$$
$$\overset{}{\underset{B_{10}H_9R}{\bigcirc}}$$

R		pK_a	Solvent[a]	Refs.
COOH		5.38	B	398
$m\text{-}C_6H_4COOH$		6.25	E	419, 420
$m\text{-}C_6H_4COOH$		7.05	D	419
$p\text{-}C_6H_4COOH$		6.26	C	419, 420
$p\text{-}C_6H_4COOH$		6.99	D	419

C-substituted m-carboranyl acids, $RCB_{10}H_{10}CR'$

R	R′	pK_a	Solvent[a]	Refs.
H	COOH	3.20	A	313
H	COOH	3.34	B	398
CH_3	COOH	3.14	B	92

(continued)

TABLE 6-2—*Continued*

R	R'	pK_a	Solvent[a]	Refs.
H	$m\text{-}C_6H_4COOH$	6.17	C	*419*
H	$m\text{-}C_6H_4COOH$	6.96	D	*419*
H	$p\text{-}C_6H_4COOH$	6.04	C	*419*
H	$p\text{-}C_6H_4COOH$	6.79	D	*419*

C-substituted *p*-carboranyl acids, $RCB_{10}H_{10}CR'$

R	R'	pK_a	Solvent[a]	Refs.
H	COOH	3.64	B	*417*

[a] Code: A = H_2O; B = 50% C_2H_5OH in H_2O; C = 75% C_2H_5OH in H_2O; D = 70% dioxane in H_2O; E = 20% C_2H_5OH in H_2O.

o-Carboranyl carboxylic acids are often obtained by hydrolysis of the corresponding esters, which unlike the acids may be prepared directly from acetylenic esters and a bis-ligand decaborane derivative (*69, 74, 91, 142, 143, 304, 317, 371, 372, 410, 455*). For example, 1-*o*-carboranyl carboxylic acid is easily obtained from methyl 1-*o*-carboranyl carboxylate (*142*):

$$B_{10}H_{12}L_2 + HC\equiv CCOOCH_3 \longrightarrow \underset{B_{10}H_{10}}{\underset{\diagdown O \diagup}{HC\!-\!CCOOCH_3}} \xrightarrow{H^+} \underset{B_{10}H_{10}}{\underset{\diagdown O \diagup}{HC\!-\!CCOOH}}$$

$$L = CH_3CN, (C_2H_5)_2S$$

Many other esters of carborane carboxylic acids, however, are surprisingly resistant to acid hydrolysis. Dimethyl-1,2-*o*-carboranyl dicarboxylate is not affected by refluxing hydrochloric acid, trifluoroacetic acid, or other acids (it may, however, be saponified in poor yield by 50% aqueous potassium hydroxide) (*91*). Ethyl 1-*o*-carboranyl acetate is similarly unreactive in 96% H_2SO_4 at 100°, but in aqueous dioxane it is converted to the acid by 8 N H_2SO_4 (*372*).

$$HC\equiv CCH_2COOC_2H_5 + B_{10}H_{12}(CH_3CN)_2 \longrightarrow$$

$$\underset{B_{10}H_{10}}{\underset{\diagdown O \diagup}{HC\!-\!CCH_2COOC_2H_5}} \xrightarrow{H^+} \underset{B_{10}H_{10}}{\underset{\diagdown O \diagup}{HC\!-\!CCH_2COOH}}$$

Ester hydrolysis in alkaline media is often accompanied by decarboxylation; thus, KOH in ethanol reacts with ethyl 1-carboranyl carboxylate to give the parent carborane as well as the carboxylic acid salt (*372*).

$$\underset{B_{10}H_{10}}{\underset{\diagdown O \diagup}{HC\!-\!CCOOC_2H_5}} \xrightarrow[C_2H_5OH]{KOH} \underset{B_{10}H_{10}}{\underset{\diagdown O \diagup}{HC\!-\!CH}} + \underset{B_{10}H_{10}}{\underset{\diagdown O \diagup}{HC\!-\!CCOOK}}$$

Similar treatment of the diethyldicarboxylate ester affords only the salt of the carboranyl monocarboxylic acid. On the other hand, both esters are completely decarboxylated to o-carborane and diethyl carbonate by a catalytic quantity of sodium ethoxide in absolute ethanol (*364, 370, 371*). A plausible mechanism (*364, 371*) involves an ionic intermediate:

$$HC\!\!-\!\!CCOOC_2H_5 \quad \xrightarrow[20°]{C_2H_5O^-} \quad HC\!\!-\!\!C\!\!-\!\!C(OC_2H_5)_2 \quad \longrightarrow$$
$$B_{10}H_{10} \qquad\qquad\qquad B_{10}H_{10} \qquad O^-$$

$$(C_2H_5O)_2CO \quad + \quad HC\!\!-\!\!C^- \quad \xrightarrow{C_2H_5OH} \quad HC\!\!-\!\!CH \quad + \; C_2H_5O^-$$
$$B_{10}H_{10} \qquad\qquad\qquad B_{10}H_{10}$$

Cleavage of the diethyl ester occurs even at $-20°$ (*370*).

Such reactions are good evidence of the strongly electron-withdrawing character of the carborane cage system (discussed below). Accordingly, decarboxylation is inhibited or prevented when the carboxyl group is separated from the carborane cage by one or more methylene groups. Thus, the ethyl esters of carboranyl acetic acid, β-carboranyl propionic acid, and γ-carboranyl butyric acid are not cleaved by sodium ethoxide (*317*) [partial decarboxylation does occur, however, in alcoholic potassium hydroxide (*372*)]. A large steric factor is also evident (*91*), since it is found that o-carborane acids or esters are more prone to decarboxylation when a substituent group is attached to the carborane cage at C(2). For example, potassium 1-phenylcarboranyl-2-carboxylate is quantitatively converted to phenyl-o-carborane in 98 % ethanol (*364*).

$$C_6H_5C\!\!-\!\!CCOOK \quad \xrightarrow[20°]{C_2H_5OH} \quad C_6H_5C\!\!-\!\!CH \quad + \; KHCO_3$$
$$B_{10}H_{10} \qquad\qquad\qquad B_{10}H_{10}$$

The esters are also cleaved by organometallic reagents such as n-butyllithium, and the lithio carboranes obtained can be readily converted to the carboxylic acid (*436*).

$$C_6H_5C\!\!-\!\!CCOOCH_3 \quad \xrightarrow{C_4H_9Li}$$
$$B_{10}H_{10}$$

$$C_4H_9COOCH_3 \quad + \quad C_6H_5C\!\!-\!\!CLi \quad \xrightarrow[2.\ H_3O^+]{1.\ CO_2} \quad C_6H_5C\!\!-\!\!CCOOH$$
$$B_{10}H_{10} \qquad\qquad\qquad B_{10}H_{10}$$

[In contrast, treatment of these esters with Grignards with subsequent acidification yields secondary or tertiary alcohols rather than cleavage products (Section 6-6). Such behavior directly parallels that of o-carboranyl ketones, and in fact the ketones have been identified as intermediates in the action of organolithium compounds and Grignards on o-carboranyl esters (*450*)].

Acyl halide derivatives of o-carborane, which in general may be prepared directly from bisacetonitriledecaborane(14), are rapidly hydrolyzed to the corresponding acids (*372*).

$$HC\equiv C(CH_2)_2COCl + B_{10}H_{12}(CH_3CN)_2 \longrightarrow$$

Phenylpropiolic chloride apparently forms the expected phenylcarboranyl acid chloride, which, however, immediately undergoes an interesting cyclization to 2,3-benzo-4,5-carboranocyclopentanone (*372, 455, 459*).

$$C_6H_5C\equiv CCOCl + B_{10}H_{12}(CH_3CN)_2 \longrightarrow$$

Phenylcarboranyl acid chloride itself may be isolated from the reaction of phenylcarborane carboxylic acid with phosphorus pentachloride. Treatment of the product with aluminum chloride gives the above-mentioned cyclic ketone; the latter compound, on reaction with alcoholic base followed by acidification, yields o-carboranyl benzoic acid (*410*).

Certain carborane carboxylic acids have been prepared by the oxidation of o-carborane alcohols, but as the latter compounds are not directly obtainable from bis-ligand decaborane species (see above), the general utility of this approach seems limited. Chromium trioxide in acid media converts primary hydroxyalkyl-o-carboranes to the corresponding acids (142, 372, 431).

$$\text{HC}\underset{\underset{B_{10}H_{10}}{\diagdown O\diagup}}{\overline{\quad\quad}}\text{C(CH}_2)_n\text{CH}_2\text{OH} \xrightarrow[\text{H}_2\text{SO}_4]{\text{CrO}_3} \text{HC}\underset{\underset{B_{10}H_{10}}{\diagdown O\diagup}}{\overline{\quad\quad}}\text{C(CH}_2)_n\text{COOH}$$

Similarly, o-carborane dicarboxylic acid is obtained from bis(hydroxymethyl)-o-carborane, although in small yield (372). In a typical reaction sequence, acetylenic alcohol is transesterified, converted to the corresponding o-carboranyl ester, and hydrolyzed to the carborane alcohol (365). The latter compound may then be oxidized to the acid with chromium trioxide (372, 455).

$$\text{HC}\equiv\text{C(CH}_2)_n\text{OH} + (\text{CH}_3\text{CO})_2\text{O} \longrightarrow \text{HC}\equiv\text{C(CH}_2)_n\overset{\text{O}}{\overset{\|}{\text{OCCH}_3}} \xrightarrow{B_{10}H_{12}(\text{CH}_3\text{CN})_2}$$
$$+ \text{CH}_3\text{COOH}$$

$$\text{HC}\underset{\underset{B_{10}H_{10}}{\diagdown O\diagup}}{\overline{\quad\quad}}\text{C(CH}_2)_n\overset{\text{O}}{\overset{\|}{\text{OCCH}_3}} \xrightarrow{\text{H}_3\text{O}^+} \text{HC}\underset{\underset{B_{10}H_{10}}{\diagdown O\diagup}}{\overline{\quad\quad}}\text{C(CH}_2)_n\text{OH} \xrightarrow[\text{H}_2\text{SO}_4]{\text{CrO}_3} \text{HC}\underset{\underset{B_{10}H_{10}}{\diagdown O\diagup}}{\overline{\quad\quad}}\text{C(CH}_2)_{n-1}\text{COOH}$$

$$n = 1, 2, 3$$

Oxidation of 1-phenyl-2-hydroxyethyl-o-carborane under similar conditions affords 1-phenyl-2-carboxymethyl-o-carborane (361).

$$\text{C}_6\text{H}_5\text{C}\underset{\underset{B_{10}H_{10}}{\diagdown O\diagup}}{\overline{\quad\quad}}\text{CCH}_2\text{CH}_2\text{OH} \xrightarrow[\text{H}_2\text{SO}_4]{\text{CrO}_3} \text{C}_6\text{H}_5\text{C}\underset{\underset{B_{10}H_{10}}{\diagdown O\diagup}}{\overline{\quad\quad}}\text{CCH}_2\text{COOH}$$

A convenient synthesis of 1,2-bis(carboxymethyl)-o-carborane utilizes the reaction of ethylene oxide with dilithio-o-carborane to form the diol, which is then oxidized (142).

$$\text{LiC}\underset{\underset{B_{10}H_{10}}{\diagdown O\diagup}}{\overline{\quad\quad}}\text{CLi} + 2\ \text{H}_2\text{C}\underset{\underset{O}{\diagdown\diagup}}{\overline{\quad\quad}}\text{CH}_2 \longrightarrow$$

$$\text{HO(CH}_2)_2\text{C}\underset{\underset{B_{10}H_{10}}{\diagdown O\diagup}}{\overline{\quad\quad}}\text{C(CH}_2)_2\text{OH} \xrightarrow[\text{H}^+]{\text{CrO}_3} \text{HOOCCH}_2\text{C}\underset{\underset{B_{10}H_{10}}{\diagdown O\diagup}}{\overline{\quad\quad}}\text{CCH}_2\text{COOH}$$

Similarly, bis[2-(β-hydroxyethyl)-1-carboranylmethyl] ether may be prepared and then converted to the corresponding ether diacid (90, 142).

$$(\text{LiC}\underset{B_{10}H_{10}}{\overset{\diagup O \diagdown}{-}}\text{CCH}_2)_2\text{O} + 2\text{CH}_2\underset{O}{\overset{\diagup\diagdown}{-}}\text{CH}_2 \longrightarrow (\text{HOCH}_2\text{CH}_2\text{C}\underset{B_{10}H_{10}}{\overset{\diagup O \diagdown}{-}}\text{CCH}_2)_2\text{O} \xrightarrow[\substack{H_3O^+, \\ acetone}]{K_2Cr_2O_7}$$

$$(\text{HOOCCH}_2\text{C}\underset{B_{10}H_{10}}{\overset{\diagup O \diagdown}{-}}\text{CCH}_2)_2\text{O}$$

Potassium permanganate in basic media oxidizes certain carborane alcohols to salts of the acids (e.g., β-hydroxyethyl- and γ-hydroxypropylcarborane) (*372*) but with 1,2-bis(hydroxymethyl)- and 1-hydroxymethyl-*o*-carborane the result is decarboxylation to give the parent carborane (*91, 372, 455*).

$$\text{HC}\underset{B_{10}H_{10}}{\overset{\diagup O \diagdown}{-}}\text{CCH}_2\text{OH} \xrightarrow[\text{OH}^-]{\text{KMnO}_4} \text{HC}\underset{B_{10}H_{10}}{\overset{\diagup O \diagdown}{-}}\text{CH} \xleftarrow[\text{OH}^-]{\text{KMnO}_4} \text{HOCH}_2\text{C}\underset{B_{10}H_{10}}{\overset{\diagup O \diagdown}{-}}\text{CCH}_2\text{OH}$$

Evidence has been obtained suggesting that the alcohol is first oxidized to a ketone, which in turn is cleaved by alkali to give the carborane (*427*). In contrast, 1,2-*o*-carboranyl dicarboxylic acid is degraded to boric acid by permanganate (*91*).

Synthesis of B-Substituted Acids

The ozonization of 3-vinyl-*o*-carborane (Section 6-4) yields 3-formyl-*o*-carborane, which upon oxidation forms 3-carboxy-*o*-carborane in high yield (*398*).

Reactions of Acids and Esters

As might be anticipated from the electrophilic character of the *o*-carboranyl system, the acid strength of *o*-carborane carboxylic acids is moderately high, with the pK_a values shown in Table 6-2. It will be noted that the C-substituted acids of the *o*-carboranyl system are stronger than the corresponding *m*-carboranyl acids, presumably due to greater (and more localized) positive charge on the cage carbon atoms in the former (*127*). A direct comparison is provided by the ionization constants of carboranyl benzoic acid derivatives (Table 6-2). These data, combined with the results of mixed acid nitration of 1-phenyl-*o*-carborane, the ^{19}F NMR chemical shifts of some *o*-carboranyl- and *m*-carboranyl fluorobenzenes, and the determined pK_a values of a number of *o*-carboranyl anilinium ions, have led Hawthorne *et al.* (*127*) to the following conclusions: (1) the o-carboranyl system is a powerful electron acceptor by an inductive mechanism, the strength of the $-I$ effect being roughly comparable to the halogens; (2) the electron-withdrawing tendency of the m-carboranyl system is considerably weaker than that of the o-carboranyl; (3) there is no appreciable ground-state extension of electron delocalization by interaction of the aryl group with the o-carboranyl unit; (4) weak ground-state extension of electron delocalization may exist via electron donation from the m-carboranyl system to the aryl ring. In addition, Russian workers have prepared several 3-o-carboranyl acid derivatives (in which the organic ligand is bonded to boron) and have measured the acid dissociation constants (*373, 398, 419, 420*). The results, supported by competitive nitration studies, indicate that the $-I$ effect of the o-carboranyl group is far weaker when the ligand is bonded at B(3) than when the attachment is at C(1). In fact, the 3-o-carboranyl system appears to be an even weaker electron acceptor than the 1-m-carboranyl unit (*398, 420*), as shown by the relative pK_a values in Table 6-2.

Not surprisingly, the observed inductive effect of the o-carboranyl group with respect to fluorophenyl ligands bonded at C(1) or B(3) has been found to be solvent-dependent, with some evidence of hydrogen bond formation between the carborane cage and proton-accepting solvents (*422a*).

The acid strengths of 1-o-carboranyl carboxylic acids are decreased when substitutents which give rise to a $+I$ effect are bonded to the carborane cage at C(2) (*373*). The greater the $+I$ effect, the greater the decrease in the acid ionization constant; thus, alkyl groups bonded at C(2) cause a moderate increase in pK_a, whereas vinyl and phenyl substituents induce a considerable increase. Such an effect is to be expected, since a shift of electron density into the o-carborane cage diminishes its electron-withdrawing ability.

It is clear that the inductive effect of the carboranyl system is strongest in the case of substituents bonded directly to the carborane cage, and falls off sharply when the functional group is separated from the carborane by methylene units (*142, 317, 364*). This is evident in the decarboxylation of carborane esters discussed above, and is also true of the acids. Thus, 1,2-bis(carboxymethyl)-o-carborane gives the expected reactions of a dicarboxylic acid: it is converted by ammonia to the diamide, by methanol to the dimethyl ester, and by phosphorus pentachloride to the diacyl dichloride (which easily regenerates the diacid on hydrolysis). Upon heating *in vacuo*, the same acid behaves similarly to adipic acid by forming a cyclic ketone (*142*).

In contrast, the properties of *o*-carboranyl-1,2-dicarboxylic acid are highly unusual. The dimethyl ester cannot be hydrolyzed to the diacid, nor can the diacid itself be esterified. The diacid does form salts with ammonia, hydrazine, aniline, and diethylamine, but none of these salts can be converted to a di-amide. Treatment of the diacid with thionyl chloride or phosphorus penta-chloride fails to give the diacid dichloride, yielding instead the anhydride. However, either the anhydride or the diacid itself may be converted to the diacid dichloride by refluxing with phosphorus oxychloride in chlorine (*142*). The following diagram summarizes these interconversions:

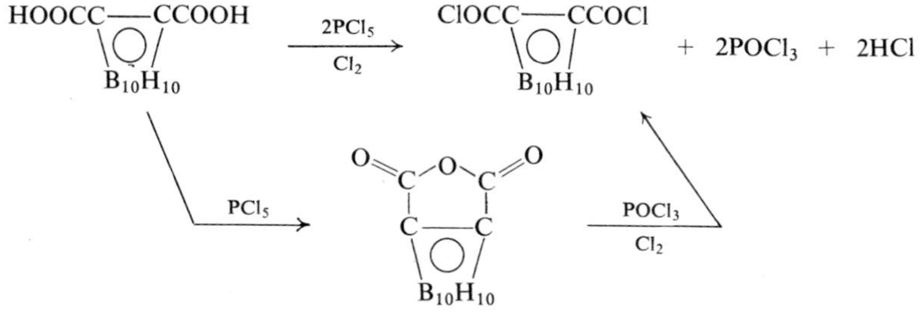

A compound closely related to the dicarboxylic acid is the ether diacid, bis(2-carboxy-1-carboranylmethyl)ether, whose preparation was given earlier. As expected from the direct carboxyl-to-carborane bonding, this compound resembles *o*-carboranyl dicarboxylic acid in that it cannot be esterified in acid media and fails to react with refluxing thionyl chloride or chlorine in phos-phorus oxychloride. It may, however, be chlorinated by phosphorus penta-chloride (*90*).

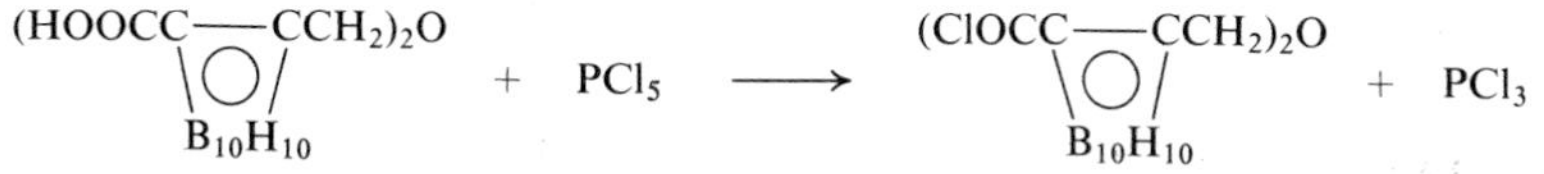

The decarboxylation of carborane carboxylic acids, as opposed to the esters (see above), has been little studied, but the acetylacetonates of beryllium and zirconium quantitatively convert 1-methyl-2-*o*-carboranyl carboxylic acid and *o*-carboranyl-1,2-dicarboxylic acid to 1-methyl-*o*-carborane and *o*-carborane, respectively (*76*). Decarboxylation also occurs in the formation of bis(1-carboranylmethyl) ether as a by-product in the pyrolysis of bis(2-carboxy-1-carboranylmethyl) ether. The main product of the reaction is a lactone (*90*).

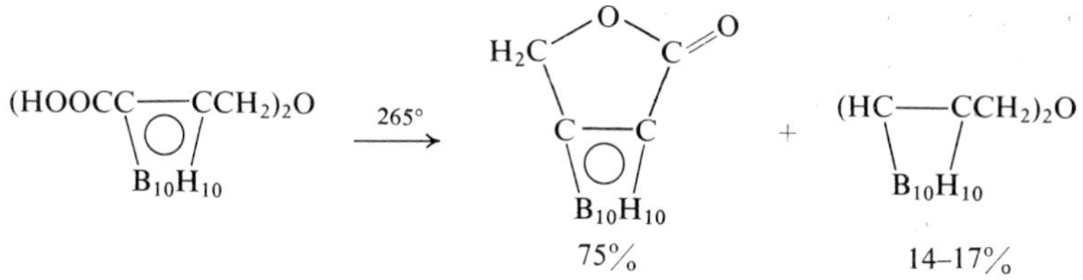

The lactone is a synthetically useful derivative, since on alkaline hydrolysis followed by acidification it is converted into the difunctional compound 1-hydroxymethyl-2-carboxy-o-carborane (90).

$$
\begin{array}{c}
\text{H}_2\text{C} \overset{\text{O}}{\diagdown}\text{C}{=}\text{O} \\
\text{C}-\text{C} \\
\diagdown\text{O}\diagup \\
\text{B}_{10}\text{H}_{10}
\end{array}
\quad
\xrightarrow[\text{2. H}_3\text{O}^+]{\text{1. NaOH}}
\quad
\begin{array}{c}
\text{HOCH}_2\text{C}-\text{CCOOH} \\
\diagdown\text{O}\diagup \\
\text{B}_{10}\text{H}_{10}
\end{array}
$$

6-6. ALCOHOLS AND ETHERS

Synthesis of Alcohols

o-Carboranyl alcohols apparently cannot be obtained directly from decaborane derivatives by reaction with acetylenic alcohols, since the borane cage is degraded by hydroxyl groups. However, if the acetylenic alcohol is first esterified and the product is allowed to react with a bis-ligand decaborane compound, the corresponding o-carboranyl ester usually forms readily.* The ester may be converted to the alcohol by transesterification (69) or by hydrolysis ($73, 91, 142, 365, 455$).

$$
\begin{array}{c}
\overset{\text{O}}{\text{CH}_3\text{COCH}_2\text{C}}-\text{CH} \\
\diagdown\text{O}\diagup \\
\text{B}_{10}\text{H}_{10}
\end{array}
+ \text{CH}_3\text{OH}
\xrightarrow{\text{HCl}}
\begin{array}{c}
\text{HOCH}_2\text{C}-\text{CH} \\
\diagdown\text{O}\diagup \\
\text{B}_{10}\text{H}_{10}
\end{array}
+ \overset{\text{O}}{\text{CH}_3\text{OCCH}_3}
$$

$$
\xrightarrow[\text{or OH}^-]{\text{LiAlH}_4}
\begin{array}{c}
\text{HC}-\text{CCH}_2\text{OH} \\
\diagdown\text{O}\diagup \\
\text{B}_{10}\text{H}_{10}
\end{array}
$$

While the hydrolysis of o-carboranyl esters has been successfully carried out in a variety of solvents and in both acidic and basic media, Zakharkin *et al.* favor acid methanolysis (365). Transesterification or hydrolysis proceeds equally readily with diesters to give the corresponding diols.

---

* The distinction between carborane esters of this type and those obtained from carboranyl carboxylic acids (Section 6-5) should be noted.

Alcohols of o-carborane may also be prepared in reactions of alkali metal carborane derivatives or carborane Grignard reagents with epoxides, aldehydes, ketones, and esters. Thus, dilithio-o-carborane reacts with ethylene oxide to yield 1,2-bis(β-hydroxyethyl)-o-carborane (Section 6-4) (90, 91, 142, 363). With α-epoxides the products are secondary alcohols (431).

$$RC\!-\!\!-\!CM \overset{\diagdown\!\!O\!\!\diagup}{B_{10}H_{10}} + \overset{CH_2\!-\!CHR'}{\diagdown_O\diagup} \longrightarrow RC\!-\!\!-\!CCH_2CHR' \overset{\diagdown\!\!O\!\!\diagup}{B_{10}H_{10}}\;\overset{|}{OH}$$

$$M = Li, Na$$
$$R = H, CH_3, CH_2\!\!=\!\!CH, C_6H_5$$
$$R' = H, CH_3, C_6H_5, CH_2Cl$$

At high temperatures the reaction with epichlorohydrin proceeds differently, yielding the epoxide and the bis(carboranyl) alcohol (431).

$$CH_3C\!-\!\!-\!CLi \overset{\diagdown\!\!O\!\!\diagup}{B_{10}H_{10}} + \overset{CH_2\!-\!CHCH_2Cl}{\diagdown_O\diagup} \longrightarrow$$

$$CH_3C\!-\!\!-\!CCH_2CH\!-\!CH_2 \overset{\diagdown\!\!O\!\!\diagup}{B_{10}H_{10}}\;\overset{\diagdown_O\diagup}{} + (CH_3C\!-\!\!-\!CCH_2)_2CHOH \overset{\diagdown\!\!O\!\!\diagup}{B_{10}H_{10}}$$

The lithium derivatives of o-carborane and of 1-methyl-, 1-phenyl-, and 1-allyl-o-carborane react with many aliphatic, aromatic, and heterocyclic aldehydes to give secondary alcohols (90, 142, 311, 363, 426). The reaction with benzaldehyde, for example, proceeds smoothly (311).

$$CH_3C\!-\!\!-\!CLi \overset{\diagdown\!\!O\!\!\diagup}{B_{10}H_{10}} + C_6H_5\overset{O}{\overset{\|}{CH}} \longrightarrow CH_3C\!-\!\!-\!C\overset{H}{\overset{|}{C}}\!-\!C_6H_5 \overset{\diagdown\!\!O\!\!\diagup}{B_{10}H_{10}}\;\overset{|}{OH}$$

Alternatively, o-carboranyl Grignard reagents may be used in place of the lithium derivatives (375, 378). Although disodio-o-carborane reacts with formaldehyde in liquid ammonia to give bis(hydroxymethyl)-o-carborane in 92% yield, the use of sodium derivatives usually leads to poor yields of carborane alcohols. This is attributed to an equilibrium in the reaction system (426):

$$RC\!-\!\!-\!CNa \overset{\diagdown\!\!O\!\!\diagup}{B_{10}H_{10}} + R'\overset{O}{\overset{\|}{CH}} \overset{NH_3\,(liq)}{\rightleftharpoons} RC\!-\!\!-\!C\overset{H}{\overset{|}{C}}\!-\!CR' \overset{\diagdown\!\!O\!\!\diagup}{B_{10}H_{10}}\;\overset{|}{ONa}$$

The products obtained in reactions of lithio- and dilithio-o-carborane with esters are not the expected tertiary alcohols, but secondary alcohols or ketones.

From the study of a few such reactions (*430, 432*), it appears that ketones are formed from methyl esters of aromatic acids and from methyl and ethyl esters of aliphatic acids, while the ethyl, propyl, and isopropyl esters of aromatic acids lead to secondary alcohols. The following examples illustrate both situations.

$$C_6H_5C\text{—}CLi\diagdown O\diagup B_{10}H_{10} + C_6H_5\overset{O}{\overset{\|}{C}}OCH_3 \longrightarrow C_6H_5C\text{—}C\overset{O}{\overset{\|}{C}}C_6H_5\diagdown O\diagup B_{10}H_{10} + LiOCH_3$$

$$C_6H_5C\text{—}CLi\diagdown O\diagup B_{10}H_{10} + C_6H_5\overset{O}{\overset{\|}{C}}OC_2H_5 \longrightarrow C_6H_5C\text{—}C\text{—}CH\text{—}C_6H_5\diagdown O\diagup B_{10}H_{10}\ \underset{OH}{} + LiOC_2H_5$$

The reaction of o-carboranyl ketones with Grignards or organolithium compounds followed by acid hydrolysis might be expected to produce tertiary alcohols, but this is frequently not the case. Such reactions often lead instead to reduction of the carbonyl group to the secondary alcohol and/or cleavage of the exopolyhedral carbon–carbon bond (*202, 317, 362, 436, 439*). For example, 1-phenyl-2-benzoyl-o-carborane reacts with ethylmagnesium bromide to give both 1-phenyl-2-(α-hydroxybenzyl)-o-carborane and 1-phenyl-o-carborane. However, n-butyllithium yields mainly the cleavage product (even at $-50°$), with very little reduction to the alcohol. Both the observed cleavage and the frequent failure to obtain tertiary alcohols in such reactions have been explained in terms of an unstable intermediate tertiary alkoxide (*436*).

$$C_6H_5C\text{—}C\overset{O}{\overset{\|}{\text{—}C}}\text{—}C_6H_5\diagdown O\diagup B_{10}H_{10} + C_4H_9Li \longrightarrow \left[C_6H_5C\text{—}C\overset{OLi}{\overset{|}{\text{—}C}}\text{—}C_6H_5\diagdown O\diagup B_{10}H_{10}\ \underset{C_4H_9}{} \right] \longrightarrow C_6H_5C\text{—}CLi\diagdown O\diagup B_{10}H_{10}$$

$$\xrightarrow{CO_2,\ H_3O^+} C_6H_5C\text{—}C\overset{O}{\overset{\|}{C}}OH\diagdown O\diagup B_{10}H_{10} \qquad \xrightarrow{H_2O} C_6H_5C\text{—}CH\diagdown O\diagup B_{10}H_{10}$$

Recent studies (*202, 450*) indicate that in general, o-carboranyl ketones are cleaved by organolithium reagents, while Grignards react with the same ketones to give secondary or tertiary alcohols. For example, a tertiary alcohol is formed in the reaction of 1-phenyl-2-(benzoylmethyl)-o-carborane with phenylmagnesium bromide (*439*).

$$\underset{B_{10}H_{10}}{C_6H_5C\overset{O}{\overset{\|}{\diagdown O \diagup}}CCH_2CC_6H_5} \quad \xrightarrow[\text{2. HCl}]{\text{1. } C_6H_5MgBr} \quad \underset{B_{10}H_{10}}{C_6H_5C\overset{OH}{\overset{|}{\diagdown O \diagup}}CCH_2C(C_6H_5)_2}$$

Zakharkin and L'vov have suggested that the intermediate tertiary alcoholates formed by the organolithium reagents are less stable than those formed by Grignards (as a consequence of the more polar O—Li bond), so that cleavage, rather than alcohol formation, is the dominant process when the lithium compounds are involved. Not surprisingly, the course of these reactions is influenced by additional factors such as temperature and the choice of solvent; thus, in hot solutions even the reactions with Grignards yield predominantly cleavage products (202).

Secondary o-carboranyl alcohols are obtained in high yield in the reactions of many organolithium and Grignard reagents with o-carboranyl aldehydes, accompanied in some cases by C—C cleavage to a minor extent (202).

Other methods which have been employed to prepare o-carborane alcohols include the reduction of o-carboranyl carboxylic acids (90) and ketones (362, 372, 439) with lithium aluminum hydride, and the acid-catalyzed opening of epoxide rings to produce diols (Section 6-4). o-Carboranyl derivatives of cellulose have also been prepared (6).

Boron-substituted hydroxy derivatives have been obtained from the treatment of o-carborane with nitric acid (Section 6-8).

Synthesis of Ethers

Aliphatic ethers containing the o-carboranyl unit are usually prepared from acetylenic ethers by reaction with a bis(ligand) decaborane species (68, 90, 101), as illustrated by the synthesis of bis(1-carboranylmethyl) ether from dipropargyl ether.

$$(HC\equiv CCH_2)_2O + 2B_{10}H_{12}(CH_3CN)_2 \longrightarrow \underset{B_{10}H_{10}}{(HC\overset{}{\overset{}{\diagdown O \diagup}}CCH_2)_2O} + 4CH_3CN + 2H_2$$

Cyclic o-carboranyl ethers are formed in a variety of reactions. The formation of epoxides from alkenylcarboranes (Section 6-4) and in the reaction of metallated carboranes with epichlorohydrin (see above) has been described. o-Carborane diols undergo condensation reactions of several types to form cyclic ethers; for example, 1,2-bis(hydroxymethyl)-o-carborane is converted to a stable, volatile ether by hot sulfuric acid (91, 142).

The same diol also reacts with formaldehyde (*99*), dibutoxymethane (*364*), and 1-epoxyisopropyl-o-carborane (*207*) to yield other cyclic ethers. In the last reaction an unusual eight-membered exocycle is formed.

Still other cyclic ethers are produced in reactions of metal derivatives of

1-bromomethyl-*o*-carborane with aldehydes and ketones, probably via intermediate alcoholates (*363*).

$$
\text{BrCH}_2\text{C}\underset{\text{B}_{10}\text{H}_{10}}{\overset{\displaystyle}{\diagdown\text{O}\diagup}}\text{CLi} + R\overset{\text{O}}{\overset{\|}{\text{C}}}R' \longrightarrow \left[\text{BrCH}_2\text{C}\underset{\text{B}_{10}\text{H}_{10}}{\overset{\displaystyle}{\diagdown\text{O}\diagup}}\text{C}\underset{R}{\overset{\text{OLi}}{\text{C}}}R' \right] \longrightarrow
$$

R = CH=CHCH$_3$, R' = H
R = CH$_3$, R' = CH$_3$
R = C$_6$H$_5$, R' = H
RR' = —(CH$_2$)$_5$—

Reactions of Alcohols and Ethers

The oxidation of primary *o*-carboranyl alcohols to the acids is discussed in Section 6-5. When carried out with chromic acid in aqueous acetone, primary β- and γ-carboranyl alcohols are easily converted to the acids, but α-carboranyl alcohols such as 1,2-bis(hydroxymethyl)-*o*-carborane give low yields of the corresponding *o*-carboranyl dicarboxylic acid (*372*). Under the same acidic conditions, secondary α- and β-*o*-carboranyl alcohols are converted to the ketones as expected (*427, 431, 439*). Oxidation of α-carboranyl alcohols by basic permanganate results in decarboxylation, as pointed out in Section 6-5 (β- and γ-carboranyl alcohols, in contrast, give the acids in good yield).

The electrophilic nature of the *o*-carboranyl cage system is apparent in some reactions of the hydroxymethyl derivatives; thus, 1-hydroxymethyl-*o*-carborane is unaffected by 48% hydrobromic acid in sulfuric acid and by sodium bromide in concentrated sulfuric acid (*364*). Replacement of the hydroxyl group by chlorine may be accomplished, however, by the action of thionyl chloride in the presence of pyridine (*364*).

The reactions of *o*-carboranyl alcohols with many other reagents proceed normally. Acylation by aliphatic acids, acid anhydrides, and acid chlorides leads to ester formation, and the action of vinyl ethers gives mixed acetals (*90, 364*). Tertiary *o*-carboranyl alcohols are cleaved by catalytic amounts of sodium hydroxide in ethanol (*202*), illustrating the instability of the alcoholates of tertiary *o*-carboranyl alcohols mentioned above.

$$
\text{HC}\underset{\text{B}_{10}\text{H}_{10}}{\overset{\displaystyle}{\diagdown\text{O}\diagup}}\text{CCH}_2\text{O}\overset{\text{O}}{\overset{\|}{\text{C}}}R \xleftarrow[\text{(RCO)}_2\text{O}]{R\overset{\text{O}}{\overset{\|}{\text{C}}}\text{Cl}} \text{HC}\underset{\text{B}_{10}\text{H}_{10}}{\overset{\displaystyle}{\diagdown\text{O}\diagup}}\text{CCH}_2\text{OH} \xrightarrow[\text{H}^+]{\text{CH}_2=\overset{\text{H}}{\overset{\|}{\text{C}}}\text{OC}_4\text{H}_9} \text{HC}\underset{\text{B}_{10}\text{H}_{10}}{\overset{\displaystyle}{\diagdown\text{O}\diagup}}\text{CCH}_2\text{O}\overset{\text{CH}_3}{\underset{\text{H}}{\text{C}}}\text{OC}_4\text{H}_9
$$

1,2-Bis(hydroxymethyl)-o-carborane reacts as expected with sodium and other active metals in inert solvents, forming dimetalloalcoholates which are unreactive except for hydrolysis (142).

The synthesis of carborane-based polymers (Chapter 8) from o-carborane alcohols has been attempted with a number of reagents, but with only limited success. The o-carboranyl system tends to resist polymerization, due to a large steric factor and possibly to electronic effects as well (97, 99). Many attempts at polymerization of o-carborane diols have led instead to the formation of exocycles, as in the formation of cyclic ethers and other exocyclic derivatives from bis(hydroxyalkyl)-o-carboranes, described earlier in this section. In addition, the reactions of 1,2-bis(hydroxymethyl)-o-carborane with trioxane and with polyphosphoric acid produce a cyclic formal and a cyclic phosphate, respectively (91). A similar reaction with formaldehyde gives a cyclic formal in high yield, as described earlier. Polyesters having molecular weights between 2,000 and 20,000 have been prepared by allowing o-carborane diols to react with organic diacids, although such polyesterifications are slow in comparison to the analogous reactions of 1,4-butanediol (99).

$$HOCH_2C\overset{\displaystyle \diagdown O \diagup}{\underset{B_{10}H_{10}}{\rule{0pt}{0pt}}}CCH_2OH \ + HOOC(CH_2)_xCOOH \longrightarrow$$

$$nH_2O + \left[OCH_2C\overset{\displaystyle \diagdown O \diagup}{\underset{B_{10}H_{10}}{\rule{0pt}{0pt}}}CCH_2OOC(CH_2)_xCO \right]_n$$

On the other hand, polymerization of o-carborane diols with o-carborane diacids takes place with difficulty if at all.

Bis(2-hydroxyethyl-1-carboranylmethyl) ether forms a polymer on reaction with formaldehyde, but the analogous hydroxymethyl derivative fails to react (98, 99).

$$(HOCH_2CH_2C\overset{\displaystyle \diagdown O \diagup}{\underset{B_{10}H_{10}}{\rule{0pt}{0pt}}}CCH_2)_2O \ + CH_2O \longrightarrow$$

$$\left[CH_2CH_2C\overset{\displaystyle \diagdown O \diagup}{\underset{B_{10}H_{10}}{\rule{0pt}{0pt}}}CCH_2OCH_2C\overset{\displaystyle \diagdown O \diagup}{\underset{B_{10}H_{10}}{\rule{0pt}{0pt}}}CCH_2CH_2OCH_2O \right]_n$$

$$(HOCH_2C\overset{\displaystyle \diagdown O \diagup}{\underset{B_{10}H_{10}}{\rule{0pt}{0pt}}}CCH_2)_2O \ + CH_2O \longrightarrow \text{No reaction}$$

Aside from the ring-opening of o-carboranyl epoxides discussed above, no characteristic reactions have been reported for o-carboranyl ethers as a class.

These compounds may be subjected to most of the usual preparative methods for o-carborane derivatives without breaking the ether linkage, as shown in many reactions cited in this chapter. Bis(1-carboranylmethyl) ether, for example, can be metallated and subsequently converted to the diacid carboranyl ether (Section 6-5).

6-7. ALDEHYDES AND KETONES

Synthesis of Aldehydes

Few methods have been reported for the preparation of o-carboranyl aldehydes. The hydrogenation of o-carboranyl acid chlorides over a palladium catalyst in boiling xylene produces the corresponding aldehyde (*303*), and the ozonization of vinyl o-carboranes forms aldehydes in nearly quantitative yield (*398, 438*).

$$n = 0,1$$

$$R = H, CH_3$$

1-Phenyl-2-formyl-o-carborane has been prepared from 1-phenyl-2-cyano-o-carborane by treatment with water and diisobutylalane (*437*).

A convenient synthesis developed by Russian workers utilizes the reaction of acetylenic aldehyde diacetates with decaborane to yield o-carboranyl aldehyde diacetates, from which the aldehydes may be obtained by acid hydrolysis. The intermediate diacetates need not be isolated (*304*). Thus, C-formyl-o-carborane is prepared in 40% yield from the diacetate of propargyl aldehyde.

A boron-substituted aldehyde, 3-formyl-o-carborane, has been prepared by the ozonization of 3-vinyl-o-carborane, which in turn is obtained from $C_2B_9H_{11}^{2-}$ ion (Chapter 9) and vinyldichloroborane (*398*).

Synthesis of Ketones

In contrast to the aldehydes, o-carboranyl ketones are produced in a wide variety of reactions reported in the literature. The chromic acid oxidation of secondary o-carboranyl alcohols to the ketones is discussed in Section 6-6. Lithium derivatives of o-carborane react easily with aliphatic, aromatic, and heterocyclic acyl chlorides to give ketones, as illustrated by the synthesis of 1-methyl-2-benzoyl-o-carborane from benzoyl chloride (*362, 435*).

The reaction is quite general, and may be used to prepare bis(o-carboranyl) ketones from o-carboranyl acid chlorides (*435*).

Similarly, the action of phosgene and of oxalyl chloride upon the appropriate lithiocarborane produces, respectively, cyclic ketones (*249*) and bis(o-carboranyl)-α-diketones (*435*).

The fact that lithiocarboranes react with acyl chlorides to give ketones rather than tertiary alcohols supports the suggestion noted earlier (Section 6-6) that the lithium alcoholates of tertiary *o*-carboranyl alcohols are unstable and decompose to the original ketone and lithium derivative. It has been shown that 1-methyl-2-lithio-*o*-carborane reacts with 1-phenyl-2-(*p*-chlorobenzoyl)-*o*-carborane, forming an equilibrium mixture which suggests a bis-(*o*-carboranyl)alcoholate intermediate (*435*).

Ketones are also formed from *o*-carboranyl Grignard reagents and acyl chlorides of aromatic, aliphatic, heterocyclic, and *o*-carboranyl carboxylic acids, in reactions similar to those of the *o*-carboranyl lithium derivatives (*378*). Alternatively, organic Grignards may be used to convert *o*-carboranyl acyl chlorides to the ketones (*435*).

The bis(*o*-carboranyl) ketone probably forms via a magnesium alcoholate which rearranges to 1-phenyl-2-chloromagnesium-*o*-carborane, the latter compound then combining with the original acyl chloride to split out magnesium dichloride (*435*). A somewhat related process involves the condensation of *o*-carboranyl acid chlorides with benzene over aluminum chloride (*317, 435*).

$$RC\!\!-\!\!\overset{\diagdown O \diagup}{\underset{B_{10}H_{10}}{C}}\!\!(CH_2)_n\overset{O}{\overset{\|}{C}}\!\!-\!\!Cl \quad \xrightarrow[AlCl_3]{C_6H_6} \quad RC\!\!-\!\!\overset{\diagdown O \diagup}{\underset{B_{10}H_{10}}{C}}\!\!(CH_2)_n\overset{O}{\overset{\|}{C}}\!\!-\!\!C_6H_5$$

$$n = 0, 1 \quad R = H, CH_3, C_6H_5, CH_2{=}CH, \text{iso-}C_3H_7$$

The acid chlorides are also capable of taking up acetylene to form o-carboranyl-β-chlorovinylketones (Section 6-4) (*379*).

Another approach to the synthesis of ketones utilizes the discovery that lithium, sodium, or halomagnesium derivatives of o-carborane undergo 1,4-addition to α,β-unsaturated ketones (*378, 429, 432*). Thus, 1-phenyl-2-lithio-o-carborane reacts readily with benzalacetophenone or benzalpinacoline in benzene, apparently via an unstable alcoholate intermediate which rearranges to the observed saturated o-carboranyl ketone (*432*).

$$C_6H_5C\!\!-\!\!\overset{\diagdown O \diagup}{\underset{B_{10}H_{10}}{C}}\!\!Li \; + \; C_6H_5\overset{H}{\overset{|}{C}}{=}\overset{H}{\overset{|}{C}}\!\!-\!\!\overset{O}{\overset{\|}{C}}\!\!R \longrightarrow$$

$$\left[\, C_6H_5C\!\!-\!\!\overset{\diagdown O \diagup}{\underset{B_{10}H_{10}}{C}}\!\!-\!\!\overset{H}{\overset{|}{C}}{-}\overset{C_6H_5}{\overset{|}{\underset{H}{C}}}{-}\overset{OLi}{\overset{/}{C}}{=}C\!\!-\!\!R \,\right] \longrightarrow C_6H_5C\!\!-\!\!\overset{\diagdown O \diagup}{\underset{B_{10}H_{10}}{C}}\!\!-\!\!\overset{C_6H_5}{\overset{|}{\underset{H}{C}}}{-}CH_2\!\!-\!\!\overset{O}{\overset{\|}{C}}\!\!R$$

$$R = C_6H_5, C(CH_3)_3$$

It was mentioned earlier (Section 6-6) that the reactions of lithio- and dilithio-o-carborane with esters are anomalous, yielding ketones or secondary alcohols rather than the expected tertiary alcohols. Methyl esters of both aliphatic and aromatic acids form ketones, as do the ethyl esters of aliphatic acids (*430, 432*). In contrast, esters of α,β-unsaturated acids are attacked at the carbonyl group by lithiocarboranes, forming an α,β-unsaturated o-carboranyl ketone. This intermediate species reacts further with a second mole of the lithiocarborane, which adds at the 1,4 positions to give a bis(o-carboranyl) ketone (*429, 432*).

$$RC\!\!-\!\!\overset{\diagdown O \diagup}{\underset{B_{10}H_{10}}{C}}\!\!Li \; + \; C_6H_5\overset{H}{\overset{|}{C}}{=}\overset{H}{\overset{|}{C}}\!\!-\!\!\overset{O}{\overset{\|}{C}}\!\!OCH_3 \longrightarrow$$

$$\left[\, C_6H_5\overset{H}{\overset{|}{C}}{=}\overset{H}{\overset{|}{C}}\!\!-\!\!\overset{O}{\overset{\|}{C}}\!\!-\!\!C\!\!-\!\!\overset{\diagdown O \diagup}{\underset{B_{10}H_{10}}{C}}\!\!R \,\right] \xrightarrow{RC\!\!-\!\!\overset{\diagdown O \diagup}{\underset{B_{10}H_{10}}{C}}\!\!Li} RC\!\!-\!\!\overset{\diagdown O \diagup}{\underset{B_{10}H_{10}}{C}}\!\!-\!\!\overset{C_6H_5}{\overset{|}{\underset{H}{C}}}{-}CH_2\!\!-\!\!\overset{O}{\overset{\|}{C}}\!\!-\!\!C\!\!-\!\!\overset{\diagdown O \diagup}{\underset{B_{10}H_{10}}{C}}\!\!R$$

$$R = CH_3, C_6H_5$$

o-Carboranyl ketones have also been prepared from acid anhydrides by reaction with lithiocarboranes. When cyclic anhydrides are used the products are *o*-carboranyl keto acids (*432*).

Cyclic ketones are formed in the pyrolysis of 1,2-bis(carboxymethyl)-*o*-carborane (*142*) (Section 6-5), the reaction of 1-phenyl-2-carboxymethyl-*o*-carborane with polyphosphoric acid (*361*), and by treatment of bis(acetonitrile) decaborane with phenyl propiolyl chloride (*372, 455, 459*).

Reactions of Aldehydes and Ketones

o-Carboranyl aldehydes are converted to secondary alcohols by organolithium compounds; thus, phenyllithium reacts with 1-methyl-2-formyl-*o*-carborane to give 1-methyl-2-(α-hydroxybenzyl)-*o*-carborane (*202, 304, 438*).

$$CH_3C\overset{\displaystyle O}{\overset{\|}{\underset{\underset{\displaystyle B_{10}H_{10}}{\diagdown O \diagup}}{-C-CH}}} \quad \xrightarrow[\text{(C}_2\text{H}_5)_2\text{O}]{C_6H_5Li, H^+} \quad CH_3C\overset{\displaystyle OH}{\overset{|}{\underset{\underset{\displaystyle B_{10}H_{10}}{\diagdown O \diagup}}{-C-\underset{H}{C}-C_6H_5}}}$$

Grignard reagents, on the other hand, react with the aldehydes in two ways: addition to the carbonyl group and reduction to the primary alcohol occur simultaneously, the product ratio depending on the specific Grignard (*202, 304*).

$$HC\overset{\displaystyle O}{\overset{\|}{\underset{\underset{\displaystyle B_{10}H_{10}}{\diagdown O \diagup}}{-CCH}}} + RMgX \quad \xrightarrow[H^+]{(C_2H_5)_2O} \quad HC\overset{\displaystyle OH}{\overset{|}{\underset{\underset{\displaystyle B_{10}H_{10}}{\diagdown O \diagup}}{-C-\underset{H}{C}R}}} + HC\underset{\underset{\displaystyle B_{10}H_{10}}{\diagdown O \diagup}}{-CCH_2-OH}$$

$$R = CH_3, C_2H_5, C_6H_5$$

The difference in the action of organolithium and Grignard reagents apparently results from the greater nucleophilic character of the former.

The reduction of o-carboranyl ketones to secondary alcohols by the action of Grignards and organolithium reagents is discussed in the preceding section. With Grignard reagents, reduction is usually accompanied by some cleavage of the ketone–carborane link, while organolithium compounds such as n-butyllithium give cleavage almost exclusively. Secondary alcohols are also formed in the reaction of o-carboranyl ketones with lithium aluminum hydride (*362, 439*); however, bis(1-phenyl-o-carboranyl) ketone is cleaved by LiAlH$_4$, yielding 1-phenyl-o-carborane (*362*).

$$(C_6H_5C\underset{\underset{\displaystyle B_{10}H_{10}}{\diagdown O \diagup}}{-C)_2CO} \quad \xrightarrow{\text{LiAlH}_4} \quad C_6H_5C\underset{\underset{\displaystyle B_{10}H_{10}}{\diagdown O \diagup}}{-CH}$$

o-Carboranyl aldehydes and ketones tend to undergo similar cleavage in basic media. This has been observed in aqueous or alcoholic ammonia (*440*), with alkali metal ethoxides or sodamide in ethanol (*303, 304, 310, 362, 432, 439*), and (in the case of ketones) during thin-layer chromatography on basic alumina (*312, 317, 435, 439*). The cleavage mechanism is not simple, and some conflicting data and interpretations exist in the literature (*312, 317, 449*). Catalytic quantities of sodium ethoxide are evidently sufficient to cause cleavage:

$$RC\overset{\displaystyle O}{\overset{\|}{\underset{\underset{\displaystyle B_{10}H_{10}}{\diagdown O \diagup}}{-C-CR'}}} \quad \xrightarrow[C_2H_5OH]{C_2H_5ONa} \quad RC\underset{\underset{\displaystyle B_{10}H_{10}}{\diagdown O \diagup}}{-CH} + R'C\overset{\displaystyle O}{\overset{\|}{OC_2H_5}}$$

$$R = H, C_6H_5, CH_3, (CH_3)_2CH$$
$$R' = H, CH_3, C_6H_5$$

In strongly basic solutions, the noncarborane cleavage product is a carboxylic acid salt rather than an ester, and the salts themselves are reportedly capable of inducing cleavage. Thus, it is found that potassium acetate in aqueous alcohol cleaves 1-methyl-2-benzoyl-*o*-carborane, yielding 1-methyl-*o*-carborane, acetic acid, and potassium benzoate (*449*).

In all known cases, the splitting of *o*-carboranyl ketones proceeds as expected by analogy with unsymmetrical diaryl ketones, in that the hydrocarbon product is formed by the more electrophilic group of the ketone; in most cases this is the *o*-carboranyl cage, so that the carborane product is a hydrocarbon rather than a carborane carboxylic acid. An interesting illustration of this effect is the cleavage of a bis(*o*-carboranyl) ketone as shown (*429, 432*).

$$CH_3C \overset{\displaystyle C_6H_5}{\underset{\displaystyle B_{10}H_{10}}{\diamond}} \overset{|}{\underset{H}{C}} - CH_2 - \overset{O}{\overset{\|}{C}} - \overset{\diamond}{\underset{B_{10}H_{10}}{C}} - CCH_3 \xrightarrow{C_2H_5ONa}$$

$$CH_3C \overset{\diamond}{\underset{B_{10}H_{10}}{}} CH \quad + \quad CH_3C \overset{\diamond}{\underset{B_{10}H_{10}}{}} \overset{C_6H_5}{\overset{|}{C}} - CHCH_2COONa$$

The facile cleavage of *o*-carboranyl ketones in basic media effectively prevents investigation of the base-catalyzed replacement of α-hydrogen atoms in these compounds, and such reactions can ordinarily be carried out only in neutral or acid solution. In acid solution, however, the electrophilic *o*-carborane cage lowers the basicity of the ketonic oxygen atom, thereby making enolization more difficult and reducing the rate of α-hydrogen replacement. Studies of the bromination of *o*-carboranyl ketones have shown that the α-hydrogen atoms have very low proton mobility in neutral and weakly acid media; thus, 1-acetyl-2-methyl-*o*-carborane is unaffected by bromine in boiling CCl_4, although bromination does take place in acetic acid solution at 115° (the carborane cage itself is not attacked under these conditions) (*442*).

$$CH_3C \overset{\diamond}{\underset{B_{10}H_{10}}{}} \overset{O}{\overset{\|}{C}} - CCH_3 \xrightarrow[CH_3COOH]{Br_2, 115°}$$

$$CH_3C \overset{\diamond}{\underset{B_{10}H_{10}}{}} \overset{O}{\overset{\|}{C}} - CCH_2Br \xrightarrow[CH_3COOH]{Br_2, 115°} CH_3C \overset{\diamond}{\underset{B_{10}H_{10}}{}} \overset{O}{\overset{\|}{C}} - CCHBr_2$$

Bromination is resisted even when the *o*-carborane cage is separated from the carboxyl group by a methylene unit, as shown by the failure of 2-(1-methyl-*o*-carboranyl)-acetophenone to react with bromine or dioxane dibromide in boiling dioxane. Again, however, such ketones are readily brominated in hot acetic acid solution (*442*).

The α-hydrogen atoms also are highly resistant to deuterium exchange in D_2SO_4. o-Carboranyl ketones are weaker bases than benzophenone and acetophenone; as expected, m-carboranyl ketones are stronger bases than o-carboranyl ketones due to the weaker $-I$ effect of the m-carborane cage (*202*) (see also Section 7-8).

The properties of many o-carboranyl aldehydes and ketones clearly suggest steric hindrance by the carborane cage. For example, although methyl ketones ordinarily form 2,4-dinitrophenylhydrazones easily, the formation of these derivatives from methyl-o-carboranyl ketones is very slow (*439*). In ethanol–sulfuric acid solution at 20°, 1-acetyl-o-carborane requires 3 days for reaction with 2,4-dinitrophenylhydrazine, while 1-acetyl-2-phenyl-o-carborane reacts in 10 days and 1-benzoyl-o-carborane fails to react at all. Steric hindrance is also evident in the reactions of α-(o-carboranyl) aldehydes, as in their failure to form bisulfite derivatives. In contrast, β-(o-carboranyl) aldehydes form such derivatives easily (*440*).

$$CH_3C\!-\!CCH_2\overset{O}{\overset{\|}{C}H} \quad \xrightarrow{\text{NaHSO}_3} \quad CH_3C\!-\!CCH_2\overset{OH}{\overset{|}{C}H}\!-\!SO_3Na$$
$$\underset{B_{10}H_{10}}{\diagdown O \diagup} \qquad\qquad\qquad \underset{B_{10}H_{10}}{\diagdown O \diagup}$$

Other characteristic reactions of o-carboranyl aldehydes include the slow formation of acetals and cyanohydrins, and the reduction of silver oxide with concomitant production of o-carboranyl carboxylic acid and o-carborane itself (*440*).

$$HC\!-\!C\overset{O}{\overset{\|}{C}H} \quad \xrightarrow[\text{NH}_4\text{NO}_3,\ \text{C}_2\text{H}_5\text{OH}]{\text{HC(OC}_2\text{H}_5)_3} \quad HC\!-\!CCH(OC_2H_5)_2$$
$$\underset{B_{10}H_{10}}{\diagdown O \diagup} \qquad\qquad\qquad\qquad \underset{B_{10}H_{10}}{\diagdown O \diagup}$$

$$\xrightarrow[(\text{C}_2\text{H}_5)_2\text{O}]{\text{HCN}} \quad HC\!-\!CC\overset{CN}{\underset{OH}{\overset{H}{<}}}$$
$$\underset{B_{10}H_{10}}{\diagdown O \diagup}$$

$$\xrightarrow[\text{H}_2\text{O}]{\text{Ag}_2\text{O}} \quad HC\!-\!CH \quad + \quad HC\!-\!CCOOH$$
$$\underset{B_{10}H_{10}}{\diagdown O \diagup} \qquad\qquad \underset{B_{10}H_{10}}{\diagdown O \diagup}$$
$$40\%$$

Aldehyde derivatives of *o*-carborane are not only cleaved by primary aliphatic amines and piperidine, but are also degraded to the dicarbaundecaborate(1−) ion. Aniline, on the other hand, readily forms Schiff bases and gives no cleavage (*440*).

$$
\underset{B_{10}H_{10}}{HC\!\!-\!\!C\overset{O}{\overset{\|}{C}}H} \xrightarrow{C_4H_9NH_2} C_2B_9H_{12}^- + HC\overset{O}{\underset{}{\overset{\|}{}}}\!\!-\!\!N\overset{H}{\underset{}{}}C_4H_9
$$

$$
\xrightarrow{H^+,\ C_6H_5NH_2} \underset{B_{10}H_{10}}{HC\!\!-\!\!CCH\!\!=\!\!NC_6H_5} \xrightarrow{LiAlH_4} \underset{B_{10}H_{10}}{HC\!\!-\!\!CCH_2NHC_6H_5}
$$

6-8. NITROGEN DERIVATIVES

Nitrates and Related Compounds

The treatment of *o*-carborane with 100 % nitric acid at 20° is reported to yield a B-hydroxy and a B-nitrato-*o*-carborane, the latter in larger quantity (*411, 418*). The nitrate, which is extremely unstable and detonates on heating, is easily converted into B-hydroxy-*o*-carborane (identical with the product formed from nitric acid) by reduction with tin and hydrochloric acid (*411, 418*).

$$
\underset{B_{10}H_{10}}{HC\!\!-\!\!CH} + HNO_3 \longrightarrow \underset{B_{10}H_9OH}{HC\!\!-\!\!CH} + \underset{B_{10}H_9ONO_2}{HC\!\!-\!\!CH}
$$

1-Methyl- and 1,2-dimethyl-*o*-carborane give analogous reactions, except that the former compound yields two isomeric nitrates as well as the B-hydroxy derivative. Although the positions of substitution have not been established, the evidence suggests that attack occurs at the 9,12 boron atoms, as is observed in electrophilic halogenation (Section 6-13). This view is supported by the isolation of two isomeric 1-methyl-B-nitrato-*o*-carboranes [in which B(9) and B(12) are nonequivalent] and by the probability that the nitration is an electrophilic process involving substitution at the most negative boron atoms (*418*).

The attack of nitric acid on phenyl- or hydroxymethyl-o-carborane deriva-tives takes place on the substituent group and not on the carborane cage itself. Thus, 1-hydroxymethyl- and 1,2-bis(hydroxymethyl)-o-carborane are con-verted to the highly explosive 1-nitrato- and 1,2-dinitrato-o-carborane, respectively, by cold nitric acid (*365*). 1-Phenyl-o-carborane reacts with 100% nitric acid or with a mixture of nitric and sulfuric acids in an inert solvent, forming 1-(m-nitrophenyl)-o-carborane and 1-(p-nitrophenyl)-o-carborane with the latter predominating (*300, 382, 388, 462*).

The ratio of isomers obtained varies somewhat with reaction conditions, but in all published results the p-nitrophenyl derivative is the main product. [The nitration of 1-phenyl-2-carboxy-o-carborane, on the other hand, pro-duces the corresponding m-nitrophenyl derivative in larger yield than the p-nitrophenyl isomer (*382*). An o-nitrophenyl derivative has been observed in only one study, as a very minor product (*127*). The mixed acid nitration of 3-phenyl-o-carborane has also been examined (*420*), with the resulting forma-tion of 3-(o-nitrophenyl)-, 3-(m-nitrophenyl)-, and 3-(p-nitrophenyl)-o-carborane in an o:m:p ratio of 3:4:3.

All of these observations on the nitration of o-carborane derivatives have been cited as evidence of the strong electron-withdrawing power of the 1-o-carbor-anyl group and the somewhat weaker effect of the 3-o-carboranyl unit.

An obvious route to nitrophenyl-o-carboranes is the oxidation of amino-phenyl derivatives. It is found that 1-(m-aminophenyl)-o-carborane is converted to 1-(m-nitrophenyl)-o-carborane by hydrogen peroxide in trifluoroacetic acid (*382*).

The 1-nitrophenyl derivatives, in turn, are easily reduced to the corresponding amines (*286, 382, 388*).

o-Carborane derivatives containing nitro or nitrito groups have also been prepared by the reaction of alkenylcarboranes with dinitrogen tetroxide (Section 6-4) and by the interaction of *p*-nitrobenzoyl chloride with lithio-carboranes (*435*).

$$RC\underset{\underset{B_{10}H_{10}}{}}{\overset{}{\diagdown O\diagup}}CLi \ + \ NO_2C_6H_4\overset{O}{\overset{\|}{C}}Cl \ \longrightarrow \ RC\underset{\underset{B_{10}H_{10}}{}}{\overset{}{\diagdown O\diagup}}C\overset{O}{\overset{\|}{C}}C_6H_4NO_2$$

However, attempts to synthesize the ethyl ester of 1-(*p*-nitrophenyl)-2-carboxyl-*o*-carborane from bis(acetonitrile)decaborane were unsuccessful due to reduction of the nitro ligand to an amino group (*372*).

$$NO_2C_6H_4\overset{O}{\overset{\|}{C}}\!\!\equiv\!\!CCOC_2H_5 + B_{10}H_{12}(CH_3CN)_2 \ \longrightarrow \ NH_2C_6H_4C\underset{\underset{B_{10}H_{10}}{}}{\overset{}{\diagdown O\diagup}}\overset{O}{\overset{\|}{C}}COC_2H_5$$

Nitroso derivatives of *o*-carborane have been synthesized from lithio-carboranes and nitrosyl chloride at very low temperatures (*169, 382*). Hydrogenation of 1-nitroso-2-phenyl-*o*-carborane proceeds readily to yield 1-hydroxylamino-2-phenyl-*o*-carborane (*382*).

$$RC\underset{\underset{B_{10}H_{10}}{}}{\overset{}{\diagdown O\diagup}}CLi \ \xrightarrow[-75 \text{ to } -125°]{NOCl} \ RC\underset{\underset{B_{10}H_{10}}{}}{\overset{}{\diagdown O\diagup}}CNO \ \xrightarrow[(R=C_6H_5)]{Sn+HCl} \ RC\underset{\underset{B_{10}H_{10}}{}}{\overset{}{\diagdown O\diagup}}C\overset{H}{\overset{|}{N}}\!-\!OH$$

$$R = H, \ CH_3, \ C_6H_5$$

Amines, Azides, and Diazonium Salts

C-Amino-*o*-carboranes have been prepared by direct reaction of amino-acetylenes with decaborane(14) (*143, 455*) and by reduction of nitrocarboranes as discussed earlier (the synthesis of B-amino-*o*-carboranes is described below). Amines are also formed in the reduction of *o*-carboranyl cyanides and some amides with lithium aluminum hydride (*364, 370, 371, 437*).

$$C_6H_5C\underset{\underset{B_{10}H_{10}}{}}{\overset{}{\diagdown O\diagup}}CCN \ \xrightarrow{LiAlH_4} \ C_6H_5C\underset{\underset{B_{10}H_{10}}{}}{\overset{}{\diagdown O\diagup}}CCH_2NH_2 \ + \ C_6H_5C\underset{\underset{B_{10}H_{10}}{}}{\overset{}{\diagdown O\diagup}}CH$$

$$HC\underset{\underset{B_{10}H_{10}}{}}{\overset{}{\diagdown O\diagup}}CCH_2\!-\!\overset{O}{\overset{\|}{C}}N(C_2H_5)_2 \ \xrightarrow{LiAlH_4} \ HC\underset{\underset{B_{10}H_{10}}{}}{\overset{}{\diagdown O\diagup}}C(CH_2)_2N(C_2H_5)_2$$

The reaction of α-amides, however, leads to decarboxylation rather than reduction to the amine, as is discussed below.

The azides of o-carboranyl carboxylic acids are converted first to isocyanates and then to the amines by 94% sulfuric acid. The azides may be obtained from o-carboranyl acyl chlorides by reaction with sodium azide in aqueous acetone (402, 410).

The dicarboxyl chloride is similarly converted to the diisocyanate; this product reacts with methanol in benzene to give a high yield of the bis(methylurethane) derivative (142).

An unusual aminoborane derivative, bis(dimethylamino)-o-carboranyl-borane, has been prepared from 1-lithio-o-carborane and chlorobis(dimethyl-amino)borane (23). This compound is deaminated by methanol, forming o-carborane, trimethyl borate, and diethylamine.

An interesting synthesis of a nitrogen mustard, 4-[bis(2-chloroethyl)amino]-phenyl-o-carborane, from 1-(p-aminophenyl)-o-carborane, has been reported (286). The latter compound, which is obtained by nitration of 1-phenyl-o-carborane followed by reduction over platinum oxide, reacts with ethylene oxide in acetic acid to produce the bis(hydroxyethyl) derivative which is then chlorinated with thionyl chloride.

$$\underset{B_{10}H_{10}}{HC\overset{\diagdown O \diagup}{\text{———}}CC_6H_4NH_2} \quad \underset{CH_3COOH}{\overset{\overset{CH_2\text{——}CH_2}{\diagdown O \diagup}}{\longrightarrow}}$$

$$\underset{B_{10}H_{10}}{HC\overset{\diagdown O \diagup}{\text{———}}CC_6H_4N(C_2H_4OH)_2} \quad \overset{SO_2Cl_2}{\longrightarrow} \quad \underset{B_{10}H_{10}}{HC\overset{\diagdown O \diagup}{\text{———}}CC_6H_4N(C_2H_4Cl)_2}$$

Other o-carborane derivatives containing C-amino functional groups are described elsewhere, viz., the synthesis of bis(difluoramino)- and difluoramino-nitrile-o-carboranes from alkenylcarboranes (Section 6-4) and the preparation of o-carboranyl disilazane ring compounds (Section 6-10).

In general, C-amino-o-carboranes are very weak bases, dissolving in concentrated sulfuric acid but precipitating out of solution on dilution (382, 388, 402, 410). Their oxidation to the corresponding nitro derivatives proceeds normally, as indicated above. Treatment of either 1-(p-aminophenyl)-o-carborane or the m-aminophenyl isomer with acetic anhydride yields the N-acetyl derivative, which is easily diazotized by nitrosylsulfuric acid in glacial acetic acid (382, 388). The diazonium salts react normally on treatment with copper(I) chloride or bromide, giving the respective halophenyl derivatives.

Azo dyes form readily from both the m- and p-diazonium salts, as illustrated by the reaction with β-naphthol in basic solution (382, 388).

The direct interaction between amines and o-carborane or its derivatives usually results in degradation of the carborane cage to an open-faced anionic species, as discussed in Chapter 9. An early report (91) of the preparation of amino-o-carborane adducts in such solutions is probably erroneous, in view of later work which indicated the ionic nature of the products (400).

B-Amino-o-carboranes have been prepared from the (3)-1,2- and (3)-1,7-dicarbollide ions in "insertion" reactions with aminodichloroboranes (Section 9-2). Zakharkin and Kalinin (396, 407) have reported the stepwise synthesis of several 3-amino-o-carboranes by reduction of o-carboranes to the dinegative ions (Section 6-2) with alkali metals in liquid ammonia, followed by reoxidation.

$$\underset{B_{10}H_{10}}{RC\!\!-\!\!CR'} \xrightarrow[\text{NH}_3\text{ (liq)}]{2M} \left[\underset{B_{10}H_{10}}{RC\!\!-\!\!CR'}\right]^{2-} \xrightarrow[-H_2]{NH_3}$$

$$\left[\underset{3\text{-}H_2NB_{10}H_9}{RC\!\!-\!\!CR'}\right]^{2-} \xrightarrow[KMnO_4]{-2e^-} \underset{3\text{-}H_2NB_{10}H_9}{RC\!\!-\!\!CR'}$$

$$R = H, CH_3, C_2H_5, CH(CH_3)_2, C_6H_5$$
$$R' = H, CH_3$$

The reaction rate with ammonia is enhanced when the ligands R and R′ are small electron-releasing groups, but steric hindrance is observed with large substituents such as phenyl. Amine substitution on 9-halo-o-carboranes proceeds similarly, but the halogen atom is removed in the process. The carborane dianions also react with primary and secondary amines, such as piperidine, forming the corresponding N-3-amino-o-carborane (*407*).

$$\left[\underset{B_{10}H_{10}}{HC\!\!-\!\!CH}\right]^{2-} \xrightarrow[\substack{THF\\-60°}]{C_5H_{10}NH} \left[\underset{3\text{-}C_5H_{10}NB_{10}H_9}{HC\!\!-\!\!CH}\right]^{2-} \xrightarrow[KMnO_4]{NH_3\text{ (liq)}} \underset{3\text{-}C_5H_{10}NB_{10}H_9}{HC\!\!-\!\!CH}$$

The 3-amino-o-carboranes are characteristic of aliphatic and aromatic amines, and may be alkylated, acylated, and diazotized. Probably because of the weak inductive electron-attracting effect at the B(3) position, these compounds are relatively basic and readily form salts with acids, in this respect contrasting sharply with the C-amino-o-carboranes discussed above (*407*).

The diazonium salts of 3-amino-o-carboranes slowly decompose to yield the corresponding 3-hydroxy derivatives, but if the diazonium salt solution is quickly added to a solution of a cuprous halide in the corresponding hydrohalic acid, the respective 3-halocarborane is formed in good yield (*397, 407, 408*).

$$\underset{3\text{-}H_2NB_{10}H_9}{RC\!\!-\!\!CR'} \xrightarrow[CH_3COOH, 5°]{NaNO_2, H_2SO_4} \underset{3\text{-}N_2^+B_{10}H_9}{RC\!\!-\!\!CR'} \xrightarrow{H_2O} \underset{3\text{-}HOB_{10}H_9}{RC\!\!-\!\!CR'}$$

$$\xrightarrow[HX]{CuX} \underset{3\text{-}XB_{10}H_9}{RC\!\!-\!\!CR'}$$

$$X = Cl, Br, F$$
$$R = H, CH_3$$
$$R' = H, CH_3$$

Amides

The amides of *o*-carboranyl carboxylic acids are prepared by direct combination of the acid chloride with ammonia or the appropriate amine (*142, 312, 317, 364, 371, 375*), or by reaction of decaborane(14) complexes with acetylenic amides (*455*). Decarboxylation of *o*-carboranyl amides is reported to occur upon treatment with sodium in liquid ammonia (*449*), although previous investigators had evidently failed to note the cleavage (*312, 317*).

As in the cleavage of ketones and esters mentioned earlier, the reactions are presumed to involve an alcoholate intermediate (*449*).

In contrast to esters and ketones, however, the cleavage of amides has not been observed with sodium ethoxide. It seems probable that the electron-withdrawing (−I) effect of the *o*-carboranyl unit is partially offset by amide groups, thus requiring a stronger attacking base for the cleavage of *o*-carboranyl amides as compared with esters and ketones. It is notable that lithium aluminum hydride decarboxylates α-carboranyl amides, but merely reduces amide groups which are not directly bonded to the carborane cage (see above) (*364, 370, 371*).

Reactions of lithiocarboranes with phenyl isocyanate or phenyl isothiocyanate yield the anilides of *o*-carborane carboxylic acids and *o*-carborane thiocarboxylic acids, respectively (*432*).

$$R = CH_3, C_6H_5$$

Cyano Derivatives

o-Carboranyl cyano compounds have been prepared from fluoramino derivatives (Section 6-4) and by the action of cyanogen chloride with lithiocarboranes in ether (*437*). The cyano- and chloro-substituted carboranes are formed in equal amounts.

$$C_6H_5C\text{---}CLi \atop B_{10}H_{10} \quad + \text{ ClCN} \quad \xrightarrow[10°]{(C_2H_5)_2O} \quad C_6H_5C\text{---}CCN \atop B_{10}H_{10} \quad + \quad C_6H_5C\text{---}CCl \atop B_{10}H_{10}$$

Lithium aluminum hydride reduces 1-phenyl-2-cyano-o-carborane to the amine, and the same cyano derivative is converted to the aldehyde by treatment with water and diisobutylalane (*437*). This carborane nitrile also reacts with methanol or ethanol, forming a cyano-substituted dicarbaundecaborane; the latter compound readily forms a methylpyridinium salt (*437*).

$$C_6H_5C\text{---}CCN \atop B_{10}H_{10} \quad \xrightarrow[20°]{CH_3OH}$$

$$(C_6H_5)(CN)C_2B_9H_{11} \quad \xrightarrow[NH_3]{CH_3C_5H_4NH^+I^-}$$

Other Nitrogen-Containing Derivatives

Pyridyl o-carborane alcohols have been prepared from lithiocarboranes and carboranyl Grignards, and by reaction with β- and γ-pyridyl aldehydes (*378, 426*).

$$RC\text{---}CLi \atop B_{10}H_{10} \quad + \quad HC(\bigcirc N)\!\!\!\!\!\!\!\!\overset{O}{\parallel} \quad \xrightarrow{H^+} \quad RC\text{---}C\text{--}\underset{H}{\overset{OH}{\overset{|}{C}}}(\bigcirc N) \atop B_{10}H_{10}$$

$$R = CH_3, C_6H_5$$

An interesting linkage of two large areas of boron chemistry has been achieved in the synthesis of an o-carboranyl derivative of borazine (*23*). The reaction of 1-lithio-o-carborane with B-trichloroborazine leads only to the formation of o-carborane, and N-trimethyl-B-trichloroborazine gives essentially the same result. However, reaction of the latter compound with 1-lithio-2-n-butyl-o-carborane forms B-tris(n-butyl-o-carboranyl)-N-trimethylborazine in high yield.

$$3C_4H_9C\text{---}CLi \atop B_{10}H_{10} \quad + \quad \begin{matrix} Cl \\ H_3C\text{--}N\overset{B}{\diagdown}N\text{--}CH_3 \\ \parallel \qquad \mid \\ Cl\text{--}B\diagdown_N\diagup B\text{--}Cl \\ \mid \\ CH_3 \end{matrix} \quad \longrightarrow \quad 3LiCl \quad + \quad \begin{matrix} R \\ \mid \\ H_3C\text{--}N\overset{B}{\diagdown}N\text{--}CH_3 \\ \parallel \qquad \mid \\ R\text{--}B\diagdown_N\diagup B\text{--}R \\ \mid \\ CH_3 \end{matrix}$$

$$R = \quad C_4H_9C\text{---}C \atop B_{10}H_{10}$$

A number of o-carboranyl acridines have been prepared for studies of tumor localization in mice. Thus, 9-chloracridine condenses with 1-(p-aminophenyl)-o-carborane to give the compound shown (*41*):

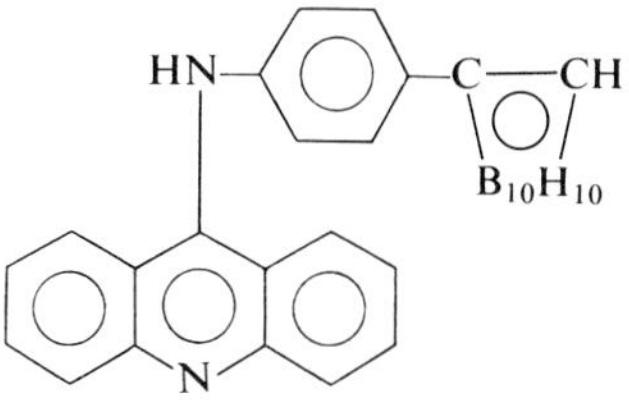

The dinitrophenylhydrazones of numerous o-carboranyl aldehydes and ketones have been prepared (*303, 431, 437–439*) as an aid in structural characterization, as have the phenylcarbamic esters of o-carboranyl secondary alcohols (*431*).

6-9. PHOSPHORUS, ARSENIC, AND ANTIMONY DERIVATIVES

The reaction of mono- or dilithio-o-carborane with chlorophosphines produces, respectively, mono- or bifunctional o-carboranyl phosphine derivatives (*7, 366, 367*). For example, with diphenylchlorophosphine the product is bis(diphenylphosphino)-o-carborane.

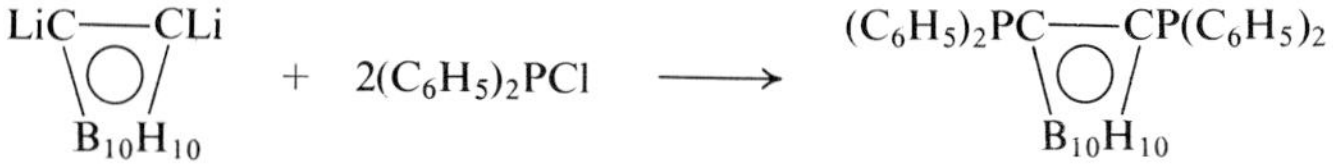

The same reaction is observed with B-brominated lithiocarboranes (*281*). Dilithio-o-carborane gives an analogous reaction with dimethyl- and diphenyl-bromoarsine, forming the 1,2-bis(dimethylarsino)- and 1,2-bis(diphenyl-arsino)-o-carboranes (*278, 358*).

Phenyldichlorophosphine reacts with dilithio-o-carborane to give the bis-(phenylchlorophosphino) derivative, which in turn is converted by ammonia to a cyclic diphospha(III) azane in 70% yield (*7*).

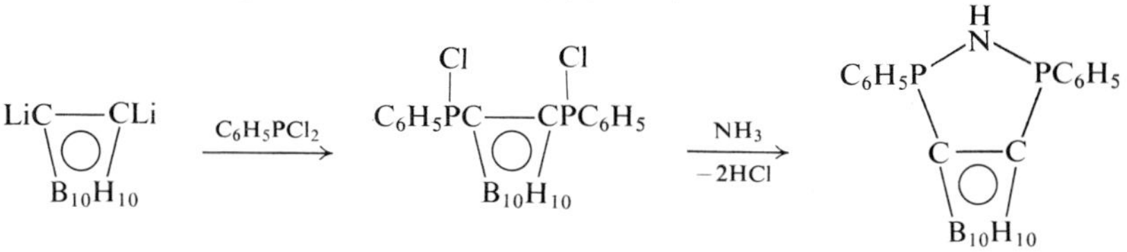

The bis(phenylchlorophosphino) derivative reacts readily with sodium azide, forming an unstable diazide, and with dilithio-o-carborane to give a cyclic bis(phenylphosphino) derivative (7).

Similar cyclic structures are produced in the reactions of dilithio-o-carborane with phosphorus trichloride (7) and with methyldibromoarsine (358).

The cyclic chlorophosphine derivative reacts with ammonia to give the

corresponding diamine, and with sodium azide to give the diazide; the latter compound is stable in refluxing benzene, and reacts with triphenylphosphine to form a phosphineimino derivative. Upon treatment with an equimolar quantity of p-bis(diphenylphosphino) benzene, the azide forms a polymer containing P—N—P linkages (7).

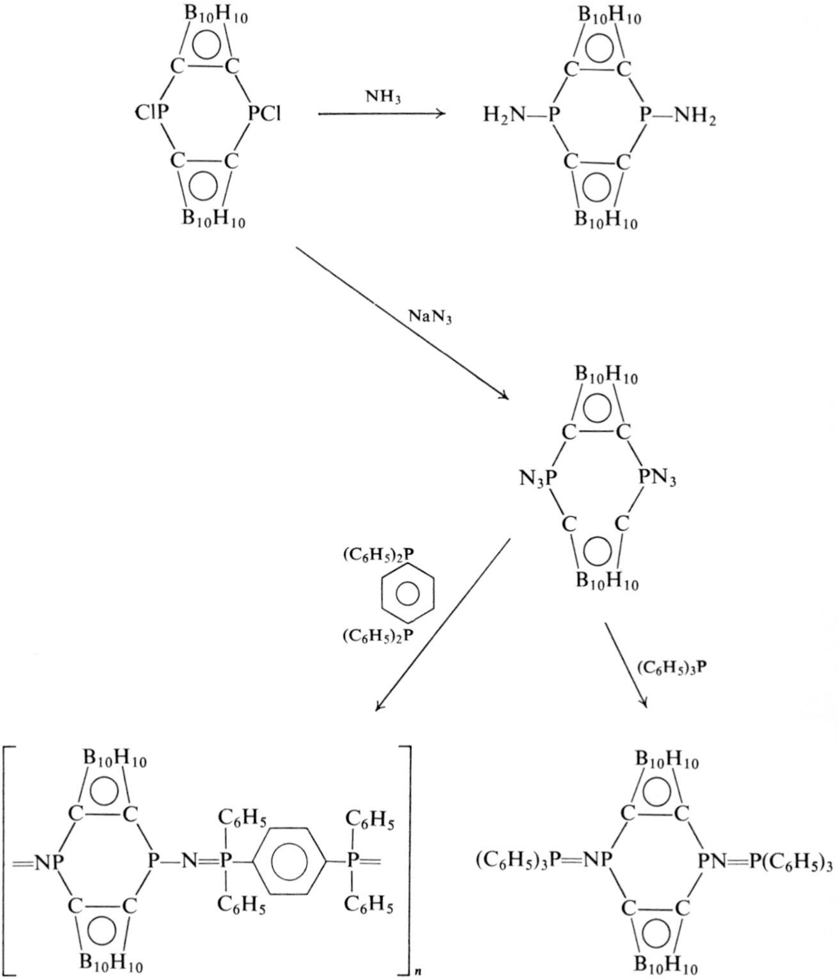

$B_{10}H_{10}$

ClP PCl $\xrightarrow{NH_3}$ H_2N-P $P-NH_2$

$B_{10}H_{10}$

$B_{10}H_{10}$

$\searrow$ NaN$_3$

$B_{10}H_{10}$

N_3P PN_3

$B_{10}H_{10}$

$(C_6H_5)_2P$

$(C_6H_5)_2P$

$(C_6H_5)_3P$

$\left[\ =NP \quad P-N=P-\underset{C_6H_5}{\overset{C_6H_5}{|}}-\underset{C_6H_5}{\overset{C_6H_5}{|}}P= \ \right]_n$

$B_{10}H_{10}$ ... $B_{10}H_{10}$

MW 2400 to 10,000

$(C_6H_5)_3P=NP$ $PN=P(C_6H_5)_3$

$B_{10}H_{10}$

Many reactions analogous to the above have been carried out using dilithio-*o*-carboranes containing one to three bromine atoms bound to the carborane cage (*9, 281*). The halogenated derivatives thus formed are generally similar to the analogous nonhalogenated species, although B-dibromo-1,2-bis(phenyl-azidophosphino)-*o*-carborane [$(N_3C_6H_5P)_2C_2B_{10}H_8Br_2$] explodes on impact or upon heating to 125° (*281*).

Monolithio-*o*-carborane derivatives react as expected with monochloro-phosphines to give the corresponding phosphinocarborane. With phosphorus trichloride, the product is a bis(*o*-carboranyl)chlorophosphine (*367*).

$$C_6H_5C\!\!-\!\!CLi \;\; \underset{B_{10}H_{10}}{\diagdown\!\!O\!\!\diagup} \;\; + \;\; (n\text{-}C_6H_{13})_2PCl \;\;\longrightarrow\;\; C_6H_5C\!\!-\!\!CP(C_6H_{13})_2 \;\; \underset{B_{10}H_{10}}{\diagdown\!\!O\!\!\diagup}$$

$$\Big\downarrow PCl_3$$

$$\left[\; C_6H_5C\!\!-\!\!C\!\!-\!\! \;\; \underset{B_{10}H_{10}}{\diagdown\!\!O\!\!\diagup}\; \right]_2 PCl$$

Attempts to synthesize tris(o-carboranyl)phosphine species have been unsuccessful, evidently due to steric hindrance about the small phosphorus atom. The arsenic and antimony analogs, however, are easily prepared (*367*):

$$C_6H_5C\!\!-\!\!CLi \;\; \underset{B_{10}H_{10}}{\diagdown\!\!O\!\!\diagup} \;\; + \;\; MCl_3 \;\;\longrightarrow\;\; \left[\; C_6H_5C\!\!-\!\!C\!\!-\!\! \;\; \underset{B_{10}H_{10}}{\diagdown\!\!O\!\!\diagup}\; \right]_3 M$$

$$M = As, Sb$$

Phosphorus trichloride does react with bromomagnesiomethyl-o-carborane, forming tris(o-carboranylmethyl)phosphine (*415*).

$$PCl_3 + 3 \;\; HC\!\!-\!\!CCH_2MgBr \;\; \underset{B_{10}H_{10}}{\diagdown\!\!O\!\!\diagup} \;\;\longrightarrow\;\; P(CH_2C\!\!-\!\!CH)_3 \;\; \underset{B_{10}H_{10}}{\diagdown\!\!O\!\!\diagup} \;\; + \;\; 3MgBrCl$$

The bis(phosphino) and bis(arsino) derivatives of o-carborane readily form chelate complexes with transition metals. Thus nickel(II) chloride combines with bis(diphenylphosphino)-o-carborane (BDC) in methanol or ethyl acetate solution to give a complex containing two BDC units (*279*). An intermediate 1:1 complex has also been isolated from the room temperature reaction; on refluxing, this is converted to the 2:1 complex.

$$(C_6H_5)_2PC\!\!-\!\!CP(C_6H_5)_2 \;\; \underset{B_{10}H_{10}}{\diagdown\!\!O\!\!\diagup} \;\; \xrightarrow[25^\circ]{NiCl_2 \cdot 6H_2O} \;\; \begin{array}{c} Cl \;\; Cl \\ \diagdown\!\!Ni\!\!\diagup \\ (C_6H_5)_2P\diagup \quad \diagdown P(C_6H_5)_2 \\ \diagdown \qquad \diagup \\ C\!\!-\!\!C \\ \underset{B_{10}H_{10}}{\diagdown\!\!O\!\!\diagup} \end{array}$$

Nickel(II) chloride also forms 2:1 (but not 1:1) complexes with the mono-, di-, and tri-B-bromo derivatives of BDC.

All of these nickel–BDC complexes are diamagnetic, and this observation, together with electronic spectra and conductance measurements, suggests a square planar configuration about the nickel atom (*279*).

Russian workers (*464*) have reported the preparation of additional 1:1 complexes of nickel, iron, and palladium with bis(diphenylphosphino)-*o*-carborane and also with its B-halo derivatives (*466*).

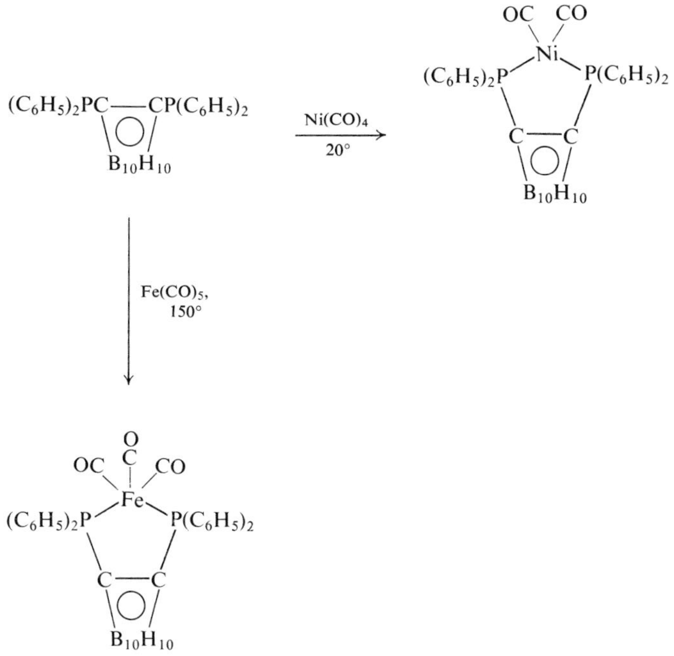

Bis(diphenylarsino)-*o*-carborane forms 1:1 complexes with $PdCl_2$ and PdI_2 (*278*), and bis(dimethylarsino)-*o*-carborane replaces carbonyl groups in metal carbonyls (*358*), forming 4-, 5-, and 6- coordinate complexes.

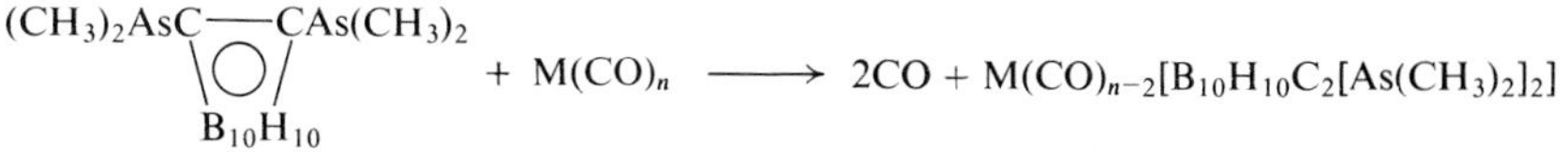

$$M = Ni, n = 4$$
$$M = Fe, n = 5$$
$$M = Mo, n = 6$$

Carboranes containing group V atoms in the cage itself are described in Chapter 9.

6-10. SILICON DERIVATIVES

o-Carboranyl Silanes

Derivatives containing silyl or alkylsilyl groups bonded directly to the o-carborane cage have been prepared by the action of alkylchlorosilanes on lithiocarboranes (*142, 173, 231, 252, 367*) and, in small yields, by the reaction of alkynylsilanes with bis(acetonitrile)decaborane (*265, 366*).

$$R_3SiCl \longrightarrow R_3SiC\underset{B_{10}H_{10}}{\diagdown O \diagup}CH$$

$$LiC\underset{B_{10}H_{10}}{\diagdown O \diagup}CH$$

$$R_2SiCl_2 \longrightarrow R_2Si\left[C\underset{B_{10}H_{10}}{\diagdown O \diagup}CH\right]_2$$

$$R_3SiCH_2Cl \longrightarrow R_3SiCH_2C\underset{B_{10}H_{10}}{\diagdown O \diagup}CH$$

$$C_6H_5C\underset{B_{10}H_{10}}{\diagdown O \diagup}CLi$$

$$MCl_4 \longrightarrow (C_6H_5C\underset{B_{10}H_{10}}{\diagdown O \diagup}C)_2MCl_2$$

$$M = Si, Ge; \quad R = CH_3, C_6H_5$$

$$R_2Si(CH_2Cl)_2$$

$$R_2SiCl_2$$

$$RSiCl_3$$

$$SiCl_4$$

$$SiHCl(CH_3)_2$$

$$R = CH_3, C_6H_5$$

$$(CH_3CN)_2B_{10}H_{12} + HC{\equiv}CSi(CH_3)_3 \longrightarrow \underset{B_{10}H_{10}}{HC{-}CSi(CH_3)_3} + 2CH_3CN + H_2$$

2.5%

The compound 1,2-bis(chlorodimethylsilyl)-*o*-carborane, prepared as indicated above, was expected to form a silicone, siloxane, or dihydroxy derivative upon hydrolysis. Interestingly, however, the cyclic tetramethyldisilaoxane was found to form quantitatively (*231*). An apparently analogous reaction occurs with ammonia, methylamine, and hydrazine to give the cyclic tetramethyldisilaazane (*231, 236*).

$$R = H, CH_3, NH_2$$

The same starting compound reacts with dilithio-o-carborane to give a six-membered ring compound containing two o-carborane units (*231*).

All of the above cyclic derivatives are reported to be thermally stable to well above *400°*. Somewhat surprisingly, they also fail to react with water, although the cyclic tetramethyldisilaoxane is converted by inorganic and organic bases to o-carborane itself and a methyl silicone polymer (*236*).

The pronounced tendency of silyl-substituted carboranes to form exo-polyhedral rings is illustrated by numerous other reactions (*236*):

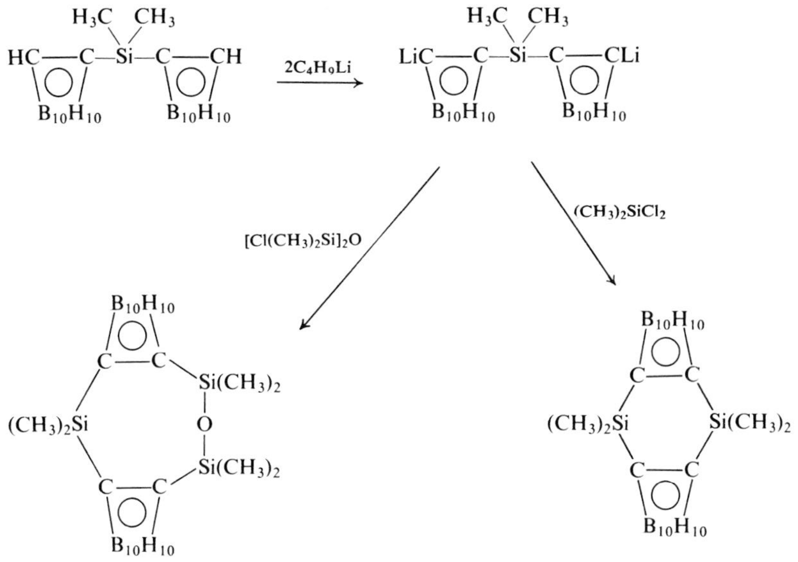

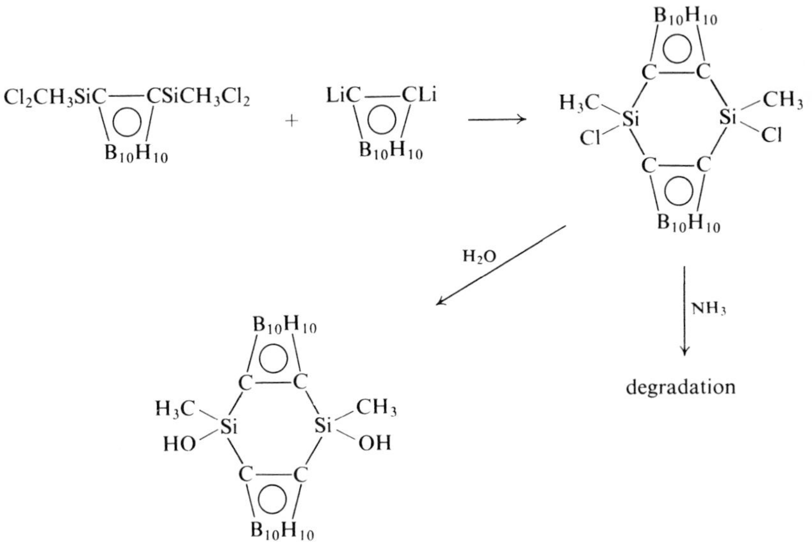

In contrast to the dimethyldichloro ring compound shown above, the analogous tetrachloro compound fails to react with water; ammonia, however, replaces all four chlorine atoms to give the tetraamino derivative (236).

o-Carboranyl Alkoxysilanes

The reaction of hydroxymethyl-*o*-carboranes with alkylchlorosilanes produces the corresponding alkoxysilane derivatives in good yield (*75, 208, 265*).

However, 1,2-bis(hydroxymethyl)-*o*-carborane interacts with dimethyl-dichlorosilane to give a seven-membered ring system:

Catalytic amounts of acid or base convert the cyclic product to the original diol (*265*).

Dilithio-*o*-carborane reacts with alkylchlorosiloxanes to form disiloxane and trisiloxane ring compounds. A 1:1 carborane–siloxane ratio yields the smaller five-membered ring, while a ratio of 1:2 produces the seven-membered ring. The chain length of the siloxane reactant apparently has little effect on the ring size (*174*).

$$n = 2, 3, 4$$

Neither product undergoes polymerization under the influence of acidic or basic catalysts.

At least two additional methods for preparing *o*-carboranyl alkoxysilanes have been reported: reaction of *o*-carboranyl alkyl Grignards with alkoxy-silanes, and addition of silanes to alkenylcarboranes (see below). The Grignard reactions are complicated by rearrangements similar to those noted previously (Section 6-2), leading in the present case to direct silicon–carborane bonds (*265*).

Attempts to polymerize the products formed in such reactions have not been notably successful, since the carborane–silicon bond is cleaved by both acids and bases. Thus, 1-methyl-2-(methyldiethoxysilyl)-*o*-carborane on treatment with acid or basic reagents yields 1-methyl-*o*-carborane and a boron-free methylsilicone polymer (*265*) (Chapter 8).

The Grignard rearrangement described above can be avoided by the use of C,C'-disubstituted carborane derivatives, so that no carborane C—H is present. In this manner, products having one or more methylene groups between the silicon atom and the carborane cage may be prepared (*265*).

$$CH_3C\!-\!CCH_2MgBr \xrightarrow{\;CH_3Si(OC_2H_5)_3\;} CH_3C\!-\!CCH_2\overset{\overset{\displaystyle CH_3}{\displaystyle |}}{Si}(OC_2H_5)_2$$
$$\underset{B_{10}H_{10}}{\diagdown\!O\!\diagup} \qquad\qquad\qquad \underset{B_{10}H_{10}}{\diagdown\!O\!\diagup}$$

The addition of silanes or their derivatives to alkenylcarboranes in the presence of platinized carbon yields alkoxysilyl carborane derivatives (*70, 71, 208, 265*). Such additions are hindered by increasing proximity of the carborane cage to the double bond; thus, no reaction is observed between methyldichlorosilane and 1-vinylcarborane, whereas the same silane reacts readily with 1-allyl- and 1-(3-butenyl)-*o*-carborane, the latter reaction being faster.

$$HC\!-\!C(CH_2)_nCH\!=\!CH_2 \;+\; \overset{\overset{\displaystyle CH_3}{\displaystyle |}}{H}Si(OC_2H_5)_2 \xrightarrow[\Delta]{Pt\text{-}C} HC\!-\!C(CH_2)_{n+2}\overset{\overset{\displaystyle CH_3}{\displaystyle |}}{Si}(OC_2H_5)_2$$
$$\underset{B_{10}H_{10}}{\diagdown\!O\!\diagup} \qquad\qquad n=1,2 \qquad\qquad \underset{B_{10}H_{10}}{\diagdown\!O\!\diagup}$$

The reactions of several 1-alkenyl-*o*-carboranes with trichlorosilane and a number of alkyldichlorosilanes have been shown to proceed as additions to the double bond, with the silicon atom adding at the position furthest from the carboranyl group (*271*).

$$HSiCl_3 + \; HC\!-\!CCH\!=\!CH_2 \xrightarrow{\;>200°\;} HC\!-\!CCH_2CH_2SiCl_3$$
$$\underset{B_{10}H_{10}}{\diagdown\!O\!\diagup} \qquad\qquad \underset{B_{10}H_{10}}{\diagdown\!O\!\diagup}$$

Alkenylcarboranes have also been used in the preparation of a series of carborane-substituted siloxanes (*70*). The products have the general structure shown, in which R, R′, and R″ are H, alkyl, or aryl groups and Q is either an *o*-carboranyl or *m*-carboranyl group.

$$R_3SiO\!\left[\!\begin{array}{c} R' \\ | \\ SiO \\ | \\ (CH_2)_xQR'' \end{array}\!\right]_n\!\!SiR_3$$

$$HC\!-\!C(CH_2)_2C\!=\!CH_2 \xrightarrow{\;CH_3SiCl_2H\;}$$
$$\underset{B_{10}H_{10}}{\diagdown\!O\!\diagup}$$

$$HC\!-\!C(CH_2)_4SiCl_2CH_3 \xrightarrow[\;(CH_3)_3SiCl\;]{H_2O,\,(C_2H_5)_2O}$$
$$\underset{B_{10}H_{10}}{\diagdown\!O\!\diagup}$$

$$(CH_3)_3SiO\!-\!\overset{\overset{\displaystyle CH_3}{\displaystyle |}}{Si}\!-\!OSi(CH_3)_3$$
$$\underset{\underset{B_{10}H_{10}}{\diagdown\!O\!\diagup}}{(CH_2)_4C\!-\!CH}$$

The synthesis of carborane–silicone polymers, of which many have been prepared, is described in Chapter 8.

6-11. GERMANIUM AND TIN DERIVATIVES

The reaction of 1,2-dilithio-*o*-carborane with dimethyldichlorogermane parallels that of the corresponding silane, and the reactions of the product, 1,2-bis(dimethylchlorogermyl)-*o*-carborane, with water and ammonia are similar to those of the analogous silicon system (*261*).

The properties of *o*-carboranyl tin compounds resemble those of the corresponding mercury compounds in many respects. Lithiocarboranes react easily with chlorostannanes as shown (*261, 366, 367*).

$$R = H, C_6H_5$$
$$R' = n\text{-}C_3H_7, C_6H_5$$

$$R = n\text{-}C_4H_9, CH_3$$

The formation of the above six-membered ring species from dialkyldichloro-stannanes is in contrast to the reaction of dilithio-o-carborane with dialkyl-dichlorosilanes, in which only nonbridged products are obtained (Section 6-10). The reaction of 1-phenyl-2-lithio-o-carborane with tin tetrahalides does not yield tetrakis(phenyl-o-carboranyl)tin, evidently due to steric hindrance, but the bis- and tris-substituted compounds are formed (*29*, *367*).

$$C_6H_5C\overset{\displaystyle\frown}{\underset{\displaystyle B_{10}H_{10}}{\smile}}CLi + SnCl_4 \longrightarrow (C_6H_5C\overset{\displaystyle\frown}{\underset{\displaystyle B_{10}H_{10}}{\smile}}C)_2SnCl_2 + (C_6H_5C\overset{\displaystyle\frown}{\underset{\displaystyle B_{10}H_{10}}{\smile}}C)_3SnCl$$

In comparison, the corresponding reactions involving the tetrachlorides of silicon and germanium yield only the bis(phenylcarboranyl) derivatives, presumably because of the smaller radius of the central atom (*367*). Tetrakis(o-carboranylmethyl)tin, in which crowding is relieved by methylene groups between the carborane units and the metal atom, has been prepared by the reaction of tin tetrachloride with bromomagnesiomethyl-o-carborane (*415*).

The tin–carborane bond, like the mercury–carborane bond, is labile with respect to nucleophilic attack. Alcoholic potassium hydroxide generates the original carborane nearly quantitatively (*29*).

$$RC\overset{\displaystyle\frown}{\underset{\displaystyle B_{10}H_{10}}{\smile}}CSnR'_3 \xrightarrow[C_2H_5OH]{KOH} RC\overset{\displaystyle\frown}{\underset{\displaystyle B_{10}H_{10}}{\smile}}CH + R'SnOH$$

$$R = C_6H_5$$
$$R' = C_3H_7, C_6H_5$$

$$(C_6H_5C\overset{\displaystyle\frown}{\underset{\displaystyle B_{10}H_{10}}{\smile}}C)_2SnCl_2 \xrightarrow[C_2H_5OH]{KOH} C_6H_5C\overset{\displaystyle\frown}{\underset{\displaystyle B_{10}H_{10}}{\smile}}CH$$

When the reaction is carried out by treatment of a benzene solution of the carborane with aqueous alkali, cleavage is accompanied by the formation of bis(phenylcarboranyl)tin oxide (*29*).

$$(C_6H_5C\overset{\displaystyle\frown}{\underset{\displaystyle B_{10}H_{10}}{\smile}}C)_2SnX_2 \xrightarrow[C_6H_6]{KOH, H_2O} (C_6H_5C\overset{\displaystyle\frown}{\underset{\displaystyle B_{10}H_{10}}{\smile}}C)_2SnO + C_6H_5C\overset{\displaystyle\frown}{\underset{\displaystyle B_{10}H_{10}}{\smile}}CH$$

$$X = Cl, n\text{-}C_4H_9 \qquad 62\% \text{ (for } X = Cl) \qquad 30\% \text{ (for } X = Cl)$$

Bis(phenylcarboranyl)tin oxide is attacked by dilute aqueous hydrobromic acid to give bis(phenylcarboranyl)tin dibromide.

The tin–carborane bonds in bis(phenylcarboranyl)tin dichloride are cleaved even by chromatography on basic alumina, although cleavage is prevented by prior acid treatment of the alumina. Germanium–carborane bonds are similarly broken on basic alumina, but silicon–carborane cleavage has evidently not been observed under these conditions (*29*).

Carboranes containing germanium, tin, and lead heteroatoms in the cage system are discussed in Chapter 9.

6-12. SULFUR DERIVATIVES

Compounds containing direct carborane–sulfur bonds may be prepared from C-metallo-*o*-carboranes. The reaction of dilithio-*o*-carborane with organic disulfides, or with sulfur followed by treatment with an alkyl halide, yields bis(thioethers). The intermediate formed in the sulfur reaction reacts with water to form 1,2-bis(mercapto)-*o*-carborane (*282*).

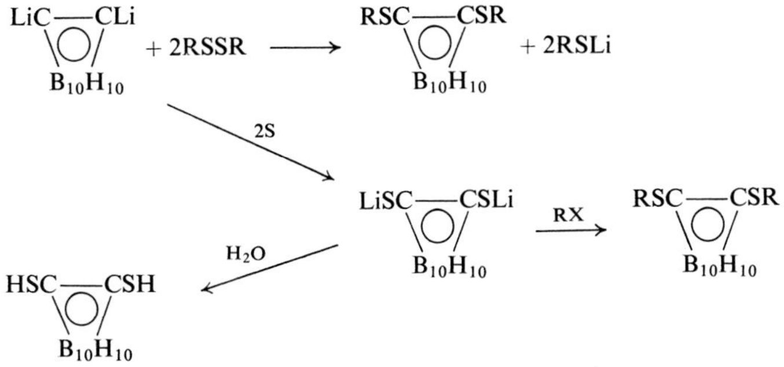

$$
\begin{array}{c}
\underset{B_{10}H_{10}}{LiC\!\!-\!\!CLi} + 2RSSR \longrightarrow \underset{B_{10}H_{10}}{RSC\!\!-\!\!CSR} + 2RSLi \\
\end{array}
$$

$$
\xrightarrow{\ \ 2S\ \ }
$$

$$
\underset{B_{10}H_{10}}{LiSC\!\!-\!\!CSLi} \xrightarrow{\ RX\ } \underset{B_{10}H_{10}}{RSC\!\!-\!\!CSR}
$$

$$
\underset{B_{10}H_{10}}{HSC\!\!-\!\!CSH} \xleftarrow{\ H_2O\ }
$$

R = H, CH$_3$, C$_2$H$_5$, CH$_2$C$_6$H$_5$, C$_6$H$_5$
X = halogen

Alternatively, the reaction between sulfur and a metallocarborane may be carried out in liquid ammonia; removal of the ammonia and treatment with dilute HCl affords the mercapto-*o*-carborane (*465*). Both mono- and dimercapto derivatives have been prepared by this route.

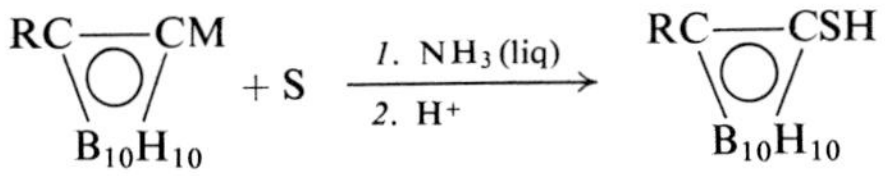

$$
\underset{B_{10}H_{10}}{RC\!\!-\!\!CM} + S \xrightarrow[\text{2. H}^+]{\textit{1.}\ NH_3\,(liq)} \underset{B_{10}H_{10}}{RC\!\!-\!\!CSH}
$$

M = Li, Na, K
R = CH$_3$, C$_6$H$_5$, H

Dilithio carborane derivatives containing one to three bromine atoms react similarly (282). Sulfur dichloride, however, yields no monomeric products on reaction with dilithio-o-carborane.

Sulfur dioxide attacks 1-phenyl-2-lithiocarborane to form a sulfinic acid, but the C—S bond in this compound is readily cleaved by gentle heating in aqueous solution (465). o-Carboranyl sulfinic acids may be oxidized to the acids and converted to the sulfonyl halides and sulfanilides (465a).

$$C_6H_5C\!\!-\!\!CLi \underset{B_{10}H_{10}}{\diagdown\!O\!\diagup} + SO_2 \xrightarrow[H^+]{H_2O} C_6H_5C\!\!-\!\!\overset{O}{\underset{B_{10}H_{10}}{\underset{\diagdown\!O\!\diagup}{\overset{\|}{C}}}}\!\!-\!\!\overset{}{\underset{O}{\overset{\|}{S}}}H \xrightarrow{\Delta} C_6H_5C\!\!-\!\!CH \underset{B_{10}H_{10}}{\diagdown\!O\!\diagup} + SO_2$$

The effect of the electron-withdrawing o-carboranyl group on the carborane–sulfur bond has been intensively studied and compared with the chemistry of conventional organosulfur compounds. For example, in contrast to the ease of metalation of alkyl or alkyl–aryl sulfides by butyllithium, this reagent cleaves 1,2-bis(methylthio)-o-carborane (282).

$$CH_3SC\!\!-\!\!CSCH_3 \underset{B_{10}H_{10}}{\diagdown\!O\!\diagup} \xrightarrow[(C_2H_5)_2O]{2C_4H_9Li} LiC\!\!-\!\!CLi \underset{B_{10}H_{10}}{\diagdown\!O\!\diagup} + 2C_4H_9SCH_3$$

Further, unlike typical organosulfides which readily form addition compounds with chlorine or bromine, no reaction is observed between bromine and 1,2-bis(methylthio)-o-carborane in refluxing carbon tetrachloride or carbon disulfide. In the presence of aluminum chloride, only bromination of the carborane cage is observed (282).

$$CH_3SC\!\!-\!\!CSCH_3 \underset{B_{10}H_{10}}{\diagdown\!O\!\diagup} \xrightarrow[AlCl_3]{2Br_2} CH_3SC\!\!-\!\!CSCH_3 \underset{B_{10}H_8Br_2}{\diagdown\!O\!\diagup} + 2HBr$$

The mercapto-o-carboranes are soluble in aqueous alkali; the resulting salts react with methyl iodide to form carboranyl sulfides, and with iodine to produce carboranyl disulfides (465).

$$RC\!\!-\!\!CSM \underset{B_{10}H_{10}}{\diagdown\!O\!\diagup} \xrightarrow[C_6H_6]{I_2} RC\!\!-\!\!C\!\!-\!\!S\!\!-\!\!S\!\!-\!\!C\!\!-\!\!CR \underset{B_{10}H_{10}}{\diagdown\!O\!\diagup}\;\underset{B_{10}H_{10}}{\diagdown\!O\!\diagup}$$

$$\xrightarrow[NH_3\,(liq)]{CH_3I} RC\!\!-\!\!CSCH_3 \underset{B_{10}H_{10}}{\diagdown\!O\!\diagup}$$

$$R = CH_3, C_6H_5$$

The S—S bond in the carboranyl disulfides is very resistant to attack by halogens or hot acidic hydrogen peroxide, probably as a consequence of the electrophilic carborane cage system (465). In contrast, 1-phenyl-2-(methylthio)-o-carborane is completely degraded to boric acid by hydrogen peroxide in acetic acid (465).

The ligand protons in 1,2-bis(mercapto)-o-carborane are easily displaced, and this compound acts as a bidentate chelating agent with both metals and nonmetals. This is illustrated by a variety of nickel(II) and cobalt(II) complexes (282, 283, 467):

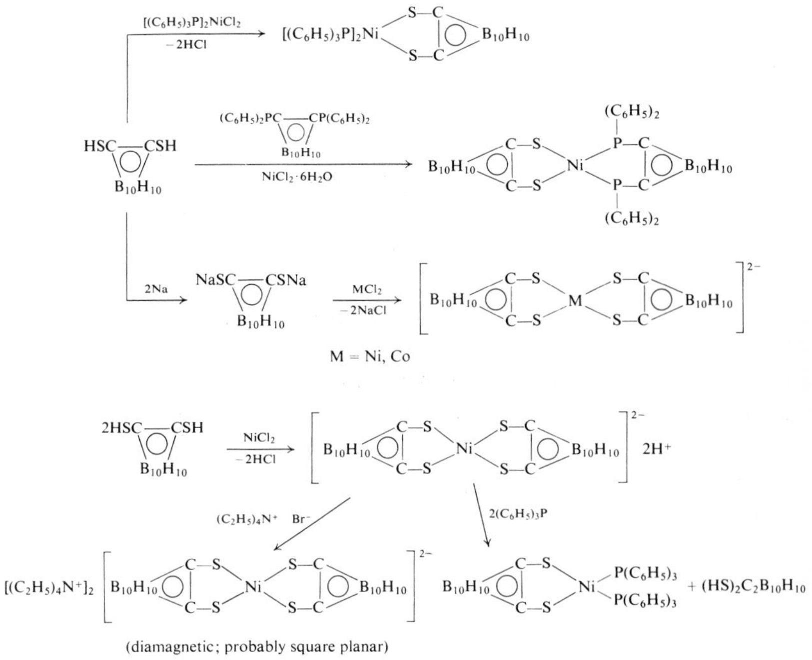

(diamagnetic; probably square planar)

The cobalt complex shown above (containing four metal–sulfur bonds) has been found to contain three unpaired electrons. It has been proposed (283) that steric requirements probably force a square-planar geometry for complexes containing two bidentate carboranyl ligands, so that this species may be a high-spin square-planar cobalt(II) complex. In comparison, the cobalt(II) analogs of the other nickel species are diamagnetic. The assumption of square-planar geometry for all of these complexes is supported by electronic spectra (283).

While π-interaction involving the metal, sulfur, and carborane cage is at least conceivable in complexes of the above type, detailed spectral studies have given no evidence for such effects, and have indicated that the metal–sulfur bonds are primarily σ in character (*283*).

Bidentate coordination of 1,2-bis(mercapto)-o-carborane to nonmetal atoms such as boron (*280*) and phosphorus (*282*) has also been observed.

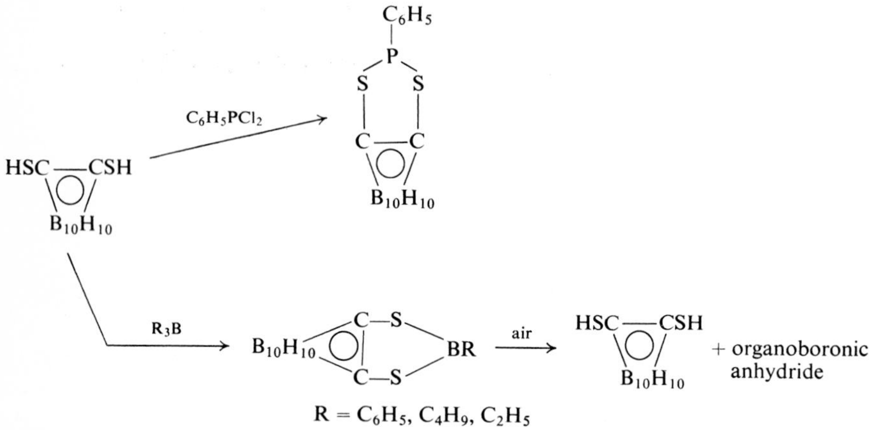

$$R = C_6H_5, C_4H_9, C_2H_5$$

Boron trichloride fails to react with 1,2-bis(mercapto)-o-carborane even in hot benzene. In the presence of acetonitrile, however, a highly reactive B-chlorinated ring compound is formed, which is converted in air to the bis(mercapto)carborane (*280*). The presence of the bis(mercapto)carborane

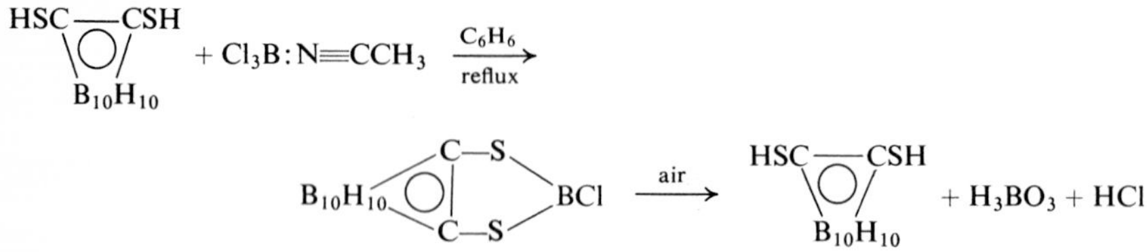

in 2:1 excess leads to an ionic product which may be precipitated as the tetraethylammonium salt. The structure shown has been postulated but not proved.

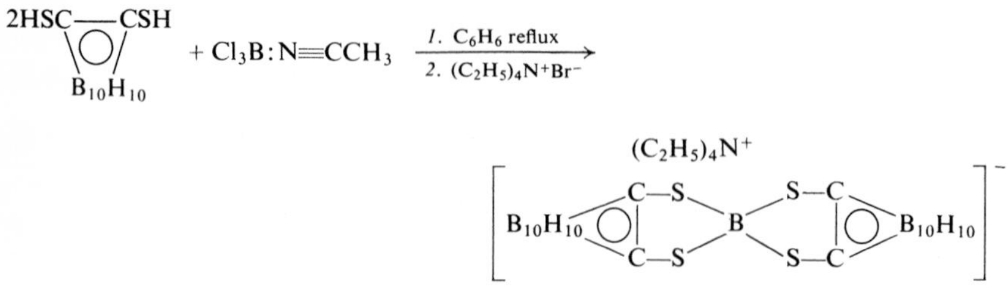

Similar boron-coordinated derivatives are obtained from 1,2-bis(mercapto)-o-carborane in reactions with aminoboranes and diborane (280).

$$\text{HSC———CSH} \overset{\text{(CH}_3)_2\text{NBH}_2}{\underset{\Delta}{\longrightarrow}} B_{10}H_{10}\underset{C—S}{\overset{C—S}{\diagdown}} BN(CH_3)_2 + 2H_2$$

$$\underset{B_{10}H_{10}}{\text{HSC———CSH}} \overset{H_2NB(C_4H_9)_2}{\longrightarrow} B_{10}H_{10}\underset{C—S}{\overset{C—S}{\diagdown}} BC_4H_9 + \text{residue}$$

$$\overset{B_2H_6}{\longrightarrow} B_{10}H_{10}\underset{C—S}{\overset{C—S}{\diagdown}} BH \text{ (proposed structure)}$$

The previously noted B-chloro ring compound reacts with phenyl isocyanate in refluxing carbon tetrachloride; subsequent treatment with water yields a final product which is probably 1,2-bis(N-phenylthiocarbamate)-o-carborane (280).

$$B_{10}H_{10}\underset{C—S}{\overset{C—S}{\diagdown}} BCl \overset{\textit{1. }2C_6H_5NCO}{\underset{\textit{2. }H_2O}{\longrightarrow}} \underset{B_{10}H_{10}}{HN—\overset{C_6H_5}{\underset{}{C}}—\overset{O}{\underset{}{C}}—SC———CS—\overset{O}{\underset{}{C}}—\overset{C_6H_5}{\underset{}{NH}}}$$

Sulfur-containing o-carborane derivatives having no direct carborane–sulfur bonds should be obtainable, in many cases, by one or more of the general routes to C-substituted o-carboranes described earlier. Although few such syntheses have been reported, 1,2-bis(methylsulfonyloxymethyl)-o-carborane has been prepared directly from decaborane (120).

$$CH_3SO_3CH_2C{\equiv}CCH_2SO_3CH_3 + B_{10}H_{14} \longrightarrow \underset{B_{10}H_{10}}{CH_3SO_3CH_2C———CCH_2SO_3CH_3}$$

6-13. HALOGEN DERIVATIVES

Nearly all of the chemistry of o-carborane presented up to this point has been concerned with substitution at the cage carbon atoms, due largely to the ease of formation of C-metallated species from which many other derivatives may be prepared by replacement of the metal atoms. In contrast, the direct halogenation of o-carborane itself leads only to boron-substituted products. Derivatives containing carbon–halogen bonds have, however, been prepared by several indirect methods which are described later in this section. The formation and properties of o-carboranyl halogen derivatives provide useful information on charge distribution, cage rearrangement mechanisms, and other properties of the icosahedral framework, and consequently much of the research on o-carborane has focused on its B- and C-halogenated derivatives.

Electrophilic Halogenation

The direct attack of chlorine, bromine, or iodine on the o-carborane system (fluorination is discussed below) has been shown in numerous studies to be highly stereospecific, starting at the boron atoms furthest removed from carbon (positions 9 and 12) followed by the adjacent borons (8 and 10). The carbon atoms and those borons directly linked to them (3,6,4,5,7,11) do not appear susceptible to Friedel-Crafts halogenation. This observed sequence of substitution correlates well with calculations of ground-state charge distribution by Potenza et al. (242, 247), who predicted on the basis of a nonempirical molecular orbital method that boron atoms 8, 9, 10, and 12 are most negative. The order of decreasing negative charge in the ground state, according to these workers, is B(8,10) > B(9,12) > B(4,5,7,11) > B(3,6) > C(1,2) (chemically equivalent atoms are grouped in parentheses). In these results atoms 8, 10, 9, and 12 have nearly equal framework charges, and although electrophilic halogenation actually occurs first at the 9, 12 positions, it has been pointed out (242, 247) that the five highest filled (and most polarizable) molecular orbitals place a higher negative charge on atoms 9 and 12 than on atoms 8 and 10. While the stoichiometry and reaction conditions may affect the product distribution, the sequence of substitution appears nearly invariant. Treatment of o-carborane with chlorine over $FeCl_3$, $AlBr_3$, or $AlCl_3$ yields mainly mono-, di-, and tri-B-chloro derivatives (essentially one isomer of each) and a small quantity of a tetrachloro product (318, 323, 384, 387, 389, 409, 461). Recent detailed studies have shown that the monochloro isomer is 9-chloro-o-carborane, with a small (~8%) yield of the 8-chloro derivative (409). Chlorination of 1-methyl-o-carborane under similar conditions yields two monochloro derivatives as well as one dichloro-, one trichloro-, and one tetrachloro compound (316, 384, 387, 409). Given the symmetry of 1-methyl-o-carborane, in which B(9) and B(12) are chemically nonequivalent while B(8) and B(10) are equivalent, and assuming that the relatively positive boron atoms 3 and 6 are not attacked (247), Zakharkin and Kalinin (387, 409) deduced that the products are, respectively, 1-methyl-9-chloro-, 1-methyl-12-

chloro-, 1-methyl-9,12-dichloro, 1-methyl-9,12,8(10)-trichloro-, and 1-methyl-9,12,8,10-tetrachloro-o-carborane. The NMR and ^{35}Cl NQR spectra (*36, 237, 323*) of these derivatives support the structural assignments given, as do the X-ray studies of the corresponding bromo and iodo derivatives discussed below. Further support is given by dipole moment measurements (*62, 205, 307*) of a number of o-carboranyl halogen derivatives, particularly those of o-carborane (4.31 D), 1,2-dibromo-o-carborane (2.82), and 9,12-dibromo-o-carborane (7.21) (*205*). The fact that the o-carborane cage is made more polar by bromination at the 9 and 12 boron atoms, and less polar by bromination at carbon, is in agreement with the view that the carbon atoms are at the positive end of the cage dipole, whereas those borons furthest from carbon constitute the negative pole (*205*).

The electrophilic bromination (*247, 281, 318, 322, 323, 384, 387, 389, 461*) and iodination (*305, 318, 323, 384, 389, 461*) of o-carborane appear to proceed similarly, except that iodination requires elevated temperatures. The substitution sequence is evidently the same as in chlorination, X-ray studies having established the structures of 9,12-$Br_2C_2B_{10}H_{10}$ (*242*), 9,12,8(10)-$Br_3C_2B_{10}H_9$ (*243*), 9-$IC_2B_{10}H_{11}$ (*10, 321*) and 1,2-$(CH_3)_2$-9,12,8,10-$Br_4C_2B_{10}H_6$ (*244*). The last-mentioned derivative is produced by bromination of 1,2-$(CH_3)_2$-$C_2B_{10}H_{10}$ at room temperature (*281*); since o-carborane itself can only be tribrominated under these conditions, the tetrabromination of the C,C'-dimethyl derivative suggests (*244*) an inductive transfer of negative charge from the methyl groups to the cage system (the methyl groups do not affect the order of halogen substitution). These results are in qualitative agreement with nonempirical molecular orbital calculations (*244*) for 1-methyl- and 1,2-dimethyl-o-carborane, which indicate the same order of charge distribution as in o-carborane itself but with increased negative charge on all framework atoms. Further, studies of the relative rates of chlorination, bromination and iodination of o-carborane and 1-methyl-o-carborane (*404*) have given evidence of a +I effect by the methyl group, which increases the rate of electrophilic substitution on the cage.

The Friedel-Crafts halogenation of 1-phenyl-o-carborane has yielded some interesting results. In the presence of $AlCl_3$ or iron filings the o-carborane cage is attacked preferentially, and it is possible to quench the reaction before halogenation of the phenyl ring begins. The electron-withdrawing carborane cage apparently renders the phenyl group passive with respect to electrophilic halogenation, since competitive bromination of equimolar amounts of benzene and 1-phenyl-o-carborane yields perbromobenzene after 30 minutes while much of the carborane is recovered unchanged (*461*). However, exhaustive electrophilic chlorination or bromination over $AlCl_3$ gives 1-pentachlorophenyl-8,9,10,12-tetrachloro-o-carborane and 1-(m,p,m-tribromophenyl)-8,9,10,12-tetrabromo-o-carborane, respectively (*384, 389, 461*).

In contrast to the above, the chlorination and bromination of 1-vinyl-o-carborane occurs on the vinyl group prior to cage substitution; iodine, however, reportedly enters the o-carborane cage only, leaving the vinyl group unaffected (see Section 6-4) (*394*).

o-Carboranyl compounds may also be halogenated by halomethanes, the substitution sequence being the same as with elemental halogens (*294, 295, 318, 406, 446*). For example, refluxing o-carborane in CCl_4 or $CHCl_3$ over $AlCl_3$ yields 9,12-dichloro-o-carborane (79%) plus traces of mono-, tri-, and tetrachloro derivatives (*406*). Iodobenzene halogenates o-carborane at 300°, giving primarily the 9- and 8-iodo derivatives (*424*).

Although the order of electrophilic substitution on the o-carborane cage appears nearly invariant, there is evidence that the choice of catalyst may alter the course of reaction detectably. Thus, the chlorination or bromination of o-carborane in the presence of iron metal (which forms $Fe(FeX_3)$ under the reaction conditions) appears less selective than is the case with $AlCl_3$ or $AlBr_3$ catalysts, yielding 8(10),12-dihalo-o-carboranes (*318*). The major dihalo products are still 9,12-substituted, however. Other catalytic effects have been noted, including the fact that CCl_4 and CH_2Cl_2 halogenate o-carborane in the presence of $AlCl_3$ but not with $Fe(FeCl_3)$ (*318*).

Photochemical Halogenation

Chlorine and bromine attack o-carborane under ultraviolet light with initial substitution occurring at boron atoms 9 and 12, followed by 8 and 10, as in Friedel-Crafts halogenation (*281, 322, 385, 460*). The photochemical reaction is somewhat less selective, however (*385*), and in the case of chlorine it proceeds until all ten borons have been halogenated. As predicted from the calculated charge distribution, atoms 3 and 6 are the last to be chlorinated (*247*). Varying the reaction time and conditions has led to the isolation of B-chlorinated derivatives having from one to ten chlorine atoms, the major products being those with four, five, eight, and ten chlorines (*241, 245, 260, 281, 385, 460, 461*). Contrary to earlier reports (*260, 262*), it is now established (*444, 461*) that the ultimate product of photochemical chlorination of o-carborane is B-decachloro-o-carborane.* Substitution on the carbon atoms does not occur even on prolonged treatment under ultraviolet irradiation (the carbons may, however, be chlorinated by other methods described below). The photochemical chlorination of 1-phenyl-o-carborane proceeds essentially identically to the

---

* The compound described as "undecachloro-o-carborane" by Schroeder *et al.* (*260*) is in fact B-decachloro-o-carborane (*257a*).

parent compound; the phenyl ring is not affected and the final product is 1-phenyl-B-decachloro-o-carborane (*385, 406*).

The bromination of *o*-carborane under ultraviolet light, unlike the chlorine reaction, takes place only with difficulty (*385, 461*). When carried out in refluxing CCl_4 the major product is B-monobromo-*o*-carborane, with 35% of the original compound unchanged (*461*). Further bromination has been achieved in the presence of iron filings, with temperatures of 120–140° required for synthesis of the B-tetrabromo derivative (it is not clear whether this reaction involves electrophilic or free radical attack, or both) (*461*). The introduction, under any conditions, of more than four bromine or iodine atoms into the *o*-carborane cage has not been reported. Ultraviolet-irradiated bromination of 1-phenyl-*o*-carborane, as with chlorination, leaves the phenyl group unchanged; the major product is 1-phenyl-12-bromo-*o*-carborane (*385, 461*).

Fluorination

The only reported synthesis of polyfluoro-*o*-carboranes is that by Kongpricha and Schroeder (*182*), who have found that elemental fluorine, unlike the other halogens, attacks *o*-carborane in a largely nonselective manner to yield B-substituted derivatives containing one to ten fluorine atoms. Excess fluorine forms B-decafluoro-*o*-carborane, but no C-substitution occurs.

$$\begin{array}{c} HC\text{---}CH \\ \diagdown O \diagup \\ B_{10}H_{10} \end{array} + 10F_2 \xrightarrow[0°]{HF\ (liq)} 10HF + \begin{array}{c} HC\text{---}CH \\ \diagdown O \diagup \\ B_{10}F_{10} \end{array}.$$

No fluorination results from the reaction of *o*-carborane with chlorine monofluoride, which instead yields B-decachloro-*o*-carborane.

3-fluoro-*o*-carborane has been prepared, as described on p. 138.

Synthesis of Halo-*o*-Carboranes from Decaborane(14) Halogen Derivatives

The reaction of structurally characterized monohalodecaboranes with acetylene or higher alkynes yields B-monohalo-*o*-carboranes in which the halogen location is established (*390, 391*).

$$B_{10}H_{13}X + RC\!\!\equiv\!\!CH \xrightarrow[\text{toluene}]{CH_3CN} \begin{array}{c} RC\!-\!\!-CH \\ \diagdown O \diagup \\ B_{10}H_9X \end{array} + 2H_2$$

$$X = Cl, Br, I;$$
$$R = H, C_6H_5, CH_2Br$$

In all reported reactions of this type, the carborane products are as expected from the location of the halogen atoms on the original decaborane framework. Thus, 1-halodecaboranes yield 9(12)-halo-o-carboranes uniquely, and 2-halodecaboranes produce only the corresponding 8(10)-halo-o-carboranes.*

The reaction of an α-acetylene with a 1-halodecaborane forms two isomeric o-carborane derivatives as expected; for example, phenylacetylene and 1-iododecaborane yield 1-phenyl-9-iodo- and 1-phenyl-12-iodo-o-carborane in equal amounts.

This type of reaction provides a convenient synthesis of B-halo-o-carborane species, such as 8-halo-o-carboranes, which are not obtained in large yield in the direct halogenation of o-carborane. Further halogenation of the carborane derivatives prepared from decaborane follows the usual substitution sequence. Thus, reaction of bromine with 1-methyl-8-bromo-o-carborane over $AlCl_3$ gives a mixture of 1-methyl-8,9-dibromo- and 1-methyl-8,12-dibromo-o-carborane. Continued bromination of the mixture yields a single isomer, 1-methyl-8,9,12-tribromo-o-carborane (*390, 391*).

Synthesis of 3-Halo-o-Carboranes

The conversion of 3-amino-o-carboranes to the 3-halo-derivatives via diazotization followed by treatment with a cuprous halide is described in Section 6-8.

The insertion reactions of the (3)-1,2-dicarbollide ion (Chapter 9) with boron trifluoride etherate and a boron trihalide afford the respective 3-halo-o-carboranes (*250*).

$$[C_2B_9H_{11}]^{2-} \xrightarrow{BX_3} \begin{array}{c} HC\!-\!\!-CH \\ \diagdown O \diagup \\ B_{10}H_9X \end{array}$$

$$X = F, Br$$

---

* Numbering of the boron atoms in $B_{10}H_{14}$ begins opposite the open side of the framework, so that borons numbered 1, 2, 3, and 4 in $B_{10}H_{14}$ become, respectively, 9, 8, 12, and 10 in the completed o-$C_2B_{10}H_{12}$ icosahedron.

Synthesis of C-Halo-*o*-Carboranes

Halogen substitution at carbon does not occur in direct reactions of *o*-carborane with fluorine, chlorine, bromine, or iodine, but a variety of C-halo-*o*-carboranes has been prepared by utilizing standard organometallic methods for the replacement of positive C-bonded hydrogen atoms with halogen. B-decachloro-*o*-carborane is easily converted to the C,C'-dilithio derivative, which reacts with chlorine forming perchloro-*o*-carborane (*262*). The same compound may be obtained directly from B-decachloro-*o*-carborane by re-action with *N*-chlorosuccinimide, or by formation of the triethylammonium salt followed by treatment with chlorine (*262, 444*). Alternatively, the triethyl-ammonium salt may be brominated to form the C,C'-dibromo-B-decachloro derivative (*444*).

B-decachloro-*o*-carborane is a moderately strong acid (see below) and is easily soluble in aqueous alkali; treatment of the resulting salt with chlorine yields perchloro-*o*-carborane (*460, 461*).

Derivatives of *o*-carborane which are halogenated *only* at carbon may be obtained from alkali metal or Grignard intermediates, or from reactions of haloalkynes with decaborane. The reaction of 1-lithio-2-phenyl-*o*-carborane with chlorine, bromine, or iodine yields the corresponding 1-halo-2-phenyl derivative (*447*).

X = Cl, Br, I

o-Carboranyl Grignard reagents (Section 6-2) undergo halogen substitution by halomethanes or elemental halogens (*142, 294, 295, 299a, 468*).

$$\underset{B_{10}H_{10}}{\overset{RC\!-\!CMgBr}{\diagdown O \diagup}} + X_2 \xrightarrow[-MgBrX]{C_6H_6 \text{ or } CCl_4} \underset{B_{10}H_{10}}{\overset{RC\!-\!CX}{\diagdown O \diagup}}$$

$$X = Cl, Br, I$$
$$R = CH_3, C_6H_5, CH\!=\!CH_2$$

It has been found, however, that the chlorination of 1-methyl-o-carboranyl magnesium bromide in the presence of ethylmagnesium bromide takes place on both carbon and boron, forming a 1-methyl-2,B-dichloro-o-carborane. It has been suggested (*294*) that the $MgCl_2$ formed under the reaction conditions acts as a Friedel-Crafts catalyst, inducing chlorination at a boron cage position. Carbon tetrachloride in tetrahydrofuran reacts with o-carboranyl magnesium bromide to form two C,B-dichloro derivatives.

$$\underset{B_{10}H_{10}}{\overset{HC\!-\!CMgBr}{\diagdown O \diagup}} \xrightarrow[THF]{CCl_4, \, 0\text{--}5°} \underset{B_{10}H_9Cl}{\overset{HC\!-\!CCl}{\diagdown O \diagup}} \quad \text{(2 isomers)}$$

The analogous reaction of 1-methyl-o-carborane gives only a single B,C-dichloro-1-methyl-o-carborane (*294, 295*).

   The preparation of 1-chloro-o-carborane has been achieved via the reaction of 1-bromomagnesio-o-carborane with p-toluenesulfonyl chloride (*296*). An equilibrium exists between 1-chloro- and 1,2-dichloro-o-carborane and o-carborane itself (the product mixture contains these compounds in amounts of 60, 10, and 30%, respectively); even 1,2-bis(bromomagnesio)-o-carborane, in reaction with excess p-toluenesulfonyl chloride, yields 50% of the 1-chloro derivative plus 31% 1,2-dichloro-o-carborane and 18% o-carborane.

   C-halo-o-carboranes seem to have only rarely been synthesized directly from haloalkynes, but the 1-n-propyl-2-bromo- and 1-n-butyl-2-bromo-derivatives have been prepared from bis(acetonitrile) decaborane and the appropriate alkyne (*143*).

$$RC\!\equiv\!CBr + B_{10}H_{12}(CH_3CN)_2 \longrightarrow \underset{B_{10}H_{10}}{\overset{RC\!-\!CBr}{\diagdown O \diagup}}$$

$$R = n\text{-}C_3H_7, \, n\text{-}C_4H_9$$

The patent literature contains a report of the preparation of 1,2-dichloro-o-carborane from dichloroacetylene and bis(acetonitrile)decaborane (*67*).

Reactions of Halogen Derivatives

Boron–halogen bonds in o-carborane derivatives are in general remarkably stable. B-decachloro-o-carborane, for example, is inert to attack by SbF_3Cl_2 at 240°, and a B-octachloro derivative does not undergo Friedel-Crafts reaction in refluxing benzene over $AlCl_3$ (260). B-hexachloro- and B-tetrachloro-o-carboranes fail to react with ammonia or amines at 20°. B-decachloro-o-carborane, the most stable of the B-chlorinated species, is not hydrolyzed by hot aqueous NaOH (443, 461), but the chlorine atoms in all of the chloro-carboranes are removed by 50% aqueous KOH containing H_2O_2 (260). B-decachloro-o-carborane is completely degraded to boron oxychlorides by dimethylsulfoxide (445).

B-decafluoro-o-carborane readily undergoes nucleophilic displacement, and in contrast to the other B-polyhalo-o-carboranes is easily hydrolyzed by water or moist air (182).

The replacement of the boron-bonded hydrogen atoms in o-carborane by halogen, particularly chlorine, increases the acidic character of the C—H bonds (192, 262, 443, 444, 460, 461). The effect is greatest in the more highly halogenated species, B-decachloro-o-carborane being comparable in strength to carboxylic acids. Zakharkin and Ogorodnikova (444) have measured the pK_a values of a number of halo-o-carboranes and compared them with those of common organic acids (Table 6-3). (The acidity of o-carboranylmercury

TABLE 6-3

ACID STRENGTH OF HALOCARBORANES IN 50% ETHANOL[a]

Compound	pK_a	Compound	pK_a
o-$C_2H_2B_{10}Cl_{10}$	6.89	p-NH_2—o-CH_3—C_6H_3COOH	6.83
o-$C_2H_2B_{10}Cl_8Br_2$	6.95	p-NO_2—C_6H_4COOH	7.68
o-$C_2H(CH_3)B_{10}Cl_{10}$	6.90	C_6H_5SH	7.78
m-$C_2H_2B_{10}Cl_{10}$	9.19	p-ClC_6H_4SH	7.06
		$CH_3(CH{=}CH)_2COOH$	7.00

[a] From Zakharkin and Ogorodnikova (444).

derivatives has been measured indirectly by determination of polarographic reduction potentials (306), and here also the chlorinated species are found to be significantly stronger acids than the nonhalogen derivatives). B-decachloro-o-carborane titrates as a *monoprotic* acid [not diprotic as reported earlier (260)] in aqueous ethanol and forms stable 1:1 adducts or salts with triethyl-amine, trimethylamine, dimethyl sulfoxide, and dimethylformamide (444), in which hydrogen bridge bonding is believed to be involved (*192*). However, B-decachloro-o-carborane evolves two moles of methane in the reaction with CH_3MgI in ether (*461*).

$$HC\!\!-\!\!CH \quad + \quad 2CH_3MgI \quad \longrightarrow \quad IMgC\!\!-\!\!CMgI \quad + \quad 2CH_4$$
$$B_{10}Cl_{10} \qquad\qquad\qquad\qquad\qquad B_{10}Cl_{10}$$

B-decachloro-o-carborane dissolves in aqueous NaOH or KOH to give both mono- and dialkali metal salts (*443, 460*). These salts, as well as those formed by C-alkyl-o-carboranes, are not hydrolyzed in aqueous solution but do react with alkyl halides in aqueous alcohol. Methyl iodide yields only 1,2-dimethyl-B-decachloro-o-carborane, regardless of whether the mono- or disodium salt is used. The monomethyl derivative, however, is easily prepared from the monoiodomagnesium salt in tetrahydrofuran.

$$HC\!\!-\!\!CNa \xrightarrow[C_2H_5OH]{2CH_3I} H_3CC\!\!-\!\!CCH_3 \quad + \quad NaI$$
$$B_{10}Cl_{10} \qquad\qquad\qquad B_{10}Cl_{10}$$

$$HC\!\!-\!\!CH + CH_3MgI \xrightarrow[-CH_4]{THF} HC\!\!-\!\!CMgI \xrightarrow[-MgI_2]{CH_3I} HC\!\!-\!\!CCH_3$$
$$B_{10}Cl_{10} \qquad\qquad\qquad B_{10}Cl_{10} \qquad\qquad B_{10}Cl_{10}$$

In contrast to the methyl iodide reaction, ethyl iodide forms only the monoethyl derivative in reactions with the mono- and disodium B-decachloro-o-carborane salts.

$$HC\!\!-\!\!CNa \xrightarrow[C_2H_5OH]{C_2H_5I} HC\!\!-\!\!CC_2H_5 \quad + \quad NaI$$
$$B_{10}Cl_{10} \qquad\qquad\qquad B_{10}Cl_{10}$$

Similar reactions of the disodium salt with allyl bromide and benzyl chloride have been used to prepare the 1-allyl, 1-benzyl, and 1,2-dibenzyl derivatives of B-decachloro-o-carborane (*443*).

Chloro-o-carboranes containing fewer than ten chlorine atoms are significantly less acidic than the B-decachloro compound; at least one octachloro-o-carborane forms a stable triethylamine adduct and may be titrated in ethanol–water solution, but derivatives having six or fewer halogens cannot be titrated with strong base (*260*).

B-bromo-o-carboranes undergo metallation, alkylation and halogenation reactions essentially analogous to those of o-carborane and its B-chloro derivatives (*281*). (However, o-carborane species having more than four bromine atoms are unknown at present.)

$$\underset{B_{10}H_{10-n}Br_n}{HC\!\!-\!\!CH} \xrightarrow{2C_4H_9Li} \underset{B_{10}H_{10-n}Br_n}{LiC\!\!-\!\!CLi} \xrightarrow{2CH_3I} \underset{B_{10}H_{10-n}Br_n}{H_3CC\!\!-\!\!CCH_3}$$

$$n = 1, 2, 3$$

$$\xrightarrow{Cl_2} \underset{B_{10}Br_nCl_mH_{10-n-m}}{HC\!\!-\!\!CH}$$

$$n = 2, m = 8;$$
$$n = 3, m = 5$$

An important and distinctive property of B-halogenated *o*-carboranes is their participation in nucleophilic substitution reactions with copper(I) chloride. As the preceding sections of this chapter indicate, electrophilic attack on the *o*-carborane cage has been commonly observed but evidence for nucleophilic exchange has rarely been cited. It has been found, however, that all halogen atoms in 9-bromo-, 9-iodo-, 9,12-dibromo-, and 9,12-diiodo-*o*-carborane are replaced by chlorine when these compounds are treated with CuCl at 250–350° (*393*, *395*). Moreover, under the same conditions 1-methyl-12-iodo- and 1-phenyl-12-iodo-*o*-carborane are smoothly converted to the corresponding chlorinated species with no side products and no effect on the alkyl or aryl group.

$$\underset{B_{10}H_{10-n}X_n}{HC\!\!-\!\!CH} \xrightarrow{CuCl} \underset{B_{10}H_{10-n}Cl_n}{HC\!\!-\!\!CH}$$

$$\underset{B_{10}H_9I}{C_6H_5C\!\!-\!\!CH} \xrightarrow{CuCl} \underset{B_{10}H_9Cl}{C_6H_5C\!\!-\!\!CH}$$

$$X = Br, I;$$
$$n = 1, 2$$

Curiously, the reaction of 8,9,10,12-tetraiodo-*o*-carborane with excess CuCl results in the replacement of only three iodine atoms by chlorine (*395*). The mechanism for these substitution reactions has not been established, although Zakharkin and Kalinin (*395*) suggest a four-center transition state involving the attack of CuCl upon the B—Br(I) bond, followed by the splitting out of CuBr.

A related discovery (*308*) of synthetic importance is the Ullman reaction of 9-iodo-*o*-carborane with copper powder, yielding 9,9'-bis(*o*-carboranyl).*

* Attempts by other workers to reproduce this reaction have failed to give even traces of biscarboranes (*275a*).

The analogous reaction of 8-iodo-o-carborane requires more severe conditions.

$$2HC\text{---}CH\underset{B_{10}H_9I}{\diagdown O\diagup} \xrightarrow[THF]{Cu} HC\text{---}CH\underset{B_{10}H_9\text{---}}{\diagdown O\diagup} \quad HC\text{---}CH\underset{\text{---}B_{10}H_9}{\diagdown O\diagup}$$

In comparison with halogen derivatives of benzene and the metallocenes, boron–halogen bonds in o-carborane derivatives are much less reactive toward nucleophilic substitution [in this connection, the reported (*309*) isotopic exchange of iodine atoms in B-iodo-o-carboranes could not be repeated and is probably incorrect (*395*)].

TABLE 6-4

POLAROGRAPHIC REDUCTION POTENTIALS OF o-CARBORANYL
HALOGEN DERIVATIVES[a]

Substituents	$-E_{1/2}$, v.	Substituents	$-E_{1/2}$, v.
Parent	2.51	8,9,12-Br_3	1.66
9-Cl	2.34	8,9,12-I_3	1.41
9-Br	2.21	1-CH=CH_2-12-I	1.52
9-I	2.13	1-CH=CH_2-9-I	1.52
9,12-Cl_2	2.03	1,2-$(CH_3)_2$-9-I	2.01
9,12-Br_2	1.90	1-C_6H_5-12-Br	1.63
9,12-I_2	1.81	1-C_6H_5-2-Br	2.01
8,9,12-Cl_3	1.71		

[a] Measured in 2×10^{-3} M dimethylformamide solution containing 0.1 M $(C_2H_5)_4NClO_4$ (from Zakharkin et al., *422*).

The halogen atoms in B-halo-o-carboranes may be replaced by hydrogen via the action of alkali metals in liquid ammonia (*421*). Thus, both 9-iodo- and 9,12-diiodo-o-carborane are easily converted to o-carborane.

$$HC\text{---}CH\underset{B_{10}H_8I_2}{\diagdown O\diagup} + 2Na \xrightarrow{NH_3 \text{ (liq)}} HC\text{---}CH\underset{B_{10}H_{10}}{\diagdown O\diagup} + 2NaI$$

Analogous reactions with 9-bromo- and 9-chloro-o-carborane give only low yields of o-carborane, evidently due to the competing formation of the o-carboranyl disodium adduct, $Na_2C_2B_{10}H_{10}$. These observations are consistent with the polarographic reduction (*422*) of o-carborane and its B-halogen derivatives, for which the half-wave potentials are given in Table 6-4. It will be noted that the ease of reduction decreases in the order Cl > Br > I and is greatest in the more highly halogenated species. Clearly, the replacement of hydrogen atoms by halogen in the 9,12 and 8(10) positions increases the electron affinity of the *o*-carborane cage.

The significant acidity of the C-bonded hydrogen atoms in *o*-carboranyl halogen derivatives led Zakharkin and Podvisotskaya (*447*) to suggest that halogen atoms bonded to *o*-carboranyl carbon atoms may have a positive character, as is the case in 1-haloalkynes. This has been borne out for bromo and iodo derivatives by reactions of 1-phenyl-2-halo-*o*-carboranes with hot alcoholic base, in which the halogen atom is replaced by hydrogen to give 1-phenyl-*o*-carborane (*447*).

$$\underset{\underset{B_{10}H_{10}}{}}{C_6H_5C\!\!-\!\!CX} \; + \; NaOH \; \xrightarrow{C_2H_5OH} \; \underset{\underset{B_{10}H_{10}}{}}{C_6H_5C\!\!-\!\!CH} \; + \; NaOX$$

$$X = Br, I$$

C-chloro-*o*-carboranes fail to give such reactions, since they are cleaved even by weak bases to yield C-chloro derivatives of dicarbaundecaborane (Section 9-1). The action of KCN and Na_2S on the bromo and iodo compounds also leads to 1-phenyl-*o*-carborane, and no cyano or thio derivatives are obtained. On the other hand, 1-phenyl-2-chloro-*o*-carborane is degraded by sodium sulfide, forming a phenylchlorodicarbaundecaborane salt.

$$\underset{\underset{B_{10}H_{10}}{}}{C_6H_5C\!\!-\!\!CCl} \; + \; Na_2S \; \xrightarrow[NaOH]{C_2H_5OH} \; C_6H_5C_2ClB_9H_{10}^-Na^+ + H_2$$

The C-bromo- and C-iodo-*o*-carboranes also undergo metal–halogen exchange in reactions with butyllithium and ethylmagnesium bromide (*299, 423*).

$$\underset{\underset{B_{10}H_{10}}{}}{RC\!\!-\!\!CX} \; + \; C_2H_5MgBr \; \xrightarrow{THF} \; \underset{\underset{B_{10}H_{10}}{}}{RC\!\!-\!\!CMgBr} \; + \; C_2H_5X$$

C-bromo- or C-chloro-*o*-carboranes in which the halogen is either on a side alkyl group or on the carborane cage itself display halogen migration at moderate temperatures (200–300°), as shown by the conversion of 1-chloromethyl- and 1-bromomethyl-*o*-carborane to the respective 1-methyl-B-halo derivatives in which the 9, 12, 8, 4, and 7 positions are halogenated to a significant extent (*424*). The low temperatures involved indicate that thermal rearrangement of the cage itself (Chapter 7) is not occurring in these reactions, and it is clear that the process is of an intermolecular rather than intramolecular nature. The same B-halo-o-carboranes are obtained when 1-methyl-2-chloro-o-carborane and 1-methyl-2-bromo-o-carborane are heated at 250–300° (*424*).

Few reactions of C,C'-dihalo-o-carboranes have been reported; Zakharkin and Podvisotskaya have found that 1,2-dichloro-, 1,2-dibromo-, and 1,2-diiodo-o-carborane undergo cage degradation by both methanol and ethanol at 20° forming C,C'-dihalodicarbaundecaboranes (Section 9-1) (*299a, 448*).

A number of halogenated o-carborane derivatives has been described elsewhere in this chapter, e.g., the synthesis of silicon- and phosphorus-containing halo-o-carboranes (Section 6-9 and 6-10).

6-14. SIGMA-BONDED TRANSITION METAL DERIVATIVES

Bis(π-cyclopentadienyl) titanium dichloride reacts easily with 1-bromomagnesiomethyl-o-carborane to form bis(π-cyclopentadienyl)-bis(σ-o-carboranylmethyl)titanium (*415*).

$$\begin{array}{c}\text{HC}\!-\!\text{CCH}_2\text{MgBr}\\ \diagdown\bigcirc\diagup\\ B_{10}H_{10}\end{array} + (C_5H_5)_2\text{TiCl}_2 \longrightarrow \begin{array}{c}(\text{HC}\!-\!\text{CCH}_2)_2\text{Ti}(C_5H_5)_2\\ \diagdown\bigcirc\diagup\\ B_{10}H_{10}\end{array}$$

The latter compound is stable in air, unlike the analogous σ-alkyl and σ-aryl titanium cyclopentadienyl derivatives. The titanium–methylene sigma bond is relatively unreactive and is cleaved by hydrochloric acid in ethanol only on prolonged heating, forming 1-methyl-o-carborane and $(C_5H_5)\text{TiCl}_2$. The reaction of 1-methyl-2-lithio-o-carborane with $(\pi\text{-}C_5H_5)\text{Fe(CO)}_2\text{I}$ yields a complex believed to contain a direct iron–carborane σ bond (Fig. 6-1) (*277*).

$$\begin{array}{c}\text{CH}_3\text{C}\!-\!\text{CLi}\\ \diagdown\bigcirc\diagup\\ B_{10}H_{10}\end{array} + (C_5H_5)\text{Fe(CO)}_2\text{I} \xrightarrow{\ 1,2\text{-}(CH_3O)_2C_2H_4\ }$$

$$1\text{-}[(\pi\text{-}C_5H_5)\text{Fe(CO)}_2]\text{-}2\text{-}CH_3\text{-}\sigma\text{-}1,2\text{-}C_2B_{10}H_{10}$$

A related species of similar structure has been prepared from $1,10\text{-}Li_2C_2B_8H_8$ (Section 5-2) (*277*).

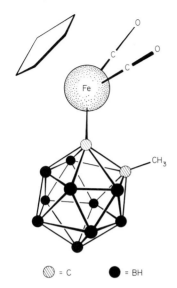

FIG. 6-1. Proposed structure of
$1\text{-}(\pi\text{-}C_5H_5)Fe(CO)_2\text{-}2\text{-}CH_3\text{-}\sigma\text{-}1,2\text{-}$
$C_2B_{10}H_{10}$ (277).

⬓ = C   ● = BH

Lithio-*o*-carboranes react with *trans*-$[(C_2H_5)_3P]_2PtCl_2$ to form diamagnetic σ-bonded carborane-platinum complexes for which the structure indicated has been deduced from spectroscopic, magnetic, and chemical properties (*31*).

$$R = CH_3, C_6H_5$$

Chelate complexes in which the *o*-carborane system is linked to the metal through phosphorus or sulfur atoms are described in Sections 6-9 and 6-12. The π-bonded dicarbollyl complexes and related species are treated in Chapter 9.

6-15. MERCURY DERIVATIVES

The chemical properties of mercury–carborane bonds are unusual in comparison with other organomercury species and, as a result, a large number of mercury derivatives of *o*-carborane has been prepared and studied. In general, these may be obtained directly from a mercury dihalide and a lithiocarborane or o-carboranyl Grignard. The product almost invariably is a symmetrical bis(carboranyl) derivative, even when the mercury dihalide is present in large excess (*366–369, 415*). Use of an alkyl mercuric halide, however, leads to the expected monocarborane derivative.

$$RC\!-\!CLi \quad \xrightarrow{HgX_2} \quad RC\!-\!C\!-\!Hg\!-\!C\!-\!CR$$
$$B_{10}H_{10} \qquad\qquad B_{10}H_{10} \qquad B_{10}H_{10}$$

$$\xrightarrow{CH_3HgX} \quad RC\!-\!C\!-\!Hg\!-\!CH_3$$
$$B_{10}H_{10}$$

$$R = H, C_6H_5, CH_3, CH_2\!\!=\!\!CH$$
$$X = Cl, Br$$

The direct mercuration of o-carborane and several of its derivatives has been achieved by reaction with organomercury hydroxides (*416*).

$$RC\!-\!CH \quad + \quad R'HgOH \quad \longrightarrow \quad RC\!-\!CHgR' \quad + \quad H_2O$$
$$B_{10}H_{10-n}X_n \qquad\qquad\qquad\qquad B_{10}H_{10-n}X_n$$

$$X = Br, Cl; \qquad n = 0, 2, 4; \qquad R = CH_3, C_6H_5, H; \qquad R' = CH_3, C_6H_5$$

Both symmetrical and unsymmetrical o-carboranyl mercury compounds are in general exceptionally stable. Methyl(phenylcarboranyl)mercury and phenyl(phenylcarboranyl)mercury are cleaved by hydrogen chloride in alcohol, reflecting the electron-withdrawing power of the carborane cage (*367, 368*). On the other hand, bis(phenylcarboranyl)mercury fails to react with hydrogen chloride even in boiling alcohol (*369*).

$$C_6H_5C\!-\!C\!-\!Hg\!-\!CH_3 \quad \xrightarrow[C_2H_5OH]{HCl}$$
$$B_{10}H_{10}$$

$$C_6H_5C\!-\!CH \quad + \quad CH_3HgCl \quad + \quad C_6H_5C\!-\!C\!-\!HgCl \quad + \quad C_6H_6$$
$$B_{10}H_{10} \qquad\qquad\qquad\qquad B_{10}H_{10}$$

Unlike most organomercuric compounds, unsymmetrical o-carboranyl mercury derivatives do not disproportionate even at the melting point; as a result the synthesis of stable $RHgR'$ compounds is straightforward, as shown by the preparation of mixed ferrocenyl-o-carboranyl derivatives (*369*).

The thermal stability of symmetrical bis(*o*-carboranyl)mercury species and their resistance to electrophilic attack is quite striking. For example, bis(phenyl-carboranyl)mercury is unreactive with bromine in refluxing dichloromethane–bromobenzene solvent (*369*). The same compound, on reaction with mercuric chloride at 210°, forms phenylcarboranyl mercuric chloride only with difficulty and over long periods (*368*). Dimethylmercury, however, does react to give methyl(phenylcarboranyl) mercury in 70% yield (*28*).

Symmetrical bis(*o*-carboranyl)mercury compounds have been found to give characteristic reactions with nucleophilic reagents, usually resulting in cleavage of the mercury–carbon bond. Thus, they may be reduced to the carborane or converted to lithiocarboranes; with pyridine they form adducts. Alcoholic base cleaves the Hg—C bond and destroys the carborane cage (*28*).

Polarographic measurements of a number of symmetrical bis(*o*-carboranyl) mercury derivatives show that such compounds are reduced at substantially less negative potentials than conventional organomercuric species (*306, 368, 415*). These data, as well as the chemical properties, are explainable in terms of the characteristic electrophilicity of the o-carboranyl system. In addition, Zakharkin *et al.* (*368*), have suggested two other factors which may contribute to the unorthodox properties of mercury–carborane bonds. Steric hindrance may be significant due to the bulkiness of the carborane cage, although such effects would tend to be minimized by the linear geometry of RHgR systems. Secondly, these authors have pointed out that the hexacoordinated carborane carbon atoms are unlikely to participate in transition states similar to those that are assumed in cleavage reactions of conventional organomercury species, since this would require an even higher valency state for carbon. (Cleavage by nucleophilic attack, on the other hand, need not involve any increase in carbon valency.)

Unlike most alkylmercuric halides, o-carboranyl mercuric halides do not form symmetrical complexes with nitrogen-containing molecules such as o-phenanthroline, but instead react to form stable 1:1 adducts of the type $RCB_{10}H_{10}CHgX \cdot C_{12}N_{12}H_8$. Analogous complexes are obtained from symmetrical bis-o-carboranyl mercury species and o-phenanthroline (*451*).

7

The $1,7\text{-}C_2B_{10}H_{12}$ and $1,12\text{-}C_2B_{10}H_{12}$ (*m*- and *p*-Carborane) Systems

7-1. COMPARISON OF *o*- AND *m*-CARBORANE

The chemistry of *m*-carborane (called neocarborane in the older literature) has been much less explored than that of the *ortho* isomer ($1,2\text{-}C_2B_{10}H_{12}$) but some general comparisons are available. To a considerable extent the reactions and properties of *m*-carborane parallel those of *o*-carborane. For example, *m*-carborane is easily metallated at carbon, and the resulting metallo-*m*-carboranes, like their *o*-carborane analogs, are convenient precursors to a large number of C-substituted derivatives. There are, however, significant differences in the derivative chemistry of the two isomers. These are usually attributable to one or more of the following: (1) the separation of cage carbon atoms in the *meta* isomer, which effectively prevents the formation of small exopolyhedral rings; (2) the generally weaker electron-attracting power of the *m*-carboranyl system; and (3) the greater thermal stability of the *m*-carboranyl cage. In general, derivatives of *m*-carborane are less polar, more volatile, and lower melting than their *o*-carborane analogs (*92, 314*). Comparison of the polarographic reduction potentials of a variety of *o*- and *m*-carboranyl compounds indicates that the *o*-carboranyl species are more easily reduced and therefore possess higher electron affinities (*306, 422*). This conclusion is consistent with the assumption of a comparatively weak *m*-carboranyl inductive effect, for which there is substantial chemical evidence to be outlined in this chapter.

Table 7-1 lists a number of characterized derivatives of *m*-carborane.

TABLE 7-1

SELECTED *m*-CARBORANE DERIVATIVES

Substituent groups[a]	mp (°C)	Other data[b]	References
Parent	272–273		233
		B	264
	263.5–265	MS, IR	92
		IR	182, 193
		DM	62, 190, 205
		Raman	37
Alkyl derivatives			
$1\text{-}CH_3$	208–210		92, 93
	202–204		428, 433
		IR	193
$1,7\text{-}(CH_3)_2$	169–170		428, 433
$1,7\text{-}(n\text{-}C_3H_7)_2$	47–48		433
Aryl derivatives			
$1\text{-}C_6H_5$	55.1–55.6		84
	54–55		441
$1\text{-}C_6H_4CH_3$	76–77		419
$1\text{-}p\text{-}C_6H_4F$	61–62		127
$1\text{-}m\text{-}C_6H_4F$	45–46		127
$1\text{-}p\text{-}C_6H_4Cl$	61–62		388
$1\text{-}p\text{-}C_6H_4Br$	71–73		388
$1\text{-}p\text{-}C_6H_4I$	77–78		388
$1\text{-}m\text{-}C_6H_4COOH$	259–260.5	pK_a (Table 6-2)	419
$1\text{-}p\text{-}C_6H_4COOH$	194–194.5	pK_a (Table 6-2)	419
$1\text{-}p\text{-}C_6H_4NH_2$	111–112		388
$1\text{-}p\text{-}C_6H_4NO_2$	148–149		382
$1\text{-}p\text{-}C_6H_4NHCOCH_3$	159–160		388
$1\text{-}CH_3\text{-}7\text{-}C_6H_5$	58–59		428, 433
$1,7\text{-}(CH_2C_6H_5)_2$	79–80		428, 433
Alkenyl and alkynyl derivatives			
$1\text{-}CH_3\text{-}7\text{-}CH{=}CH_2$	75–76		202
$1\text{-}CH_3\text{-}7\text{-}CH_2CH{=}CH_2$		bp 142–143 (13 mm)	202
$1,7\text{-}C(CH_3){=}CH_2$	46–47		216
Carboxylic acids			
$1\text{-}COOH$	163–164	IR	376
		IR	298, 380
	163.5–164.5	IR, pK_a (Table 6-2)	313, 398
$1\text{-}CH_2COOH$	135–136		379
$1\text{-}CH_3\text{-}7\text{-}COOH$	108–111	pK_a (Table 6-2)	92, 93
		IR	380
$1\text{-}C_6H_5\text{-}7\text{-}COOH$	120–121		441
$1,7\text{-}(COOH)_2$	202–204		92, 93
$1\text{-}CH_3\text{-}7\text{-}CH_2COOH$		IR	380

(continued)

TABLE 7-1—*Continued*

Substituent groups[a]	mp (°C)	Other data[b]	References
Esters and acyl halides			
1,7-$(COOCH_3)_2$	52.5–54	bp 110 (0.7 mm)	*92*
1,7-$(COCl)_2$		bp 86 (0.5 mm)	*92*
		bp 67 (0.1 mm)	*249*
1-C_6H_5-7-COCl		bp 157–159 (4 mm)	*441*
Alcohols			
1-CH_3-7-CH_2OH	181–182	IR	*433*
1-CH_3-7-$CH(CH_3)OH$	42–43		*439*
1-CH_3-7-$C(CH_3)_2OH$	29–30	H, IR	*202*
1-CH_3-7-$CH(C_6H_5)OH$		bp 152–153 (1.5 mm), IR, H	*202*
1-CH_3-7-$C(C_6H_5)_2OH$	90–91	IR, H	*202*
1,7-$(CH_2OH)_2$	194–196		*92*
1,7-$(CH(CH_3)OH)_2$	106–107	IR	*433*
1,7-$(C(CH_3)_2OH)_2$	122–123		*216*
1,7-$(CH(C_6H_5)OH)_2$	96–97	IR	*433*
1,7-$(C(OH)(CF_3)_2)_2$	80–81		*216*
1-C_6H_5-7-$(CH—CB_{10}H_{10}C—C_6H_5)$	126–128		*441*
| OH			
Epoxides			

1-CH_3-7-CH_2CH—CH_2
           \\ /
            O
	51–52		*433*

Aldehydes			
1-CHO	213–214	IR	*304*
1-CH_3-7-CHO	143–144		*202*
1-CH_3-7-CH_2CHO		bp 98 (2 mm)	*202*
Ketones			
1-CH_3-7-$COCH_3$		bp 78–80 (3 mm)	*439*
		IR	*442*
1-CH_2Br-7-$COCH_3$		bp 98–102 (3 mm), IR	*442*
1,7-$(COC_6H_5)_2$	57–58		*433*
1-C_6H_5-7-$(CO$-*p*-$C_6H_4Cl)$	58–59		*441*
1-C_6H_5-7-$(CO—CB_{10}H_{10}CC_6H_5)$	144–145		*441*

1-C_6H_5-7-$(CO$—C——C—$C_6H_5)$
                  \\O/
                $B_{10}H_{10}$
	130–131		*441*

Amines			
1-NH_2	290–291		*402*
2-$N(C_6H_5)_2$	115–117	IR	*250*

(*continued*)

TABLE 7-1—*Continued*

Substituent groups[a]	mp (°C)	Other data[b]	References
$4\text{-}N(C_6H_5)_2{}^c$	68–70		*250*
$1\text{-}C_6H_5\text{-}7\text{-}NH_2$	87–88		*402*
Amides and azides			
$1\text{-}C_6H_5\text{-}7\text{-}CONH_2$	125–126		*441*
$1,7\text{-}(CONH_2)_2$	184–185.5		*92*
	183		*249*
$1\text{-}CONH_3$	69–70		*402*
$1\text{-}C_6H_5\text{-}7\text{-}CONH_3$	51–52		*402*
Phosphorus derivatives			
$1,7\text{-}(P(C_6H_5)_2)_2$	101		*8*
$1,7\text{-}(P(C_6H_5)Cl)_2$		bp 229 (0.35 mm)	*8*
$1,7\text{-}(PCl_2)_2$		bp 119 (0.3 mm)	*8*
$1,7\text{-}(P(OCH_3)_2)_2$	Violent dec.		*8*
$1,7\text{-}(P(C_6H_5)(OCH_3))_2$	112–114		*8*
$1,7\text{-}(P[N(CH_3)_2]_2)_2$	73–75		*8*
$1,7\text{-}(P(C_6H_5)N(CH_3)_2)_2$	100–102		*8*
Silicon derivatives			
$1,7\text{-}(Si(CH_3)_2Cl)_2$		bp 102–104 (0.1 mm)	*232*
$1,7\text{-}(Si(CH_3)Cl_2)_2$		bp 103–105 (0.15 mm)	*232*
$1,7\text{-}(Si(C_6H_5)_2Cl)_2$	131–133		*232*
	121–122		*252*
$1,7\text{-}(Si(C_6H_5)_2CH_3)_2$	138–140		*252*
$1,7\text{-}(Si(CH_3)_2C_6H_5)_2$	82–85		*252*
$1,7\text{-}(Si(CH_3)_2OH)_2$	98–99.5		*232*
$1,7\text{-}(Si(CH_3)(OH)_2)_2$	136.5–138.5		*232*
$1,7\text{-}(Si(C_6H_5)_2OH)_2$	153–155		*232*
$1,7\text{-}(Si(CH_3)_2OCH_3)_2$	36–37	bp 108–110 (0.1 mm)	*232*
$1,7\text{-}(Si(C_6H_5)_2OCH_3)_2$	151–153		*232*
$1,7\text{-}(Si(CH_3)_2NH_2)_2$	41.5–43.5		*232*
Sulfur derivatives			
$1\text{-}CH_3\text{-}7\text{-}SH$	162–164		*465*
$1,7\text{-}(SH)_2$	164–165		*282*
$1,7\text{-}(SC_6H_5)_2$		bp 114 (0.1 mm)	*282*
$1,7\text{-}(SCl)_2$		bp 97–98 (0.17 mm), MS, IR	*269*
$1,7\text{-}(S(p\text{-}C_6H_4CH_3))_2$	131–132		*282*
Monohalo derivatives[d]			
2-F	259–260	IR	*250*
$4\text{-}F^c$	263–265		*250*
9-Cl	216–217		*383*
		DM	*62, 307*
		IR	*318*

(*continued*)

TABLE 7-1—*Continued*

Substituent groups[a]	mp (°C)	Other data[b]	References
1-CH$_3$-9-Cl	150–151		*383*
1-CH$_3$-7-Br	147–148		*423*
1-Br	193–194	IR	*299a*
9-Br		IR	*193, 318*
	171–172		*383*
		DM	*62, 307*
1-CH$_3$-9-Br	102–103		*383*
1-I	105–105.5	IR	*299a*
1-C$_6$H$_5$-7-I	86–87		*441*
1-C$_6$H$_5$-B-I	89–90		*322*
9-I	109–110		*383, 389*
		DM	*62, 307*
		IR	*193, 318*
Dihalo derivatives[d]			
1,7-F$_2$	230–231	IR	*182*
1,7-Cl$_2$		IR	*193, 299a*
	123–124		*299a*
9,10-Cl$_2$	217–218		*389*
	222–223	IR	*318*
		DM	*62, 307*
1-CH$_3$-9,10-Cl$_2$	151–152		*383*
1,7-(CH$_3$)$_2$-9,10-Cl$_2$	118–119		*383*
1,7-Br$_2$		DM	*205*
		IR	*193*
9,10-Br$_2$		IR	*193, 318*
	187–188		*383*
		DM	*62, 205, 307*
1-CH$_3$-9,10-Br$_2$	119–120		*383*
1,2-(CH$_3$)$_2$-B-Br$_2$	113–114		*281*
1,7-I$_2$	74–75		*423*
		IR	*299a*
9,10-I$_2$		IR	*318*
		DM	*62, 307*
1-CH$_3$-9,10-I$_2$	128–129		*383, 389*
Trihalo derivatives[d]			
4,9,10-Cl$_3$	252.5–253	DM	*62, 307*
B-Cl$_3$	234–235		*383*
	175–176		*383*
B-Br$_3$		IR	*193*
	268–269		*383*
	135–136		*383*
1-CH$_3$-4,9,10-Br$_3$	188–189		*307*
1-CH$_3$-8,9,10-Br$_3$	221–222		*307*
1-CH$_3$-B-Br$_3$	203–205		*383*

(*continued*)

TABLE 7-1—*Continued*

Substituent groups[a]	mp (°C)	Other data[b]	References
Tetrahalo and hexahalo derivatives[d]			
4,8,9,10-Cl_4		DM	*62, 307*
B-Cl_4	277–278		*383*
	279–280		*318*
4,8,9,10-Br_4	324–325	DM	*62, 307*
		IR	*193*
4,6,8,9,10,11-Br_6	312–313		*307*
B-Br_6		IR	*193*
B-I_4	339–341		*318*
Decahalo and dodecahalo derivatives[d]			
B-F_{10}	240–242	IR	*182*
B-Cl_{10}	232–233	pK_a (Table 6-3)	*444*
	235	B	*264*
	232–233	IR	*314*
		DM	*62, 307*
B-Cl_8-B-Br_2	232–234		*281*
F_{12}		IR	*182*
Cl_{12}	443	IR, B	*262*

[a] Order of substituents is as in Table 6-1, page 56.
[b] IR = infrared spectra or band positions; B = ^{11}B NMR spectra; H = ^1H NMR spectra; MS = mass spectra or cutoff m/e value; DM = dipole moment; bp = boiling point (°C); UV = ultraviolet spectra or band positions.
[c] Tentative structural assignment.
[d] Numbering of cage positions is as shown in Fig. 2-1m (Chapter 2), regardless of systems used in individual papers in the literature.

7-2. SYNTHESIS AND STRUCTURE OF *m*-CARBORANE

The only preparative route to *m*-carborane of practical significance at present is the thermal rearrangement of *o*-carborane (*92, 93, 234, 314, 315, 403*). This occurs slowly (24–48 hr) at 425–500° in an inert atmosphere, giving nearly quantitative yields. Recent development of a flow process (*234*) involving short residence times (less than 1 min) at 600° has permitted 98 % conversion to *m*-carborane; at 700° further isomerization occurs, producing *p*-carborane in 22% yield. Mechanisms of these rearrangements are discussed in the following section. When carried out at 600° in a closed vessel, the *o*-to-*m*-carborane isomerization is accompanied by the formation of a bis(*m*-carboranyl) species of unknown structure, which is also obtained when *m*-carborane is heated at the same temperature (*315*). A different bis(*m*-carboranyl) compound, presumably 1,1′-bis(*m*-carboranyl), is produced by heating 1,1′-bis(*o*-carboranyl) (Section 6-4) at 400° (*168*).

$$\text{HC}\underset{\underset{B_{10}H_{10}}{}}{\overset{}{\diagdown\text{O}\diagup}}\text{C}\text{—}\text{C}\underset{\underset{B_{10}H_{10}}{}}{\overset{}{\diagdown\text{O}\diagup}}\text{CH} \xrightarrow{\ 400°\ } \text{HCB}_{10}\text{H}_{10}\text{C—CB}_{10}\text{H}_{10}\text{CH*}$$

* The symbol $\text{HCB}_{10}\text{H}_{10}\text{CH}$ is in common use to designate the $1,7\text{-C}_2\text{B}_{10}\text{H}_{12}$ isomer.

The rearrangement of C- and B-substituted derivatives of *o*-carborane to the corresponding *m*-carborane isomers may be carried out when thermally stable functional groups are involved, but in many cases such attempted isomerizations result in extensive degradation. For example, while 1-methyl-*o*-carborane is converted to 1-methyl-*m*-carborane in 69% yield at 400°, 1,2-bis(hydroxymethyl)-*o*-carborane decomposes at the same temperature (*92*). At 450° 1-isopropyl-*o*-carborane undergoes both dealkylation and cage isomerization, forming *m*-carborane in >90% yield (*256*).

The isomerization of *o*-carborane derivatives containing large substituted silyl groups at carbon takes place at much lower temperatures than the rearrangement of *o*-carborane itself, presumably due to relief of steric crowding in the vicinity of the cage carbon atoms (*252*). Thus, 1,2-bis(methyldiphenylsilyl)- and 1,2-bis(chlorodiphenylsilyl)-*o*-carborane are converted to the respective *m*-carborane derivatives at 260°. Halo-*o*-carboranes are generally converted to halo-*m*-carboranes with little decomposition; such isomerizations have been useful in studies of the *o*- to *m*- cage rearrangment mechanism (see below).

The only reported synthesis of *m*-carborane not involving the thermal isomerization of the *ortho* isomer is the reaction of the C,C′-dimethyl derivative of $1,6\text{-C}_2\text{B}_8\text{H}_{10}$ with B_2H_6 at 225°, which produces 1,7-dimethyl-*m*-carborane in 27% yield (Section 5-2) (*329*).

The icosahedral structure of *m*-carborane with the carbon atoms in nonadjacent 1,7 positions (Fig. 2-1m) has been established from an X-ray study of the B-decachloro derivative (*246*), supported by ^{11}B NMR spectra (*264, 337*) and dipole moment measurements (*62, 190, 205*) and by an electron diffraction study (*206*).

7-3. MECHANISMS OF ICOSAHEDRAL CARBORANE REARRANGEMENTS

ortho-meta and meta-para Isomerizations

The processes involved in the conversion of o-carborane to its *meta* and *para* isomers have been of considerable interest, the focus of attention being the geometry of the reaction intermediate. Lipscomb and associates (*150, 166, 197*) have proposed a cuboctahedral intermediate in a cooperative diamond–square–diamond (dsd) rearrangement mechanism (Fig. 7-1), whereas Grafstein

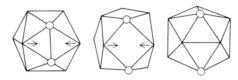

FIG. 7-1. Dsd rearrangement of o-carborane to m-carborane via cuboctahedral intermediate.

and Dvorak (*92*), and later Zakharkin and Kalinin (*383*), have favored a process involving rotation of the two halves of the icosahedron in opposite directions. Still another possibility is the rotation of triangular faces in the icosahedron, as suggested by Muetterties and Knoth (*215*, p. 70). Experimental studies (*167*) of the isomerization of halogen-labeled o-carboranes have given results suggesting that the actual mechanism is more complex than any of these, assuming that the identification of isomers is correct (see below). Thus, when 9-bromo-o-carborane is heated at 395–425° until a constant product distribution is reached, a nonrandom mixture of all possible B-monobromo-m-carborane isomers is obtained; in earlier stages of the reaction, however, all possible B-monobromo-o-carboranes are found. At these temperatures the m-carborane products are unchanged once formed. These data do not appear to be consistent with the simple cuboctahedron, rotating pyramid, or rotating triangle rearrangements when taken without modifications (*167*). However, a pathway involving the rotation of triangles in the cuboctahedral intermediates has been made to fit the observed product distribution by assuming (1) preferential rotation of those triangles furthest removed from carbon, and (2) conversion of only 10% of the cuboctahedral intermediates existing at any one time to m-carborane products (although the remaining 90% of the rearrangements would yield *ortho* isomers under this scheme, the fact that the *meta* isomers are unchanged once formed would assure complete conversion to m-carborane products if the reaction time is sufficiently long).

The same approach has been extended to the *meta–para* isomerization by heating selected monohalo-*m*-carboranes at 560–570° and examining the resulting product mixtures (*119*). Again, the data appear to support a mechanism involving rotation of triangles in the cuboctahedral intermediate,* although in this case, unlike the *ortho–meta* rearrangement, it is assumed that the rotation of B_3 and B_2C triangles is equally probable. Extensive computations (*119*) have indicated that the other suggested pathways yield isomer distributions which are not in accord with the experimental results. A special case arises in the rearrangement of fluoro-*o*-carboranes, which appears to be strongly influenced by electronic effects of some sort; certainly the final product distributions are considerably different from those observed in the isomerization of chloro- and bromo- *o*- and *m*-carboranes. Thus, 3-fluoro-*o*-carborane at 420° yields only 4-fluoro-*o*-, 2-fluoro-*m*-, and 4-fluoro-*m*-carborane (*119*, *250*).

The experimental results of Lipscomb and co-workers are based upon identification of the various carborane isomers from their chromatographic elution times, which, it is assumed, are a direct function of the dipole moments of the compounds. The moments in most cases have been estimated but not directly determined (*119*, *167*). This correlation, however, may not be completely reliable, as shown by chromatographic data on tri- and tetrachloro-*p*-carboranes, in which a few differences have been noted between the observed elution times and those calculated from dipole moments (*276*).

Zakharkin and Kalinin (*383*, *405*) have studied the isomerization of mono- and dihalo-*o*-carboranes and have reported different product distributions than those obtained by Lipscomb, but the Russian workers employed lower temperatures, and in some cases shorter reaction times, in their experiments. In addition, it seems probable that the analytical methods used in the latter investigations failed to reveal all of the isomers actually present.†

The "Reverse Isomerization" of *m*-Carborane: Conversion of *m*- to *o*-Carborane

Molecular orbital calculations by Hoffmann and Lipscomb (*150*) on the relative stabilities of *o*- and *m*-carborane indicate that the *meta* isomer is the

---

* The simple cuboctahedral mechanism (i.e., with no triangle rotations) is incapable of converting *m*- or *o*-carborane to the *para* isomer (*197*).

† The thermal isomerization of *o*- and *m*-carboranyl halogen derivatives has recently been reinvestigated by Zakharkin and Kalinin (*399a*), with results essentially in agreement with those of Lipscomb and co-workers (*119*, *167*). However, the Russian workers still favor a pentagonal bipyramidal rotation mechanism. They also present evidence for intermolecular hydrogen–iodine exchange during isomerization of iodo-*o*-carboranes (*399a*).

more stable, in agreement with the observed *o*- to *m*-carborane thermal re-arrangement. While rejecting the possibility of direct *m*- to *o*-carborane trans-formation, these authors predicted that the introduction of negative charge to the icosahedral cage would create mono- or dianions whose *meta* and *para* isomers are of comparable stability to the ortho isomer, so that *meta* → *ortho* conversion might be possible. Following tentative evidence reported by Graf-stein and Dvorak (*92*), the prediction has been conclusively verified by Zak-harkin, Kalinin, and Podvisotskaya (*412*, *414*, *453*), who found that *m*-carborane and certain of its organofunctional derivatives react with alkali metals in liquid ammonia to form *m*-dianions which can be quantitatively oxidized to the corresponding neutral *o*-carborane. The process has been postulated to involve three steps: reduction of the *m*-carboranyl species with the addition of two electrons; isomerization of the *m*-dianion to the *o*-dianion; and oxidation of the latter to the neutral *o*-carboranyl species.

$$RCB_{10}H_{10}CR' + 2e^- \xrightarrow[\text{NH}_3 \text{ (liq)}]{\text{Na}} [RCB_{10}H_{10}CR']^{2-} \longrightarrow \left[\begin{array}{c} RC-CR' \\ \diagdown O \diagup \\ B_{10}H_{10} \end{array} \right]^{2-}$$

$$\begin{array}{c} RC-CR' \\ \diagdown O \diagup \\ B_{10}H_{10} \end{array} \xleftarrow[\text{O}_2 \text{ or KMnO}_4]{-2e^-}$$

a. R = H, R′ = C_6H_5
b. R = R′ = C_6H_5
c. R = R′ = CH_3

No hydrogen is evolved in the reaction with alkali metals (which occurs rapidly even at −70°!) and the intermediate formation of a *m*-carboranyl radical-anion is suggested by an EPR spectrum of the solution. The alkali metal addition may also be carried out in tetrahydrofuran, but in this case, a small quantity of an electron-carrier such as biphenyl or naphthalene is required. Hydrolysis of the dianions yields water-soluble dicarbadodecaborane monoanions which are identical with products of the hydrolysis of *o*-carborane–alkali metal adducts (Section 6-2).

Treatment of 1,1′-bis(*m*-carboranyl) with sodium in liquid ammonia, fol-lowed by oxidation with $KMnO_4$, yields 1-*o*-carboranyl-*m*-carborane, 1,1′-bis(*o*-carboranyl), and the original compound (*414*).

$$HCB_{10}H_{10}C-CB_{10}H_{10}CH \xrightarrow{2e^-} [HCB_{10}H_{10}C-CB_{10}H_{10}CH]^{2-} \xrightarrow[\text{NH}_3 \text{ (liq)}]{\text{KMnO}_4}$$

$$\begin{array}{c} HCB_{10}H_{10}C-C-CH \\ \diagdown O \diagup \\ B_{10}H_{10} \end{array} + HCB_{10}H_{10}C-CB_{10}H_{10}CH + \begin{array}{c} HC-C-C-CH \\ \diagdown O \diagup \; \diagdown O \diagup \\ B_{10}H_{10} \quad B_{10}H_{10} \end{array}$$

Recently it has been found (*399*) that *p*-carborane may be converted to a dinegative ion which rearranges to the *m*-carborane dianion, thus allowing complete interconversion of all three $1,2\text{-}C_2B_{10}H_{12}$ carborane isomers (Section 7-15).

7-4. METALLATION OF *m*-CARBORANE

The C-metallation of *m*-carborane by organometallic reagents proceeds under conditions similar to those used for the *o*-carborane system (Section 6-2). The reaction of *n*-butyllithium with *m*-carborane is slower than that of *o*-carborane, probably due to the weaker electron-acceptor effect of the *meta* isomer (*433*); the C—H bonds are apparently less polar and less acidic in *m*-carborane than in the *ortho* species (*306, 416, 433*). As in the case of C-lithio- and C-magnesio-*o*-carboranes (Section 6-2), an equilibrium exists between the mono- and dimetallated *m*-carborane species (*377*).

$$2HCB_{10}H_{10}CLi \rightleftharpoons LiCB_{10}H_{10}CLi + HCB_{10}H_{10}CH$$

The equilibrium is shifted further to the left than in the corresponding *o*-carborane system due to the lesser acidity of the *m*-carboranyl C—H bonds.

The metallation of *m*-carborane has also been accomplished with alkali metal amides in liquid ammonia (*428, 433*).

7-5. ALKYL, ARYL, AND ALKENYL *m*-CARBORANE DERIVATIVES

Synthesis

C-Alkyl and C-aryl-*m*-carboranes are usually prepared via metallation of *m*-carborane followed by reaction with the appropriate organohalide (*67, 92, 281, 428, 433*). However, since many alkyl and aryl functional groups are capable of withstanding high temperatures, *m*-carborane derivatives of this type can often be obtained by thermal isomerization of the *o*-carborane analog (yields may be considerably less than quantitative, however) (*92, 127, 419*). An example is the synthesis of 1-(*p*-tolyl)-*m*-carborane (*419*).

$$CH_3C_6H_4C{\equiv}CH + B_{10}H_{14} \xrightarrow[\text{CH}_3\text{C}_6\text{H}_5]{\text{C}_6\text{H}_5\text{N(CH}_3)_2}$$

$$\begin{array}{c}CH_3C_6H_4C{-\!\!-\!\!-}CH \\ \diagdown\!\!\overset{}{\underset{}{O}}\!\!\diagup \\ B_{10}H_{10}\end{array} \xrightarrow{410°} CH_3C_6H_4CB_{10}H_{10}CH \quad (30\%)$$

The only reported syntheses of simple alkenyl-*m*-carboranes are those of 1,7-bis(2-isopropenyl)-*m*-carborane from 1,7-bis(dimethylcarbinol)-*m*-carborane (*216*), and of several vinyl-*m*-carboranes by pyrolysis of the acetates of the corresponding secondary alcohols (*202*).

$$\begin{array}{c}\underset{|}{\overset{CH_3}{|}} \quad \underset{|}{\overset{CH_3}{|}} \\ HOC{-}CB_{10}H_{10}C{-}OH \\ \underset{}{\overset{|}{CH_3}} \quad \underset{}{\overset{|}{CH_3}}\end{array} \xrightarrow[\text{H}_2\text{SO}_4]{\text{AlCl}_3 \ \text{or}} \begin{array}{c}\underset{|}{\overset{CH_3}{|}} \quad \underset{|}{\overset{CH_3}{|}} \\ H_2C{=}C{-}CB_{10}H_{10}C{-}C{=}CH_2\end{array}$$

Reactions

Aryl *m*-carboranes, like aryl *o*-carboranes (Section 6-8), undergo nitration to yield nitrophenyl derivatives and reduction of the latter to the aminophenyl compounds (*382, 388*). The competitive mixed acid nitration of an equimolar mixture of 1-phenyl-*o*-carborane and 1-phenyl-*m*-carborane at 20° results in nitration of the *m*-carborane derivative only (*382*), probably reflecting the relatively weak inductive electron-withdrawing character of the 1-*m*-carboranyl system [however, the 1-*m*-carboranyl group is a stronger electron-attractor than the 3-*o*-carboranyl system, as shown by competitive nitration studies of 1-phenyl-*m*-carborane and 3-phenyl-*o*-carborane in which the latter compound is preferentially attacked (*420*)]. These conclusions are in general agreement with studies of the ^{19}F NMR chemical shifts in 1-fluorophenyl- *o*- and *m*-carboranes (*127, 422a*).

The reaction of 1,7-diphenyl-*m*-carborane with anhydrous hydrazine destroys the carborane cage; in contrast, 1,2-diphenyl-*o*-carborane is degraded only to the dicarbaundecaborate salt $[N_2H_5{}^+(C_6H_5)_2C_2B_9H_{10}^-]$ with the evolution of 1 mole of H_2 (*302*). The action of hydrazine on the C-phenyl derivatives of both *o*- and *m*-carborane produces the corresponding dicarbaundecaborate salts (*302, 382, 401*), and even the reaction with 1,7-diphenyl-*m*-carborane can be stopped at the dicarbaundecaborate stage if a high-boiling inert solvent such as toluene is used (*302*). *m*-Carborane and its derivatives are degraded by a variety of other bases also, as outlined in Chapter 9.

7-6. *m*-CARBORANYL CARBOXYLIC ACIDS AND ESTERS

Synthesis

The C-substituted mono- and dicarboxylic acids may be obtained by the reaction of CO_2 with the lithio derivatives (*92, 93, 376, 441*) or with C-bromo-magnesio-*m*-carborane (*298, 313*) (obtained from the reaction of *m*-carborane with ethylmagnesium bromide); both procedures are analogous to the syntheses of *o*-carboranyl carboxylic acids described in Chapter 6. Acetic acid derivatives of *m*- and *o*-carborane have been prepared by the reaction of C-sodiocarboranes with sodium haloacetates (Section 6-5) (*379*).

Reactions

m-Carboranyl carboxylic acids are weaker than the corresponding *o*-carboranyl species (Table 6-2), a fact which is assumed to reflect the difference in electron-attracting power referred to earlier. The C,C'-dicarboxylic acid of *m*-carborane is readily esterified by alcohols under acid conditions (*92*), in contrast to the *o*-carborane diacid (Section 6-5).

$$HOOCCB_{10}H_{10}CCOOH \xrightarrow[H^+]{2CH_3OH} CH_3OOCCB_{10}H_{10}CCOOCH_3$$

As might be expected from steric considerations, the *m*-carborane diacid fails to form an exocyclic anhydride on treatment with thionyl chloride, refluxing acetic anhydride, or hot sulfuric acid (all of which readily convert the *o*-carborane diacid into the anhydride) (*92*).

The esters and salts of *m*-carborane carboxylic acids are more resistant to decarboxylation than the *o*-carborane analogs (*441*). Thus, while potassium 1-phenyl-*o*-carboranyl-2-carboxylate in boiling water is completely decarboxyl-ated to 1-phenyl-*o*-carborane, the analogous *m*-carborane derivative is stable under these conditions. Similarly, the ethyl ester of 1-phenyl-*o*-carboranyl-2-carboxylic acid is rapidly cleaved to 1-phenyl-*o*-carborane by sodium ethoxide in ethanol at 20°, but the corresponding *m*-carboranyl ester is not cleaved in 24 hr. Again, these observations have been attributed to the relative −I effects, although steric factors may also be involved (*441*).

7-7. *m*-CARBORANYL ALCOHOLS AND ETHERS

Synthesis

Many *m*-carboranyl alcohols have been prepared by the reaction of C-metallo-*m*-carboranes with aldehydes and ketones. An example is the conversion of 1,7-dilithio-*m*-carborane to the 1,7-bis(hydroxymethyl) derivative by paraformaldehyde (*92, 93*).

$$\text{LiCB}_{10}\text{H}_{10}\text{CLi} + (\text{CH}_2\text{O})_n \xrightarrow[\text{H}^+]{\text{C}_6\text{H}_6} \text{HOCH}_2\text{CB}_{10}\text{H}_{10}\text{CCH}_2\text{OH}$$

Secondary alcohols are obtained by the action of aliphatic, aromatic, and heterocyclic aldehydes on lithio-*m*-carboranes (*433, 439*); ketones react in the same manner to form tertiary alcohols (*216*).

$$\text{C}_6\text{H}_5\text{CB}_{10}\text{H}_{10}\text{CLi} + \text{CH}_3\text{OC}_6\text{H}_4\text{CHO} \xrightarrow{\text{H}^+} \overset{\overset{\displaystyle \text{OH}}{|}}{\underset{\underset{\displaystyle \text{H}}{}}{\text{CH}_3\text{OC}_6\text{H}_4\text{CCB}_{10}\text{H}_{10}\text{CC}_6\text{H}_5}}$$

$$\text{LiCB}_{10}\text{H}_{10}\text{Li} + 2\text{R}\overset{\overset{\displaystyle \text{O}}{\|}}{\text{C}}\text{R}' \xrightarrow{\text{H}^+} \overset{\displaystyle \text{R}'}{\underset{\displaystyle \text{R}}{\text{HOC}}}\text{CB}_{10}\text{H}_{10}\overset{\displaystyle \text{R}'}{\underset{\displaystyle \text{R}}{\text{COH}}}$$

a. $R = R' = CH_3$; b. $R = R' = CF_3$; c. $R = CH_3$, $R' = CF_3$; d. $R = CF_3$, $R' = CF_2Cl$.

Alternatively, *m*-carboranyl aldehydes and ketones may be treated with organolithium or organomagnesium compounds to yield, respectively, secondary and tertiary alcohols, accompanied by some cleavage of the exopolyhedral carbon–carbon bond (*202*).

$$\text{CH}_3\text{CB}_{10}\text{H}_{10}\overset{\overset{\displaystyle \text{O}}{\|}}{\text{C}}\text{H} + \text{HC}\underset{\text{B}_{10}\text{H}_{10}}{\overset{}{\diagdown}}\text{CCH}_2\text{MgBr} \xrightarrow[-\text{MgBrCl}]{\text{HCl}} \overset{\overset{\displaystyle \text{OH}}{|}}{\text{CH}_3\text{CB}_{10}\text{H}_{10}\text{C}}\underset{\text{H}}{\overset{}{}}\text{—CH}_2\text{C}\underset{\text{B}_{10}\text{H}_{10}}{\overset{}{\diagdown}}\text{CH}$$

93.5%

These results are in contrast with the behavior of *o*-carboranyl ketones (Section 6-7), which in general are cleaved by organolithium reagents but are converted to alcohols by Grignards. The alcoholate intermediates produced in reactions of *m*-carboranyl ketones are evidently much more stable than their *o*-carborane analogs, so that the predominant products are alcohols even in reactions with organolithium compounds (*202*).

Lithio-*m*-carboranes react with α-epoxides to form secondary alcohols, as do lithio-*o*-carboranes, but the *m*-carboranyl reagents are much less reactive (*433*). Epichlorohydrin reacts with 1-methyl-7-lithio-*m*-carborane to give the epoxide only (*433*), in contrast to the *o*-carboranyl analog which yields both the epoxide and the bis(*o*-carboranyl) alcohol (Section 6-6).

No general syntheses of *m*-carboranyl ethers have been reported.

Reactions

The conversion of 1,7-bis(dimethylcarbinol)-*m*-carborane to the bis(isopropenyl) derivative was mentioned in Section 7-5. The same diol reacts in unusual fashion with a bromine–sulfuric acid mixture, resulting both in cleavage of the carborane–carbon bonds and in halogenation of the borane cage. If the CH_3 groups in this diol are replaced by CF_3 groups, the acidity of the hydroxyl hydrogen atoms increases considerably (*216*).

$$\begin{array}{c} CH_3 \quad\quad CH_3 \\ | \quad\quad\quad | \\ HOCCB_{10}H_{10}CCOH \\ | \quad\quad\quad | \\ CH_3 \quad\quad CH_3 \end{array} \xrightarrow[H_2SO_4]{Br_2} BrCB_{10}H_{10-n}Br_nCBr$$

$$n = 1, 2$$

The oxidation of secondary *m*-carboranyl alcohols with chromic acid, like that of *o*-carboranyl alcohols, proceeds smoothly to the ketones (*433*, *439*). Primary and secondary *m*-carboranyl alcohols undergo oxidative cleavage

$$\begin{array}{c} OH \\ | \\ C_6H_5CB_{10}H_{10}CCR \\ | \\ H \end{array} \xrightarrow[H_2SO_4]{CrO_3} \begin{array}{c} O \\ || \\ C_6H_5CB_{10}H_{10}CCR \end{array}$$

$$R = CH_3, C_2H_5, C_6H_5$$

with $KMnO_4$ to give the parent carborane, again paralleling the behavior of *o*-carboranyl alcohols (*427*, *433*).

Unlike 1,2-bis(hydroxymethyl)-*o*-carborane, which is converted by H_2SO_4 at 140° to a cyclic ether (Section 6-6), 1,7-bis(hydroxymethyl)-*m*-carborane is unreactive under the same conditions; at 175° sulfonation occurs but no volatile products are obtained (*92*). Formation of a cyclic ether would not be expected in this case, of course, since the functional groups are attached to nonadjacent cage carbon atoms.

7-8. *m*-CARBORANYL ALDEHYDES AND KETONES

At least two of the known routes to *o*-carboranyl aldehydes—the ozonolysis of vinylcarboranes, and the hydrogenation of carboranyl acid chlorides over palladium (Section 6-7)—have been successfully employed in the preparation of *m*-carboranyl aldehydes (*202, 303, 304*). Ketone derivatives of *m*-carborane have been synthesized by the chromic acid oxidation of *m*-carboranyl secondary alcohols and by the reaction of lithio-*m*-carboranes with acyl chlorides (*441*), both methods being analogous to known syntheses of *o*-carboranyl ketones (Section 6-7). Unlike 1,2-dilithio-*o*-carborane, which forms a cyclic diketone on reaction with phosgene, 1,7-dilithio-*m*-carborane produces the diacid dichloride (*249*).

$$\text{LiCB}_{10}\text{H}_{10}\text{CLi} \xrightarrow[-2\text{LiCl}]{2\text{COCl}_2} \overset{\displaystyle O}{\overset{\|}{\text{ClCC}}}\text{B}_{10}\text{H}_{10}\overset{\displaystyle O}{\overset{\|}{\text{CCCl}}} + \text{polymer}$$
$$20\%$$

The reaction of 1-phenyl-7-lithio-*m*-carborane with methyl cinnamate (*433*) evidently parallels the reaction of this α,β-unsaturated ester with 1-phenyl-2-lithio-*o*-carborane (Section 6-7), the ultimate product in each case being the respective bis(carboranyl) ketone.

Reactions

The −I electron-acceptor effect in the *m*-carborane cage system, though weak relative to *o*-carborane, is sufficiently strong to induce cleavage in many *m*-carboranyl aldehydes and ketones in basic solution (*303, 304, 439, 441*).

$$\text{C}_6\text{H}_5\text{CB}_{10}\text{H}_{10}\overset{\displaystyle O}{\overset{\|}{\text{CC}}}\text{C}_6\text{H}_4\text{Cl} \xrightarrow[\substack{1.\ \text{C}_2\text{H}_5\text{ONa} \\ 2.\ \text{H}_3\text{O}^+}]{\text{C}_2\text{H}_5\text{OH}} \text{C}_6\text{H}_5\text{CB}_{10}\text{H}_{10}\text{CH} + \text{C}_2\text{H}_5\overset{\displaystyle O}{\overset{\|}{\text{OC}}}\text{C}_6\text{H}_4\text{Cl}$$

As a class, however, aldehydes and ketones of *m*-carborane are more resistant to cleavage than those of *o*-carborane. Thus, 1-formyl-*m*-carborane is cleaved by sodium ethoxide in alcohol only under reflux conditions (*303, 304*), but 1-formyl-*o*-carborane is attacked by this reagent at room temperature (Section 6-7). Also, while bis(1-phenyl-*o*-carboranyl) ketone is cleaved by LiAlH$_4$, the *m*-carboranyl analog is merely reduced to the secondary carbinol (*441*).

$$(\text{C}_6\text{H}_5\text{CB}_{10}\text{H}_{10}\text{C})_2\text{CO} \xrightarrow{\text{LiAlH}_4} (\text{C}_6\text{H}_5\text{CB}_{10}\text{H}_{10}\text{C})_2\underset{\underset{\displaystyle H}{|}}{\text{C}}\text{—OH}$$

The reduction of *m*-carboranyl aldehydes and ketones to the alcohols is discussed in the previous section.

The basic properties of *o*- and *m*-carboranyl ketones have been carefully studied and compared; as expected from their relative $-I$ effects, *m*-carboranyl ketones are the stronger bases (both *o*- and *m*-carboranyl ketones are far less basic than benzophenone and acetophenone) (*442*).

The α-hydrogen atoms in both *m*- and *o*-carboranyl ketones have low proton mobility. Thus, 1-acetyl-2-methyl-*o*-carborane is not attacked by bromine in boiling CCl_4, but 1-acetyl-7-methyl-*m*-carborane does react slowly under these conditions to form 1-bromoacetyl-7-methyl-*m*-carborane (*442*).

7-9. *m*-CARBORANYL NITROGEN AND PHOSPHORUS DERIVATIVES

Nitrates, Amines, and Diazonium Salts

In contrast to the behavior of *o*-carborane (Section 6-8), *m*-carborane reportedly fails to react with 100% HNO_3 at room temperature (*418*), and no cage-substituted nitrato-*m*-carboranes have been characterized. The mixed acid nitration of aryl *m*-carboranes to the nitrophenyl derivatives, and the reduction of the latter species to aminophenyl-*m*-carboranes, apparently follows the same course as the reactions of the corresponding *o*-carboranyl compounds (*382, 388*). Aryl *m*-carboranes are nitrated somewhat more readily than are their *ortho* analogs, as is noted in Section 7-5.

Amino-*m*-carboranes may be prepared via the azides of *m*-carboranyl carboxylic acids, as described earlier for *o*-carboranyl amines (*402*). The action of alkali metals on *m*-carboranes in liquid ammonia produces B-amino-*o*-carboranes (*396*), but B-amino-*m*-carboranes are formed in reactions of m-$C_2B_9H_{11}^{2-}$ salts with aminodichloroboranes (*250*) (Section 9-2). In addition, it has been found that the pyrolysis of 3-diphenylamino-*o*-carborane at 400° yields 2-diphenylamino-*m*-carborane and probably 4-diphenylamino-*m*-carborane also (*250*).

The diazotization of aminophenyl-*m*-carboranes is easily accomplished by treatment with nitrosylsulfuric acid in glacial acetic acid; the diazonium salts react in a normal fashion with KI, CuCl, and CuBr to form the respective halophenyl-*m*-carboranes, and with β-naphthol to produce azo dyes (Section 6-8) (*388*).

$$p\text{-}^+N_2C_6H_4CB_{10}H_{10}CH \xrightarrow[HCl]{CuCl} ClC_6H_4CB_{10}H_{10}CH$$

The *m*-carborane cage is considerably more resistant to attack by amines than is the *o*-carborane system; thus, *m*-carborane is unreactive with piperidine in pentane and with *n*-propylamine in refluxing hexane (*92*) (*o*-carborane is degraded to dicarbaundecaborane species by these reagents as described in Chapter 9).

Amides

The C,C'-diacid dichloride of *m*-carborane, mentioned in the preceding section, reacts with ammonia to form the diamide (*93, 249*).

$$\underset{\text{ClCCB}_{10}\text{H}_{10}\text{CCCl}}{\overset{\text{O} \quad\quad \text{O}}{\| \quad\quad \|}} \xrightarrow{\text{2NH}_3} \underset{\text{H}_2\text{NCCB}_{10}\text{H}_{10}\text{CCNH}_2}{\overset{\text{O} \quad\quad \text{O}}{\| \quad\quad \|}}$$

The synthesis of *m*-carboranyl carboxanilides and thiocarboxanilides from lithio-*m*-carboranes and phenyl isocyanate or phenyl isothiocyanate has been reported without details (*433*).

Phosphorus Derivatives

The reaction of 1,2-dilithio-*o*-carborane with PCl_3 to form a cyclic bis(*o*-carboranyl) product (Section 6-9) has no parallel in *m*-carborane chemistry. Treatment of 1,7-dilithio-*m*-carborane with PCl_3 yields the bis(dichlorophosphino) derivative and a polymer; the former product is easily converted to the corresponding methoxy and dimethylamino derivatives (*8*).

$$\text{LiCB}_{10}\text{H}_{10}\text{CLi} \xrightarrow{\text{2PCl}_3} \text{Cl}_2\text{PCB}_{10}\text{H}_{10}\text{CPCl}_2 + \left[\text{—CB}_{10}\text{H}_{10}\text{CP—} \right]_n$$

$$n \leqslant 5$$

$$\xrightarrow{\text{RH}} \text{R}_2\text{PCB}_{10}\text{H}_{10}\text{CPR}_2$$

$$R = CH_3O, (CH_3)_2N$$

Both diphenylchlorophosphine and phenyldichlorophosphine react with 1,7-dilithio-*m*-carborane to form the expected C,C'-difunctional derivatives (*8*), the *o*-carboranyl analogs of which were described in Section 6-9.

7-10. *m*-CARBORANYL SILICON DERIVATIVES

The preparation of C,C'-difunctional *m*-carboranes from alkylchlorosilanes and 1,7-dilithio-*m*-carborane is straightforward; in contrast to the silyl *o*-carboranes (Section 6-10), however, silyl derivatives of *m*-carborane have shown no tendency to form cyclic structures (*232, 252, 261*). Thus, 1,7-bis-(chlorodimethylsilyl)-*m*-carborane reacts with ammonia, water, and methanol to form the disubstituted aminodimethylsilyl, hydroxymethylsilyl, and methoxydimethylsilyl derivatives, respectively; the reaction with water is interesting in that the analogous *o*-carborane derivative has resisted all attempts at hydrolysis (*232*).

$$
\begin{array}{ccc}
& \overset{R}{\underset{R}{|}}\quad\overset{R}{\underset{R}{|}} & \overset{R}{\underset{R}{|}}\quad\overset{R}{\underset{R}{|}} \\
\text{LiCB}_{10}\text{H}_{10}\text{CLi} & \xrightarrow[\text{ether}]{R_2\text{SiCl}_2} \quad \text{ClSiCB}_{10}\text{H}_{10}\text{CSiCl} & \xrightarrow{\text{NH}_3} \quad \text{H}_2\text{NSiCB}_{10}\text{H}_{10}\text{CSiNH}_2
\end{array}
$$

$$\downarrow \begin{array}{c}\text{CH}_3\text{SiCl}_3\\ \text{ether}\end{array}$$

$$
\begin{array}{c}
\text{Cl}_2\text{SiCB}_{10}\text{H}_{10}\text{CSiCl}_2 \\
\underset{\text{CH}_3}{|} \qquad \underset{\text{CH}_3}{|}
\end{array}
$$

$$\searrow \text{H}_2\text{O} \qquad \searrow \text{CH}_3\text{OH}$$

$$
\begin{array}{c}
\overset{R}{\underset{R}{|}}\qquad\overset{R}{\underset{R}{|}} \\
\text{CH}_3\text{OSiCB}_{10}\text{H}_{10}\text{CSiOCH}_3
\end{array}
$$

$$\downarrow \text{H}_2\text{O}$$

$$
\begin{array}{c}
\overset{R}{\underset{R}{|}}\qquad\overset{R}{\underset{R}{|}} \\
\text{HOSiCB}_{10}\text{H}_{10}\text{CSiOH}
\end{array}
$$

$$
\begin{array}{c}
(\text{HO})_2\text{SiCB}_{10}\text{H}_{10}\text{CSi(OH)}_2 \\
\underset{\text{CH}_3}{|} \qquad \underset{\text{CH}_3}{|}
\end{array}
$$

$$R = \text{CH}_3, \text{C}_6\text{H}_5$$

An alternative route to silyl *m*-carboranes is the thermal isomerization of silyl *o*-carboranes, which occurs at lower temperatures than are required for the conversion of unsubstituted *o*- to *m*-carborane (*252*).

Other silicon derivatives of *m*-carborane, including alkoxysilanes and a variety of polymers, are discussed in Chapter 8.

7-11. *m*-CARBORANYL GERMANIUM, TIN, AND LEAD DERIVATIVES

Tetrachlorogermane reacts with 1,7-dilithio-*m*-carborane to form 1,7-bis(trichlorogermyl)-*m*-carborane, but the reaction of the dilithio compound with two moles of dichlorodimethylgermane produces only a small yield of the bis(chlorodimethyl)-m-carborane monomer; the major product is a polymer containing an average of six carboranyl units (261).

$$\text{LiCB}_{10}\text{H}_{10}\text{CLi} \xrightarrow[\text{ether}]{\text{2GeCl}_4} \text{Cl}_3\text{GeCB}_{10}\text{H}_{10}\text{CGeCl}_3$$

$$\xrightarrow[\text{ether}]{\text{2(CH}_3)_2\text{GeCl}_2}$$

$$\underset{\overset{|}{\text{CH}_3}}{\overset{\overset{\text{CH}_3}{|}}{\text{ClGeCB}_{10}\text{H}_{10}\text{CGeCl}}} + \underset{\overset{|}{\text{CH}_3}}{\overset{\overset{\text{CH}_3}{|}}{\text{ClGe}}} \left[\underset{\overset{|}{\text{CH}_3}}{\overset{\overset{\text{CH}_3}{|}}{\text{CB}_{10}\text{H}_{10}\text{CGe}}} \right]_{\sim 6} \text{Cl}$$

$$\qquad\qquad\qquad 20\% \qquad\qquad\qquad 80\%$$

Similar polymeric products are obtained when 1,7-dilithio-m-carborane is treated with germanium, tin, and lead compounds of type R$_2$MCl$_2$ in tetrahydrofuran (33, 35).

$$\text{LiCB}_{10}\text{H}_{10}\text{CLi} \xrightarrow[\text{THF, 0°}]{\text{R}_2\text{MCl}_2} \left[\underset{\overset{|}{\text{R}}}{\overset{\overset{\text{R}}{|}}{\text{CB}_{10}\text{H}_{10}\text{CM}}} \right]_n$$

$$M = \text{Si, Ge, Sn, Pb}$$
$$R = \text{CH}_3, \text{C}_2\text{H}_5, n\text{-C}_4\text{H}_9, \text{C}_6\text{H}_5$$

Dichlorodiphenylstannane in ether yields only the monofunctional derivative, but in refluxing toluene some polymeric material is obtained (261).

$$\text{LiCB}_{10}\text{H}_{10}\text{CLi} \xrightarrow[25°, \text{ether}]{(\text{C}_6\text{H}_5)_2\text{SnCl}_2} \underset{\overset{|}{\text{C}_6\text{H}_5}}{\overset{\overset{\text{C}_6\text{H}_5}{|}}{\text{HCB}_{10}\text{H}_{10}\text{CCl}}}$$

A series of monomeric m-carboranyl tin derivatives has been prepared from the reaction of 1,7-dilithio-m-carborane with organohalostannanes (32).

$$\text{LiCB}_{10}\text{H}_{10}\text{CLi} + 2\text{ClSnR}_3 \xrightarrow{\text{ether, N}_2} \text{R}_3\text{SnCB}_{10}\text{H}_{10}\text{CSnR}_3$$

$$R = \text{C}_4\text{H}_9, \text{C}_6\text{H}_5$$

$$\text{R}'\text{CB}_{10}\text{H}_{10}\text{CLi} + \text{Cl}_2\text{SnR}_2 \longrightarrow (\text{R}'\text{CB}_{10}\text{H}_{10}\text{C})_2\text{SnR}_2$$

$$R = \text{CH}_3, \text{C}_6\text{H}_5;$$
$$R' = \text{CH}_3, \text{C}_6\text{H}_5$$

7-12. *m*-CARBORANYL SULFUR DERIVATIVES

The methods outlined for the synthesis of *o*-carboranyl sulfur derivatives (Section 6-12) may be applied to the preparation of the *m*-carboranyl analogs. These include the reactions of alkali metal derivatives with organic disulfides (*282*) or with elemental sulfur in liquid ammonia (*465*).

$$\text{LiCB}_{10}\text{H}_{10}\text{CLi} + 2\text{RSSR} \longrightarrow \text{RSCB}_{10}\text{H}_{10}\text{CSR} + 2\text{RSLi}$$

$$R = H, C_2H_5, CH_3C_6H_4$$

$$\text{RCB}_{10}\text{H}_{10}\text{CM} + \text{S} \xrightarrow[\text{2. H}^+]{\text{1. NH}_3 \text{ (liq)}} \text{RCB}_{10}\text{H}_{10}\text{CSH}$$

$$R = CH_3$$
$$M = \text{Li, Na, K}$$

Direct chlorination of 1,7-bis(mercapto)-*m*-carborane produces the bis(chlorosulfenyl) derivative, whose chlorine atoms may be displaced by a variety of nucleophilic reagents. Oxidation with NaOCl converts it to 1,7-bis(chlorosulfonyl)-*m*-carborane (*269*).

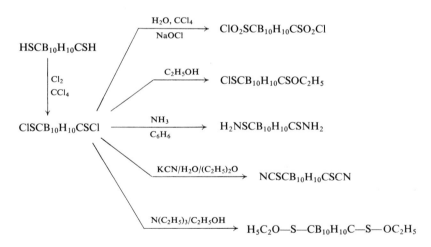

Reaction of the bis(chlorosulfenyl) derivative with 1,7-dilithio-*m*-carborane forms a linear polymer (*269*).

$$\text{ClSCB}_{10}\text{H}_{10}\text{CSCl} \xrightarrow[-\text{LiCl}]{\text{LiCB}_{10}\text{H}_{10}\text{CLi}} \left[\text{SCB}_{10}\text{H}_{10}\text{CS}\right]_{12-30}$$

7-13. *m*-CARBORANYL HALOGEN DERIVATIVES

Synthesis from Halo-*o*-Carboranes

The perhalo- and B-decahalo-*m*-carboranes may be prepared by the high-temperature isomerization of the corresponding *o*-carborane derivatives (*67, 314, 444, 461*). Thermal rearrangements of partially halogenated *o*-carboranes, however, are often complex (Section 7-3) and seem less useful as a route to specific halo-*m*-carboranes than direct halogenation of *m*-carborane, described below. An exception to this observation may be the rearrangement of mono-halo-*o*-carboranes such as 3-fluoro-*o*-carborane, which on pyrolysis yields only the 2-fluoro- and 4-fluoro-*m*-carboranes as final products (*119, 250*).

Electrophilic Halogenation

Calculations of the charge distribution in *m*-carborane place the highest negative charge on atoms 9 and 10, which are the only boron atoms not adjacent to carbon (*152*). As in *o*-carborane, the carbon atoms (1 and 7) are found to be the most positive in the cage system, with negative charge increasing in the order $(1,7) < (2,3) < (5,12) < (4,6,8,11) < (9,10)$. In agreement with this ordering, electrophilic halogenation of *m*-carborane takes place first at the 9,10 boron atoms in reactions of elemental halogens (*11, 216, 281, 305, 314, 318, 322, 323, 389, 409*) and halomethanes (*383, 406*); further halogenation is not observed under most conditions studied. This is consistent with the electrophilic halogenation of *o*-carborane, in which only the four boron atoms not bonded to carbon are susceptible to halogen substitution (Section 6-13).

Russian workers have reported that the exhaustive chlorination and iodination of *m*-carborane over AlCl$_3$ at 80° yields tri- and tetrahalo derivatives (*318, 319, 389*), while bromination under these conditions proceeds as far as the hexahalo compound (*307, 383, 389*). The latter result appears to conflict with the results of Smith *et al.* (*281*), who obtained only a dibromo derivative as the ultimate product of the bromination of *m*-carborane over AlCl$_3$ and AlBr$_3$. As with *o*-carborane, such reactions seem highly sensitive to the choice of catalyst; thus, the halogenation of *m*-carborane over iron filings in CCl$_4$ yields only mono- and dihalo derivatives (*314, 318*). Similar results have been obtained in the iodination and bromination of *m*-carborane in aqueous media (*305*).

The fact that most halo-*m*-carboranes obtained in these reactions are found in only one structural isomer (*322, 323, 383, 409*) is evidence that the electrophilic halogenation of *m*-carborane, like that of *o*-carborane, is highly stereospecific under most conditions. This observation, supported by dipole moment measurements (*62, 307*), X-ray studies (*11, 338*), and ^{35}Cl NQR spectra (*36*), has established boron atoms 9 and 10 as the initial halogenation sites. Using dipole moment data (*307*) and the isomer count method (*383*), several workers have concluded that the third, and subsequent, halogen atoms enter the 4,6,8,11 group, in agreement with the charge density calculations mentioned above.

A comparative rate study (*404*) has shown that *m*-carborane is less susceptible than *o*-carborane to electrophilic halogenation, as expected from the weaker electron-withdrawing power of the *m*-carborane system. Thus, *o*-carborane is chlorinated (over AlCl$_3$ in CH$_2$Cl$_2$) 3.5 times faster than *m*-carborane, while the ratios for bromination and iodination are 7:1 and 11.5:1, respectively.

Fluorination

The reaction of *m*-carborane with elemental fluorine in liquid HF is much less stereospecific than with other halogens. In excess fluorine the ultimate product is B-decafluoro-*m*-carborane, with no C-fluorination occurring (*182*).

$$HCB_{10}H_{10}CH + 10F_2 \xrightarrow[0°]{HF \text{ (liq)}} 10HF + HCB_{10}F_{10}CH$$

Substitution at carbon has been achieved, however, by reaction of 1,7-dilithio-*m*-carborane with perchloryl fluoride; subsequent treatment with fluorine yields perfluoro-*m*-carborane (*182*).

$$LiCB_{10}H_{10}CLi \xrightarrow[-2LiClO_3]{2ClO_3F} FCB_{10}H_{10}CF \xrightarrow[HF \text{ (liq), } 0°]{10F_2} FCB_{10}F_{10}CF$$

Chlorine monofluoride effects only chlorine substitution, but chlorine trifluoride yields monofluoropolychloro derivatives (*182*).

$$HCB_{10}H_{10}CH \xrightarrow[HF \text{ (liq), } 0°]{ClF_3} HCB_{10}Cl_xH_{10-x}CH + HCB_{10}FCl_xH_{9-x}CH$$
$$x = 1–6$$

The synthesis of 2-fluoro-*m*-carborane has been achieved by the degradation of *m*-carborane to the dicarbollide ion, $1,7\text{-}C_2B_9H_{11}^{2-}$ (Chapter 9), followed by reaction with boron trifluoride etherate (*250*).

$$C_2B_9H_{11}{}^{2-} \xrightarrow{\;BF_3 \cdot (C_2H_5)_2O\;} HCB_{10}H_9FCH + 2F^- + (C_2H_5)_2O$$

Photochemical Halogenation

The ultraviolet-irradiated chlorination and bromination of *m*-carborane may be less stereospecific than electrophilic halogenation (as is the case with *o*-carborane (Section 6-13)) but this has not been clearly demonstrated. The 1:1 reaction with chlorine in CCl_4 produces two monochloro-*m*-carboranes (*385, 403*), but with excess chlorine the B-decachloro derivative is easily obtained (*264, 444, 461*).

$$HCB_{10}H_{10}CH + 10Cl_2 \xrightarrow[-10HCl]{} HCB_{10}Cl_{10}CH$$

Similarly, the chlorination of 9,10-dibromo-*m*-carborane yields $HCB_{10}Br_2\text{-}Cl_8CH$ (*281*).

Both *m*- and *o*-carborane are highly resistant to photochemical bromination, and only 9-bromo-*m*-carborane has been obtained by this route. The reaction of 1-phenyl-*m*-carborane with bromine or iodine yields only the 1-phenyl-9-bromo- and 1-phenyl-10-bromo-*m*-carboranes (as with 1-phenyl-*o*-carborane, the aryl ring is unaffected) (*385*).

C-Halo-*m*-Carboranes

The synthesis of C-halocarboranes from C-lithiocarboranes (Section 6-13) has been applied successfully to the *m*-carborane system, as shown by the chlorination of $LiCB_{10}Cl_{10}CLi$ to produce perchloro-*m*-carborane (*262*). Similarly, $LiCB_{10}H_{10}CLi$ reacts easily with F_2, Cl_2, Br_2, and I_2 to give the 1,7-dihalo derivatives (*182, 299a, 423*). Certain C-monohalo derivatives have also been prepared (Table 7-1) (*299a*).

Reactions and Properties of Halo-*m*-Carboranes

B-decachloro-*m*-carborane is a much weaker acid than B-decachloro-*o*-carborane (*444*) (Table 6-3, Section 6-13). In contrast to its *o*-carboranyl analog, B-decachloro-*m*-carborane dissolves in aqueous basic solution only when heated (*443*). However, B-bromo- and B-iodo-*m*-carboranes undergo nucleophilic exchange with CuCl in the same manner as B-halo-*o*-carboranes (*393*). Other reactions paralleling those of B-halo-*o*-carboranes are the Ullman synthesis of 9,9'-bis(*m*-carboranyl) from 9-iodo-*m*-carborane (*308*) and the dehalogenation of 9-iodo-*m*-carborane by the action of sodium in liquid ammonia (*421*).

The halogen atoms in C-bromo- and C-iodo-*m*-carboranes are easily replaced by hydrogen in hot alcoholic base (*423, 441*), but C-chloro-*m*-carboranes are degraded with evolution of H_2 (*423*). These results correspond to the behavior of C-monohalo-*o*-carboranes. However, in contrast to the degradation of C,C'-dihalo-*o*-carboranes in methanol and ethanol, C,C'-dihalo-*m*-carboranes are stable in these solvents (*423*).

$$C_6H_5CB_{10}H_{10}CI \xrightarrow[\text{C}_2\text{H}_5\text{OH}]{\text{KOH}} C_6H_5CB_{10}H_{10}CH$$

$$XCB_{10}H_{10}CX \xrightarrow[\text{C}_2\text{H}_5\text{OH}]{\text{NaOH}} XCB_{10}H_{10}CH \xrightarrow[\text{C}_2\text{H}_5\text{OH}]{\text{NaOH}} HCB_{10}H_{10}CH$$

$$X = Br, I$$

B-decafluoro-*m*-carborane is readily hydrolyzed in water and in moist air, a property also exhibited by its *ortho* and *para* isomers (*182*). No explanation has been given for this somewhat surprising behavior (B-decachloro-*o*-carboranes are hydrolytically stable). The degradation of B-decachloro-*m*-carborane by dimethyl sulfoxide has also been reported (*445*).

7-14. *m*-CARBORANYL MERCURY DERIVATIVES

The synthesis of symmetrical and mixed *m*-carboranyl mercury compounds from mercuric halides and lithio-*m*-carboranes (*452*) is analogous to the formation of the *o*-carboranyl compounds (Section 6-15).

$$RCB_{10}H_{10}CLi \xrightarrow{HgCl_2} (RCB_{10}H_{10}C)_2Hg$$

$$R = CH_3, C_6H_5$$

$$C_6H_5CB_{10}H_{10}CLi + C_6H_5HgCl \longrightarrow C_6H_5CB_{10}H_{10}CHgCl + C_6H_5Li$$

$$C_6H_5CB_{10}H_{10}CHgC\!\!-\!\!CC_6H_5 \qquad\qquad C_6H_5C\!\!-\!\!CLi$$
$$\underset{B_{10}H_{10}}{\diagdown O\diagup} \qquad\qquad\qquad \underset{B_{10}H_{10}}{\diagdown O\diagup}$$

The reaction of 1,7-dilithio-*m*-carborane with $HgCl_2$ leads to a linear polymer of molecular weight 10,000 (*34*). The direct C-mercuration of the carborane cage with organomercury hydroxides, which has been accomplished with *o*-carboranyl compounds, has not been successful in the case of the *m*-carboranyl system (*416*).

Mercury derivatives of *m*-carborane are generally similar to the corresponding *o*-carboranyl compounds in their chemical and physical properties, and resemble organomercuric compounds which contain strong electron withdrawing groups such as perfluoroalkyl, pentafluorophenyl, and ethynyl. The comparatively weak −I inductive effect of the *m*-carboranyl system is manifested in the greater reactivity of *m*-carboranyl mercury species with electrophilic reagents (HCl, $HgCl_2$, Br_2) compared to *o*-carboranyl mercury compounds; still, *m*-carboranyl mercury derivatives are less reactive with such reagents than are dialkyl- and diarylmercurials (*452*). In reactions with Br_2, only the external Hg—C bond is cleaved and the Hg-*m*-carborane links are unaffected.

$$CH_3HgCB_{10}H_{10}CHgCH_3 + 2Br_2 \rightarrow BrHgCB_{10}H_{10}CHgBr + 2CH_3Br$$

Consistent with these results, polarographic measurements indicate that the Hg—C bond polarity decreases in the order *o*-carboranyl > *m*-carboranyl > diaryl mercury compounds (*452*)

Symmetrical *m*-carboranyl mercury derivatives exhibit high thermal stability; $(CH_3CB_{10}H_{10}C)_2Hg$, for example, is unchanged on prolonged heating at 275° (*452*). Mixed derivatives, such as $CH_3HgCB_{10}H_{10}CCH_3$ and $CH_3HgCB_{10}H_{10}CCl$, are also stable but undergo complete symmetrization at 220–250°, contrasting sharply in this respect with unsymmetrical *o*-carboranyl mercury compounds (Section 6-15).

$$2CH_3CB_{10}H_{10}CHgCH_3 \xrightarrow{\Delta} (CH_3CB_{10}H_{10}C)_2Hg + CH_3HgCH_3$$

Unsymmetrical *m*-carboranyl mercuric halides, like the corresponding *o*-carboranyl compounds but unlike alkylmercuric halides, do not react with *o*-phenanthroline to give symmetrical products; instead, stable 1:1 adducts are obtained (*451*).

$$CH_3CB_{10}H_{10}CHgX + C_{12}H_8N_2 \rightarrow CH_3CB_{10}H_{10}CHgX \cdot C_{12}H_8N_2$$
$$X = Cl, Br$$

These adducts do not undergo symmetrization even on heating in benzene.

Studies of the polarographic reduction potentials of symmetrical *o*- and *m*-carboranyl mercury derivatives (*306*, *452*) have led to the conclusion, consistent with the chemical evidence, that the *o*-carboranyl species are stronger acids than their *m*-carboranyl analogs. The introduction of a methyl group has little effect on the *o*-carboranyl cage but substantially lowers the acidity of *m*-carborane. Phenyl groups actually increase the acidity of the *o*-carboranyl system while lowering that of *m*-carborane.

7-15. *p*-CARBORANE

Synthesis and Structure

Until recently *p*-carborane (1,12-$C_2B_{10}H_{12}$) was a rare compound, obtainable only as a minor product in the decomposition of *o*- or *m*-carborane above 600° (*233*). An improved synthesis (*234*, *276*), involving the continuous passage of *o*-carborane in a stream of N_2 through a converter at 700°, produces 75% *m*-carborane and 25% of the *para* isomer. Separation of the products on basic alumina gives pure *p*-carborane in decagram quantities, thus opening the way to an extensive derivative chemistry of this isomer (the known *p*-carboranyl derivatives are summarized in Table 7-2). The assignment of an icosahedral structure with the carbons at opposite vertices (Fig. 2-1n) is convincingly supported by the ^{11}B NMR spectrum, consisting of one doublet which collapses to a singlet on decoupling (*233*, *337*); the zero dipole moment (*190*); and the isomer distribution of B-halogenated species, discussed below.

TABLE 7-2

DERIVATIVES OF p-CARBORANE

Substituent groups	mp (°C)	Other data[a]	Refs.
Parent	259–261	IR, B	233
		DM	190
1-COOH	173–174	pK_a (Table 6-2)	417
1,12-(COOH)$_2$	356–357		417
1,12-(SH)$_2$	157–158		269
1,12-(SCl)$_2$	71–73		269
B-F$_{10}$	222–224		182
B-Cl	179–180	MS	276
	191–192		307a
B-Cl$_2$[b]	155–158	MS	276
	151–152		307a
B-Cl$_3$[b]	136–139	MS	276
B-Cl$_4$[b]	128–131	MS	276
B-Cl$_5$[b]	125–128	MS	276
B-Cl$_6$[b]	114–115	MS	276
B-Cl$_7$[b]	135–137	MS	276
B-Cl$_8$[b]	175–177	MS	276
B-Cl$_9$	216–218	MS	276
B-Cl$_{10}$	302–304	MS	276
B-Br	150–151	MS	233, 276
	140–141		307a
B-Br$_2$[b]	55–60	MS	276
B-I	50–52	MS	276
	60–61		307a
B-I$_2$[b]	47–48	MS	276

[a] IR = infrared spectra or band positions; B = ^{11}B NMR spectra; DM = dipole moment; MS = mass spectrum or cutoff m/e value.
[b] Mixture of isomers.

Organic and Organometallic Derivatives

The few reports of p-carborane derivative chemistry in the literature are concerned with the synthesis of polymers or polymer precursors, as discussed in Chapter 8. The preparation and properties of 1,12-dilithio-p-carborane, not surprisingly, parallel those of the m-carboranyl system.*

* The symbol $\overline{HCB_{10}H_{10}CH}$ is used here to represent $1,12$-$C_2B_{10}H_{12}$.

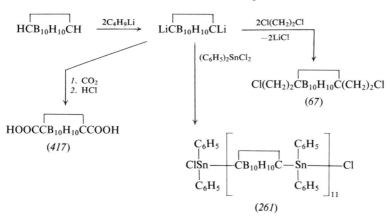

(261)

Competitive lithiation of *m*- and *p*-carborane indicates that the *para* isomer is less acidic (*417*). This result is consistent with the relatively high pK_a of 3.64 for *p*-carboranyl carboxylic acid (Table 6-2), demonstrating that the *para* isomer is a weaker electron-acceptor than the *ortho* and *meta* systems (*417*). Other differences in the behavior of *p*-carborane as compared to its *ortho* and *meta* isomers have been noted, particularly the greater tendency of *p*-carboranyl species to polymerize (*261*) and the resistance of *p*-carborane to cage degradation in basic media (*250*). Although B-decafluoro-*p*-carborane is not hydrolyzed in air (in contrast to the corresponding *o*- and *m*-carboranyl compounds), this derivative is completely degraded to boric acid on treatment with aqueous acetonitrile (*182*).

Zakharkin and Kalinin (*399*) have reported that *p*-carborane reacts with 2 moles of sodium in liquid ammonia to form the dianion $p\text{-}C_2B_{10}H_{12}^{2-}$, which at $-45°$ rearranges to the $m\text{-}C_2B_{10}H_{12}^{2-}$ ion; oxidation of the latter species yields *m*-carborane. This "reverse isomerization" is, of course, analogous to the conversion of *m*- to *o*-carborane in liquid ammonia (Section 7-3).

Halogen Derivatives

The halogenation of *p*-carborane has been studied in some detail. Electrophilic chlorination of *p*-carborane in refluxing CCl_4 over $AlCl_3$ yields B-substituted mono-, di-, tri, and tetrachloro derivatives, but bromination and iodination under the same conditions gives essentially only the mono- and dihalo species (*276, 307a, 417*). Photochemical chlorination of *p*-carborane, as with *o*- and *m*-carboranes, proceeds both faster and farther than the electrophilic reactions; B-monochloro- through B-decachloro-*p*-carboranes are produced, with complete conversion to the latter compound occurring within six hours (276). Direct fluorination of p-carborane in liquid HF at $0°$ forms B-decafluoro-p-carborane in high yield (182).

Unlike the *ortho* and *meta* isomers, all ten boron atoms in p-carborane are equivalent, so that each B-monohalo species should exist in only one isomeric form. This has been verified for the B-monofunctional fluoro-, chloro-, bromo-, and iodo-p-carboranes (276). Since all boron atoms are equivalent, B-halogenation beyond the monohalo stage would occur randomly unless a halogen substituent effect of the *ortho*- or *meta*-directing type were present. Some rather tentative evidence for such an effect has been presented by Sieckhaus *et al.* (276), on the basis of gas chromatograms of the product mixtures obtained in the photochemical chlorination of p-carborane. However, the isomer distribution is not grossly different from that predicted on a purely random statistical basis, so that the directive effect, if any, appears weak. The same authors suggest (276a) that a stronger substituent effect should be observed in electrophilic chlorination, in support of which they point to the somewhat different isomer distributions (for the mono- to tetrachloro derivatives) obtained in the electrophilic and photochemical reactions. At this writing the evidence is inconclusive, although it must be said that some directive influence by the halogen substituent atoms, however weak, is surely present.

8

Carborane Polymers

8-1. INTRODUCTION

The enormous scope of carborane chemistry practically assures significant applications of this field to human needs, no doubt in some ways not now predictable. Given the versatility of boron in forming stable aromatic-like cage structures involving many different metallic and nonmetallic elements, it is no longer rash to view the field as being as rich in synthetic possibilities as organic chemistry. It is conceivable that carborane-based fibers, oils, pigments, and even drugs may one day be important commercial products. At present, however, the practical utilization of carborane compounds revolves almost exclusively around polymers, particularly those with outstanding resistance to degradation by heat and oxidation. In fact, the bulk of the published derivative chemistry on the icosahedral o-, m-, and p-carborane systems derives from industrial research aimed at the synthesis of such polymers. The underlying rationale for this effort is that the carborane cage units not only possess considerable thermal and chemical stability, but may also act as energy sinks and thereby increase the stability of neighboring bonds in a polymeric chain. The actual properties of carborane-based polymers vary widely, but some of them are truly extraordinary materials capable of withstanding extreme conditions in which conventional organic and inorganic polymers are severely degraded.

Much of the research in this area remains in private files and has not appeared in the open literature. The published data are nonetheless extensive, and it is apparent that carborane polymer chemistry is an extremely complex field which overlaps the borders of chemistry, physics, and materials science. The following short discussion is limited to the more important synthetic routes to carborane polymers, with emphasis on the chemistry rather than on the quantitative physical-mechanical data which have accrued for these materials.

8-2. GENERAL CONSIDERATIONS

All of the known and characterized carborane polymers are based upon the icosahedral $C_2B_{10}H_{12}$ isomers; polymeric materials which may involve other carborane cage systems have been reported but none has been structurally identified. Two broad types of carborane polymers are recognizable: those having o-, m-, or p-carborane units in the polymer chain itself, and those in which carborane cages are attached to the chain as pendant groups. It is convenient to designate these as classes I and II, respectively. With few exceptions, the polymers exhibiting unusual chemical and thermal stability are in class I; the attachment of carboranyl groups to a backbone of conventional organic structure does not appear to greatly increase the resistance to initial chain degradation (it may, of course, lead to other desirable properties). Another generalization which has emerged from the considerable research in this area is that the m- and p-carborane isomers offer greater versatility than the *ortho* isomer in the synthesis of stable high polymers. Steric problems are frequently encountered in the development of class I polymers based on o-carborane, a common one being the tendency of C,C'-difunctional o-carborane derivatives to form exopolyhedral rings (Chapter 6). These points are discussed in greater detail in the following sections.

8-3. CLASS I POLYMERS

Polyesters

o-Carborane diols and diacids react with organic or fluorocarbon diols and diacids (or with each other), yielding polyesters of molecular weight 2,000 to 20,000 (Section 6-6) (97–99). The chain lengths and properties of these products are dependent upon reaction conditions and the monomer ratio, and materials ranging from hard brittle solids to viscous liquids are obtained. Fusion of bis(hydroxymethyl)-o-carborane with small perfluorinated acids yields low molecular weight (1000–3400) polyesters (472). Polyester carboranes obtained from glutaric, adipic, azelaic, and sebacic acids have been shown to be polydisperse systems (253), but none of these materials appears to have unusual heat- or oxidative resistance in comparison with conventional polyesters.

Polyformals

Although reactions of *o*-carborane glycols with formaldehyde or dipropyl formal frequently yield cyclic formals, linear carborane polyformals have been obtained from bis(2-hydroxyethyl-1-*o*-carboranylmethyl) ether (Section 6-6). Treatment of bis(hydroxymethyl)-*o*-carborane with propylene oxide yields a poly(oxypropylene)carborane diol which is thermally stable to 200° (*4*). A related reaction is the polymerization of epoxypropoxypropyltriethoxysilane with bis(hydroxymethyl)-*o*-carborane (*285*). The kinetics of these copolymerizations have been examined in detail (*5, 285*).

$$HOCH_2C\!-\!\!-\!\!CCH_2OH \quad + \quad CH_3CH\!-\!CH_2 \quad \longrightarrow$$
$$\underset{B_{10}H_{10}}{\diagdown O \diagup} \qquad\qquad\qquad \underset{O}{\diagdown \diagup}$$

$$H\!-\!\!\left[\!-OCH_2CH\underset{CH_3}{\big|}\!-\!\right]_m\!\!-\!OCH_2C\!-\!\!-\!\!CCH_2O\!-\!\!\left[\!-CH\underset{CH_3}{\big|}\!-\!CH_2O\!-\!\right]_n\!\!-\!H$$
$$\underset{B_{10}H_{10}}{\diagdown O \diagup}$$

$$m, n \cong 5$$

$$HOCH_2C\!-\!\!-\!\!CCH_2OH \quad + \quad CH_2\!-\!CH\!-\!CH_2O(CH_2)_3Si(OC_2H_5)_3 \quad \longrightarrow$$
$$\underset{B_{10}H_{10}}{\diagdown O \diagup} \qquad\qquad\qquad \underset{O}{\diagdown \diagup}$$

$$\left[\!-OCH_2CH\underset{(C_2H_5O)_3Si(CH_2)_3-O-CH_2}{\big|}\!-\!O\!-\!CH_2\!-\!C\!-\!\!-\!\!CCH_2O\!-\!CHCH_2O\underset{CH_2-O-(CH_2)_3Si(OC_2H_5)_3}{\big|}\!-\!\right]$$
$$\underset{B_{10}H_{10}}{\diagdown O \diagup}$$

The poly(oxypropylene)carborane diol described above reacts with hexamethylene diisocyanate at 100° to form a polyurethane (*4*).

Siloxanes

Early attempts to prepare carborane-siloxane polymers by well-known methods of polymerization were unsuccessful; thus, hydrolysis of 1,7-bis (chlorodimethylsilyl)-*m*-carborane produced only the bis(hydroxydimethylsilyl) derivative (Section 7-10), and the latter compound could not be polymerized (*235*). Mayes, Greene, and Cohen (*208*) synthesized a cyclic *m*-carboranesiloxane which undergoes acid-catalyzed ring-opening polymerization.

$$\underset{\overset{|}{CH_3}}{\overset{\overset{CH_3}{|}}{Cl\,Si}}-(CH_2)_3CB_{10}H_{10}C(CH_2)_3\underset{\overset{|}{CH_3}}{\overset{\overset{CH_3}{|}}{Si}}Cl \;+\; H_2O \quad \xrightarrow{\;acetone\;}$$

$$\begin{array}{c} \qquad\qquad\quad O \\ (CH_3)_2Si \qquad Si(CH_3)_2 \\ \quad | \qquad\qquad | \\ (CH_2)_3 \;\; (CH_2)_3 \\ \quad\;\; CB_{10}H_{10}C \end{array}$$

$$\xleftarrow{\;H_2SO_4\;}$$

45%

+ linear polymer,
MW 6300 (55%)

$$HO\!-\!\!\left[\,\underset{\overset{|}{CH_3}}{\overset{\overset{CH_3}{|}}{Si}}(CH_2)_3CB_{10}H_{10}C(CH_2)_3\underset{\overset{|}{CH_3}}{\overset{\overset{CH_3}{|}}{Si}}\!-\!O\,\right]_{\!n}\!\!-\!H$$

elastomeric gum, MW ~ 8500

From a practical standpoint, the most important breakthrough in the development of exceptionally heat-resistant carborane elastomers has been the synthesis of poly-*m*-carboranesiloxanes by a group of chemists at the Olin Corporation. This family of polymers, known under the trade name Dexsil, consists of dialkyl- or diarylsiloxane units linked by *m*-carborane cage structures, with a small proportion of 1-vinyl-*o*-carboranyl side groups to promote crosslinking (*144, 145, 235, 258, 259*). The formulas of Dexsil 201* (R = CH_3) and Dexsil 202 (R = C_6H_5) (an insoluble gum and a viscous liquid, respectively) are shown for illustration (*258*).

$$\left[\,\underset{\overset{|}{CH_3}}{\overset{\overset{CH_3}{|}}{-Si}}CB_{10}H_{10}C\underset{\overset{|}{CH_3}}{\overset{\overset{CH_3}{|}}{Si}}\!-\!O\,\right]_{1\cdot0}\!\left[\,\underset{\overset{|}{R}}{\overset{\overset{R}{|}}{-Si}}\!-\!O\,\right]_{0\cdot997}\!\left[\,\underset{\overset{|}{\underset{B_{10}H_{10}}{C-C-CH=CH_2}}}{\overset{\overset{CH_3}{|}}{-Si}}\!-\!O\,\right]_{0\cdot003}$$

The Dexsils are obtained by condensation polymerization of bis(methoxy-dimethylsilyl)-*m*-carborane with bis(chlorosilyl)-*m*-carborane derivatives, alkylchlorosilanes, or alkylchlorosiloxanes in the presence of a catalyst. In each case methyl chloride is produced as an elimination product (*145–147, 235, 258*).

* The first digit of the three-digit code designates the number of siloxy groups in the repeating unit.

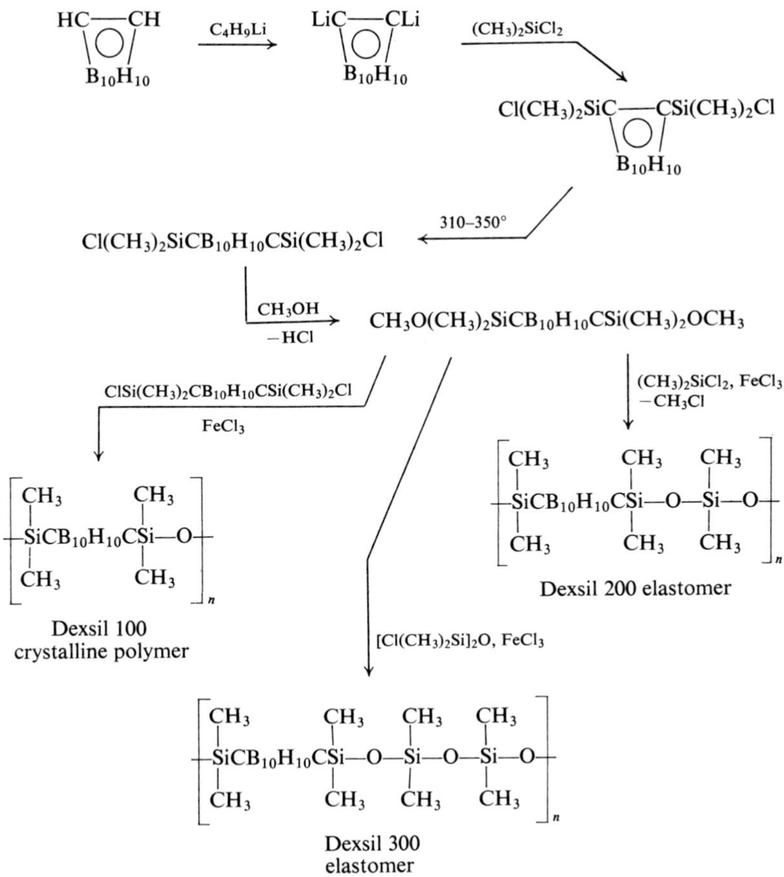

$$\underset{\underset{B_{10}H_{10}}{}}{HC\text{---}CH} \xrightarrow{C_4H_9Li} \underset{\underset{B_{10}H_{10}}{}}{LiC\text{---}CLi} \xrightarrow{(CH_3)_2SiCl_2}$$

$$\underset{\underset{B_{10}H_{10}}{}}{Cl(CH_3)_2SiC\text{---}CSi(CH_3)_2Cl}$$

$$Cl(CH_3)_2SiCB_{10}H_{10}CSi(CH_3)_2Cl \xleftarrow{310\text{--}350°}$$

$$\xrightarrow[-HCl]{CH_3OH} \quad CH_3O(CH_3)_2SiCB_{10}H_{10}CSi(CH_3)_2OCH_3$$

$$\underset{FeCl_3}{ClSi(CH_3)_2CB_{10}H_{10}CSi(CH_3)_2Cl} \qquad \xrightarrow[-CH_3Cl]{(CH_3)_2SiCl_2, FeCl_3}$$

$$\left[\begin{array}{c} CH_3 \qquad CH_3 \\ | \qquad\qquad | \\ \text{---}SiCB_{10}H_{10}CSi\text{---}O\text{---} \\ | \qquad\qquad | \\ CH_3 \qquad CH_3 \end{array}\right]_n$$

Dexsil 100
crystalline polymer

$$\left[\begin{array}{c} CH_3 \qquad CH_3 \quad CH_3 \\ | \qquad\qquad | \qquad | \\ \text{---}SiCB_{10}H_{10}CSi\text{---}O\text{---}Si\text{---}O\text{---} \\ | \qquad\qquad | \qquad | \\ CH_3 \qquad CH_3 \quad CH_3 \end{array}\right]_n$$

Dexsil 200 elastomer

$$[Cl(CH_3)_2Si]_2O, FeCl_3$$

$$\left[\begin{array}{c} CH_3 \qquad CH_3 \quad CH_3 \quad CH_3 \\ | \qquad\qquad | \qquad | \qquad | \\ \text{---}SiCB_{10}H_{10}CSi\text{---}O\text{---}Si\text{---}O\text{---}Si\text{---}O\text{---} \\ | \qquad\qquad | \qquad | \qquad | \\ CH_3 \qquad CH_3 \quad CH_3 \quad CH_3 \end{array}\right]_n$$

Dexsil 300
elastomer

Elastomers of the Dexsil 400 type (with four siloxy groups per repeating unit) have been prepared by similar reactions (258).

Some correlations of the structures of these polymers with their observed properties have been noted. For example, the thermal and oxidative stability tends to decrease with increasing number of siloxy groups not bonded to carborane, an effect that has been attributed to inductive stabilization of silyl groups by adjacent electron-withdrawing *m*-carboranyl units (44, 96, 331). Thus, polymers of the type shown below are susceptible to thermal oxidation at 240° (in a nitrogen atmosphere they are stable to 350°) (208, 331).

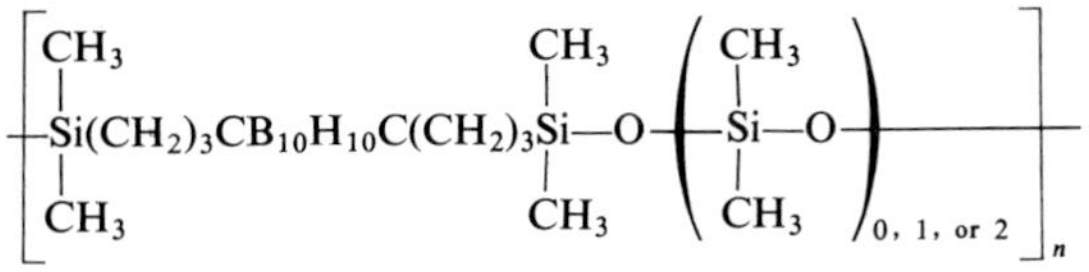

$$\left[\begin{array}{c} CH_3 \qquad\qquad\qquad CH_3 \quad \left(CH_3\right) \\ | \qquad\qquad\qquad\qquad | \qquad | \\ \text{---}Si(CH_2)_3CB_{10}H_{10}C(CH_2)_3Si\text{---}O\text{---}\left(Si\text{---}O\right) \\ | \qquad\qquad\qquad\qquad | \qquad | \\ CH_3 \qquad\qquad\qquad CH_3 \quad \left(CH_3\right)_{0,\ 1,\ or\ 2} \end{array}\right]_n$$

The replacement of methyl with phenyl substituents, as in Dexsil 202 (above), leads to viscous liquids of low molecular weight. Random copolymers have also been prepared, consisting of Dexsil 100 and 200 units in various ratios (*258*):

$$
\begin{bmatrix}
& CH_3 & & CH_3 \\
& | & & | \\
-Si & -CB_{10}H_{10}C- & Si & -O- \\
& | & & | \\
& CH_3 & & CH_3
\end{bmatrix}_a
\begin{bmatrix}
CH_3 & & CH_3 & CH_3 \\
| & & | & | \\
Si-CB_{10}H_{10}C- & Si & -O-Si-O- \\
| & & | & | \\
CH_3 & & CH_3 & CH_3
\end{bmatrix}_b
$$

Dexsil 125 ($a/b = 3.0$) is a powder, while Dexsil 150 ($a/b = 1.0$) and Dexsil 175 ($a/b = 0.33$) are rubberlike; all, however, are partly crystalline.

The elastomeric, mechanical, and high-temperature behavior of the poly-*m*-carboranesiloxanes has been thoroughly investigated (*44, 96, 287, 288, 235,*

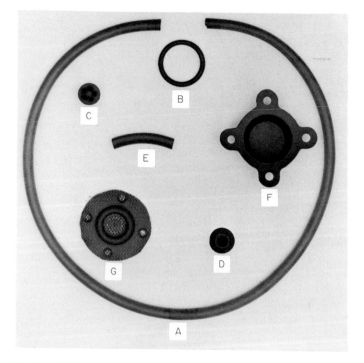

FIG. 8-1. Dexsil 201 products. (A) High-temperature wire jacketing; (B) O-ring; (C) electrical connector face seal; (D) missile nose cone seal; (E) glass fiber-wrapped door seal; (F) coated Nomex diaphragm for use as hot air control device in jet engines; (G) supporting material for glass cloth. Courtesy of Dr. H. A. Schroeder, Olin Corporation.

359). In general they display high thermal stability, and some of them have been treated by the methods of modern rubber technology to give noncrystalline elastomers with extraordinary properties not matched by other synthetic rubbers (*263*). Thus, Dexsil 201 compares favorably with silicones and fluorocarbons in retention of elongation, tensile strength, and high temperature performance. With the addition of antioxidants and fillers, and peroxide curing, carboranesiloxane elastomers which retain their flexibility at 900°F and even 1000°F have been produced in limited commercial quantities as gaskets, O-rings, and electrical connector inserts (Fig. 8-1). Dexsil 202, a short-chain polymer with no crosslinking, cannot be converted to an elastomer but it does form slightly flexible films on heating to 500–600°F. Its solubility in organic solvents makes it convenient to produce extremely heat-resistant coatings on metals and other materials. Possible applications at temperatures as high as 1500°F are under investigation (*258*).

Polymers with Single Atom Links

Chain polymers in which carborane icosahedra are joined by single atoms in place of the more flexible siloxane or ester linkages are, as expected, hard materials with thermoplastic properties. Polymers of this type with linking atoms from groups IV, V, and VI of the periodic table have been prepared and are described below.

Si-, Ge-, Sn-, OR Pb-LINKED CHAINS

The synthesis of compounds of type $+CB_{10}H_{10}CMR_2+_n$ (where M represents Si, Ge, Sn, or Pb) from $LiCB_{10}H_{10}CLi$ and R_2MCl_2 is discussed in Sections 7-10 and 7-11. The formation of such polymers is usually accompanied by monomeric products, and the tendency toward polymerization is dependent on such factors as solvent, temperature, reactant ratio, and the substituent groups on the metal. Thus, polymerization is strongly favored by $(CH_3)_2SnCl_2$ in place of $(C_6H_5)_2SnCl_2$. The chain length of the dimethyltin-*m*-carborane polymer varies from 8 to 30 repeating units, but the softening points are low (~250°) even in the long-chain products (substitution of a Br atom at B(9) on each *m*-carborane cage in the dimethyltin polymer results in a softening point nearly 100° higher) (*258*). The silicon, germanium, and lead analogs of the tin polymers also soften at low temperatures (*35, 261*).

A few *p*-carborane polymers of this type have been reported, and these remain hard up to relatively high temperatures (300–400°). Reaction of 1,2-dilithio-p-carborane with diphenyltin dichloride yields products averaging 11 repeating units, but the use of dimethyltin dichloride results in a polymer which is insoluble in ordinary organic solvents and is air-stable to 425° (*258*).

$$\overline{\text{LiCB}_{10}\text{H}_{10}\text{CLi}} + (\text{CH}_3)_2\text{SnCl}_2 \longrightarrow \text{ClSn}\left[\overset{\text{CH}_3}{\underset{\text{CH}_3}{|}}\overline{\text{CB}_{10}\text{H}_{10}\text{C}}-\overset{\text{CH}_3}{\underset{\text{CH}_3}{\overset{|}{\text{Sn}}}}\right]_n\text{Cl}$$

Heteropolymers containing both m- and p-carborane units are obtained from condensation of the respective dilithio species with dimethyltin dichloride. The linking atoms may also be randomized, as in the second reaction below (*258*).

$$\overline{\text{LiCB}_{10}\text{H}_{10}\text{CLi}} + \overline{\text{LiCB}_{10}\text{H}_{10}\text{CLi}} + 2(\text{CH}_3)_2\text{SnCl}_2 \longrightarrow$$

$$\left(\overset{\text{CH}_3}{\underset{\text{CH}_3}{\overset{|}{\text{CB}_{10}\text{H}_{10}\text{CSn}}}}\right)\left(\overline{\text{CB}_{10}\text{H}_{10}\overset{\text{CH}_3}{\underset{\text{CH}_3}{\overset{|}{\text{CSn}}}}}\right) \quad \text{mp } 370\text{–}375°$$

$$\text{LiCB}_{10}\text{H}_{10}\text{CLi} + (\text{CH}_3)_2\text{SnCl}_2 + (\text{CH}_3)_2\text{GeCl}_2 \longrightarrow$$

$$\left(\overset{\text{CH}_3}{\underset{\text{CH}_3}{\overset{|}{\text{CB}_{10}\text{H}_{10}\text{CSn}}}}\right)\left(\overset{\text{CH}_3}{\underset{\text{CH}_3}{\overset{|}{\text{CB}_{10}\text{H}_{10}\text{CGe}}}}\right) \quad \text{mp } 350\text{–}355°$$

PHOSPHORUS- AND SULFUR-LINKED CHAINS

The reaction of dilithio-m-carborane with phosphorus trichloride yields a low molecular weight linear polymer and bis(dichlorophosphino)-m-carborane, as described in Section 7-9. The corresponding reaction of dilithio-o-carborane yields only cyclic products (Section 6-9).

Some sulfur-linked chains of fairly high molecular weight have been prepared from bis(chlorosulfenyl)-m-carborane (Section 7-12) and dilithio-m-carborane (*258, 268*). Somewhat higher melting polymers based on p-carborane have been obtained (*258*).

$$\text{ClSCB}_{10}\text{H}_{10}\text{CSCl} \xrightarrow{\text{LiCB}_{10}\text{H}_{10}\text{CLi}} \text{Cl}\text{+}\text{CB}_{10}\text{H}_{10}\text{CS}\text{+}_{x}\text{—Cl} \qquad \text{mp } 231\text{–}233°$$

$$x \cong 30$$

$$\text{LiCB}_{10}\text{H}_{10}\text{CLi} \xrightarrow[\text{ether}]{S_8} \text{LiSCB}_{10}\text{H}_{10}\text{CSLi} \xrightarrow[\text{H}_2\text{O}]{\text{HCl}} \text{HSCB}_{10}\text{H}_{10}\text{CSH}$$

$$\Big\downarrow \text{Cl}_2, \text{CCl}_4$$

$$\text{ClO}_2\text{SCB}_{10}\text{H}_{10}\text{CSO}_2\text{Cl} \xleftarrow[\text{H}_2\text{O}]{\text{NaOCl}} \text{ClSCB}_{10}\text{H}_{10}\text{CSCl} \xrightarrow{\text{LiCB}_{10}\text{H}_{10}\text{CLi}}$$

$$\text{Cl}\text{+}\text{CB}_{10}\text{H}_{10}\text{C—S}\text{+}_{n}\text{OH} \qquad \text{mp} > 420°$$

MERCURY-LINKED CHAINS

A polymer consisting simply of *m*-carborane icosahedra connected by mercury atoms has been reported (*34*).

$$\text{LiCB}_{10}\text{H}_{10}\text{CLi} + \text{HgCl}_2 \;\rightarrow\; \text{+CB}_{10}\text{H}_{10}\text{C—Hg+}_{n} \qquad \begin{array}{l} \text{mp} > 300° \\ \text{MW} \sim 10{,}000 \end{array}$$

Other Class I Polymers

Long-chain compounds of curious structure involving cyclic *o*-carboranyl-phosphorus rings with P—N—P linkages have been prepared (Section 6-9) with molecular weights between 2400 and 10,000 (*7, 258*). The reaction of B-methyl-*N*-triphenylborazole with bis(aminomethyl)- or bis(hydroxymethyl)-*o*-carborane yields transparent yellow, brittle solids (MW ~ 1500) which are soluble in dioxane, acetone, and alcohols but insoluble in water, benzene, and carbon tetrachloride. The polymers prepared from the bis(hydroxymethyl) derivative are air-stable to ~500° and remain hard up to 150° (*3*).

A white nitroso rubber terpolymer of uncertain structure has been synthesized from 1-nitrosocarborane, nitrosofluoromethane, and tetrafluoroethylene (*169*). Poly(*o*-carboranyloxadiazoles) of low thermal stability (decomposing below 230°) have been prepared in a two-step synthesis: the condensation of bis(carboxymethyl)-*o*-carborane with oxalyl, succinoyl, or adipoyl hydrazide in N,N'-dimethylformamide at 100° yields the corresponding poly(o-carboranyl hydrazides), which are converted to polymers of the structure shown by heating at 110–120° (*183*).

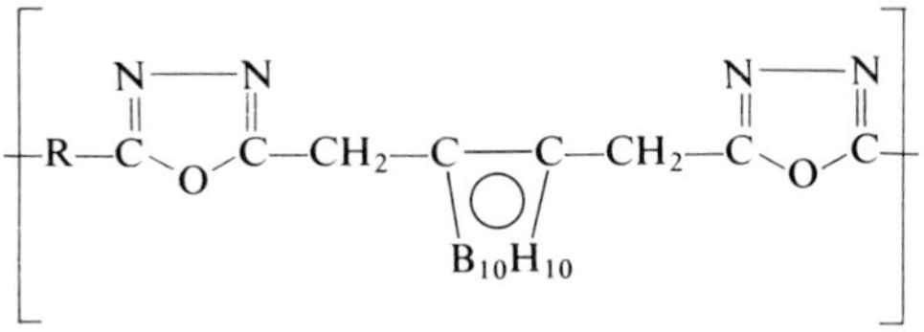

8-4. CLASS II POLYMERS

Poly(o-carboranylmethacrylates) are formed in the free radical-catalyzed polymerization of 1-methacryloyloxymethyl-o-carborane or its alkyl derivatives (*99, 102*).

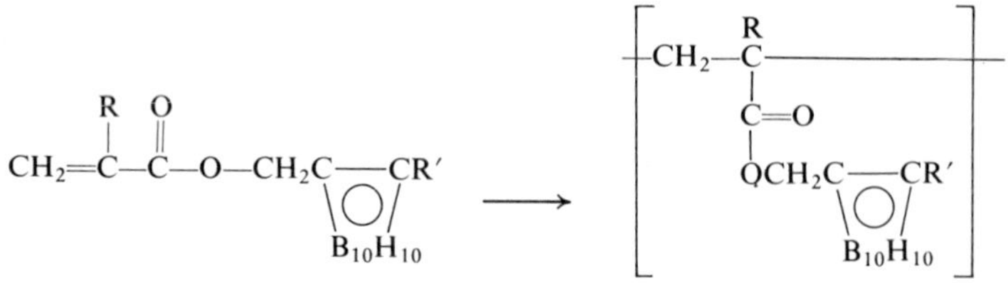

R = H, CH₃;   R' = H, CH₃

The compound in which R' = CH₃ is a brittle, colorless, linear polymer with no crosslinking. When R' = H, the product gives evidence of crosslinking involving the o-carboranyl C—H bond (*102*). The polymer in which R = R' = H is a hard brittle solid which softens to a rubber at 165°C; this unusually high softening point for an acrylate has been attributed to the large molar volume of the carborane icosahedra (*99*).

Poly(o-carboranylorganosiloxanes) are obtained by hydrolysis-condensation of o-carboranylchlorosilanes, or by their copolymerization with chlorosilanes or alkoxysilanes (*99, 208, 272*). The homopolymers are of low molecular weight and are accompanied by large quantities of cyclic products, but copolymers of molecular weights up to 14,000 have been prepared. These materials are tough, flexible solids which are stable in nitrogen or air to 400–450° (*96, 99*).

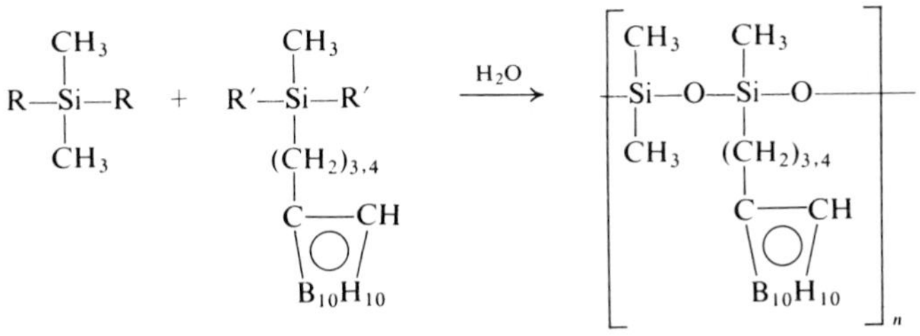

$$
\begin{array}{ccc}
\underset{\underset{CH_3}{|}}{\overset{\overset{CH_3}{|}}{R-Si-R}} & + & \underset{\underset{\underset{\underset{\underset{B_{10}H_{10}}{}}{C-CH}}{(CH_2)_{3,4}}}{|}}{\overset{\overset{CH_3}{|}}{R'-Si-R'}}
\end{array}
\xrightarrow{H_2O}
\left[\begin{array}{c}
\underset{\underset{CH_3}{|}}{\overset{\overset{CH_3}{|}}{-Si-O-}}\overset{\overset{CH_3}{|}}{\underset{\underset{\underset{\underset{B_{10}H_{10}}{}}{C-CH}}{(CH_2)_{3,4}}}{Si-O-}}
\end{array}\right]_n
$$

$$R = Cl, OC_2H_5;$$
$$R' = Cl, OC_2H_5$$

Polymers of closely related structure, prepared from alkyl- or aryl(*o*-carboranylisopropyl)chlorosilanes, are viscous liquids with freezing points as low as $-108°$ and stable to at least $350°$ (*272*). Another route to *o*-carboranyl-substituted linear polysiloxanes is the ring-opening polymerization of cyclic *o*-carboranylsiloxanes, which in turn may be prepared by the platinum-catalyzed addition of silanes to alkenylcarboranes (Section 6-10) (*208*).

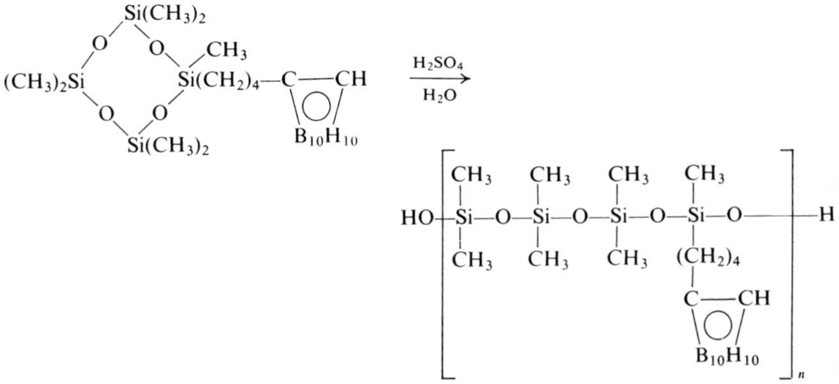

The reaction of poly(alkylhydrosiloxanes) with 1-isopropyl-*o*-carborane at 200–300° yields clear liquid polymers with the structure shown; the ethylated species are thermally more stable than the original poly(ethylhydrosiloxane) (*284*).

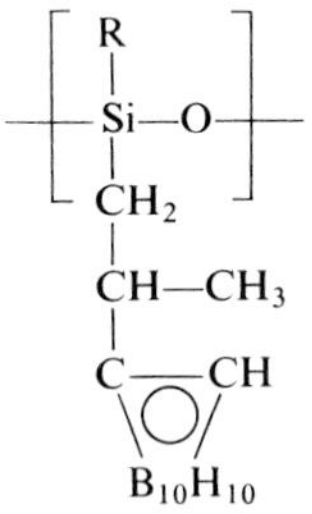

$$
\left[\begin{array}{c}
\underset{\underset{\underset{\underset{\underset{\underset{B_{10}H_{10}}{}}{C-CH}}{CH-CH_3}}{CH_2}}{|}}{\overset{\overset{R}{|}}{-Si-O-}}
\end{array}\right]
$$

$$R = CH_3, C_2H_5, C_6H_5$$

A study of the thermal behavior of copolymers of the above type has shown that thermooxidative cleavage of the *o*-carboranyl pendant groups occurs at lower temperatures than are required for breaking the Si—O and Si—C bonds in the chain (*43*). It has been suggested (*43*) that the carborane cage units do inhibit thermal decomposition, but that the effect is primarily steric rather than inductive in nature. Other workers, however, have attributed the stabilization to an *o*-carboranyl inductive effect (*96*).

Degradation of the Icosahedral Cage. Heteroatom Carboranes and Transition Metal π-Complexes

9-1. DEGRADATION OF o- AND m-CARBORANE

$C_2B_9H_{12}^-$ Ions

All three icosahedral carboranes are highly resistant to thermal and chemical attack, a fact which has permitted the development of a large derivative chemistry in which the cage system remains intact. Nevertheless, Wiesboeck and Hawthorne (*348*) discovered that o-carborane can be selectively degraded by strong bases such as methoxide ion, effectively removing one boron atom from the cage and forming the 1,2-dicarbaundecaborate(1−) ion. More recent work has established that m-carborane undergoes analogous reactions, yielding salts of 1,7-dicarbaundecaborate(1−) (*84, 136, 138, 392*), but p-carborane appears unreactive toward basic reagents (*250*).

$$1,2\text{-} \quad \text{or} \quad 1,7\text{-}C_2B_{10}H_{12} + CH_3O^- + 2CH_3OH \xrightarrow[40°]{\text{KOH, } CH_3OH}$$

$$1,2\text{-} \quad \text{or} \quad 1,7\text{-}C_2B_9H_{12}^- + B(OCH_3)_3 + H_2$$

The 1,2- or $1,7\text{-}C_2B_9H_{12}^-$ species (or their substituted derivatives) have also been obtained in reactions of the o- or m-carborane cage systems with tri-alkylamines (*356, 381*), hydrazine (*91, 301, 302, 382, 401*), ammonia (*374, 381*), and piperidine (*136, 386, 400*). The synthesis of $1,2\text{-}(CH_3)_2\text{-}1,2\text{-}C_2B_9H_{10}^-$ has been published in full detail (*126*). In addition, the $1,7\text{-}C_2B_9H_{12}^-$ ion has been prepared from the *closo*-carborane $1,8\text{-}C_2B_9H_{11}$ (Section 5-2).

As expected from charge distribution studies, the attack of bases on the icosahedral carboranes occurs at the most electronegative boron atoms, which on the *ortho* isomer are atoms 3 and 6 and in the *meta* isomer are atoms 2 and 3 (see Figs. 2-1l and 2-1m). Removal of B(3) or B(6) from *o*-carborane generates an icosahedral fragment with adjacent carbons (1 and 2) on the open face (*138, 348*); abstraction of B(2) or B(3) from *m*-carborane produces an icosahedral fragment with the carbons in nonadjacent positions on the open face (Fig. 9-1). The resulting ions are conveniently designated $(3)\text{-}1,2\text{-}C_2B_9H_{12}^-$ and $(3)\text{-}1,7\text{-}C_2B_9H_{12}^-$, respectively (*137*), thus retaining the numbering system of the original icosahedral species and indicating in parentheses the boron atom removed. While the exact location of the "extra" hydrogen atom in both $C_2B_9H_{12}^-$ isomers is uncertain, it is assumed from ^{11}B NMR spectra (*138*) to be bonded in some manner to the open face of the ion. A six-center hydrogen

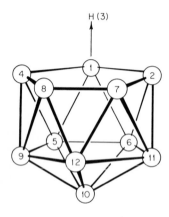

FIG. 9-1. Structure of the (3)-1,2- and $(3)\text{-}1,7\text{-}C_2B_9H_{12}^-$ ions, showing numbering of cage atoms (*138*).

bridge bond has been proposed (*138*), in which the proton occupies the position of the "missing" boron atom.

The mechanism of *o*- and *m*-carborane base degradation has not been established [although Hawthorne (*138*) has proposed a five-step sequence involving *o*-carborane and ethoxide ion], and it undoubtedly varies with the choice of basic reagent (*138, 381*). The highly selective nature of the attack at B(3) and B(6) in *o*-carborane is convincingly shown by the fact that treatment of *o*-carborane derivatives with refluxing piperidine produces the corresponding $(3)\text{-}1,2\text{-}C_2B_9H_{12}^-$ derivative *except* for 3,6-disubstituted *o*-carboranes, which are unreactive under these conditions (*135*). The piperidine reactions, of course, yield piperidinium salts of $C_2B_9H_{12}^-$, but early reports (*386, 400*) that a B—N bonded $C_2B_9H_{12}^-$· piperidine adduct is formed now appear incorrect (*136*). A second piperidine molecule is indeed present in the crystalline product, but it is weakly held [perhaps by hydrogen bonds (*136*)] and may be removed by vacuum distillation (*386, 400*).

$$\underset{B_{10}H_{10}}{\overset{HC\text{---}CH}{\diagdown O \diagup}} + 4C_5H_{10}NH \xrightarrow[C_6H_6]{20°} [C_2B_9H_{12}^-C_5H_{10}NH_2^+]\cdot C_5H_{10}NH + (C_5H_{10}N)_2BH$$

$$C_5H_{10}NH + C_5H_{10}NH_2^+C_2B_9H_{12}^- \xleftarrow{\Delta}$$

Dicarbaundecaborate(1−) adducts containing B—N, B—O, or B—S dative bonds are reportedly formed in the reaction of trimethylamine with C-bromomethyl-*o*-carboranes (*356*) and in "oxidative-substitution" reactions of $C_2B_9H_{12}^-$ salts with Lewis bases in the presence of ferric chloride (*357*). Similar reactions have been reported involving pyridine and quinoline (*25*).

$$\underset{B_{10}H_9Br}{\overset{H_3CC\text{---}CCH_2Br}{\diagdown O \diagup}} + N(CH_3)_3 + 3H_2O \longrightarrow$$
$$[(CH_3)_2C_2B_9H_8Br]\cdot N(CH_3)_3 + B(OH)_3 + H_2 + HBr$$

$$(3)\text{-}1,2\text{-}C_2B_9H_{12}^- + 2FeCl_3 + L \rightarrow (3)\text{-}1,2\text{-}C_2B_9H_{11}L \quad (2 \text{ isomers}) + 2FeCl_2 + HCl + Cl^-$$

The $C_2B_9H_{11}L$ species are isoelectronic with $C_2B_9H_{12}^-$ and, like the latter ion, contain one nonterminal hydrogen atom; the base ligand is probably bonded to a boron atom on the open face (*357*).

In general, the *o*- and *m*-carborane cage systems are stable toward alcohols (*66*), but 1,2-dihalo-*o*-carboranes are degraded by both methanol and ethanol (*448*), yielding C,C'-dihalo derivatives of dicarbaundecaborane(13) (discussed below). Treatment of 1,2-bis(halomethyl)-*o*-carboranes with aqueous am-

$$\underset{B_{10}H_{10}}{\overset{XC\text{---}CX}{\diagdown O \diagup}} \quad 3CH_3OH \xrightarrow{20°} C_2X_2B_9H_{11} + (CH_3O)_3B + H_2$$
$$X = Cl, Br, I$$

monia leads to bis(aminomethyl)dicarbaundecaborate inner salts (*374*).

$$\underset{B_{10}H_{10}}{\overset{XCH_2C\text{---}CCH_2X}{\diagdown O \diagup}} \xrightarrow[H_2O]{NH_3} NH_2CH_2C_2B_9H_{10}CH_2^-NH_3^+$$
$$X = Cl, Br$$

The (3)-1,2- and (3)-1,7-$C_2B_9H_{12}^-$ ions are colorless, air-stable, and resistant to attack by aqueous bases or nonoxidizing acids; the potassium salts, at least, show high thermal stability (*138*). However, the (3)-1,2-$C_2B_9H_{12}^-$ ion quantitatively rearranges (*84*) to the (3)-1,7 isomer at 300°, the product being identical to that produced directly from *m*-carborane by degradation with basic reagents. The cage frameworks of both ions are destroyed by certain oxidizing agents such as bromine in acid media (301) and aqueous ferric chloride (357). Iodine effects an apparent electrophilic substitution on $1,2\text{-}C_2B_9H_{12}^-$ in aqueous ethanol, forming a $C_2B_9H_{11}I^-$ in which the iodine may be attached to a boron on the open face adjacent to carbon (218). Mono- and diiodo derivatives of $C_2B_9H_{12}^-$ have also been prepared by the reaction of B-iodo-o-carboranes with piperidine (309). A claim (309) that the iodine in these ions undergoes rapid exchange with $Na^{131}I$ has been disputed (395) by others who failed to reproduce the earlier results.

$C_2B_9H_{13}$

The (3)-1,2- and $(3)\text{-}1,7\text{-}C_2B_9H_{12}^-$ ions can be reversibly protonated at room temperature to yield, respectively, (3)-1,2- and $(3)\text{-}1,7\text{-}C_2B_9H_{13}$ (dicarba-undecaborane(13)) in parent or derivative form (348).

$$C_2B_9H_{12}^- + H_3O^+ \rightleftharpoons C_2B_9H_{13} + H_2O$$

The 1,7 isomer is highly unstable and has not been characterized, but (3)-$1,2\text{-}C_2B_9H_{13}$ is a volatile, hygroscopic compound which titrates as a strong monoprotic acid in aqueous solution. Spectroscopic and chemical evidence (348) suggests that the $C_2B_9H_{13}$ isomers are icosahedral fragments with two bridge protons on the open face (Fig. 2-2g), analogous to the structure of the isoelectronic $B_{11}H_{13}^{2-}$ ion as established by an X-ray study (80) [the parent $B_{11}H_{15}$ hydride has been isolated as a bisdioxanate adduct (63)].

The $C_2B_9H_{13}$ species are thermally unstable and lose hydrogen at 75–100° to form the same closed polyhedral carborane, $1,8\text{-}C_2B_9H_{11}$ (Section 5-1) (84, 327, 329). The high-temperature protonation of either $C_2B_9H_{12}^-$ ion yields $1,8\text{-}C_2B_9H_{11}$ directly (14, 84). The following diagram summarizes these interconversions (C-phenyl derivatives of (3)-1,2- and $(3)\text{-}1,7\text{-}C_2B_9H_{12}^-$ give analogous reactions):

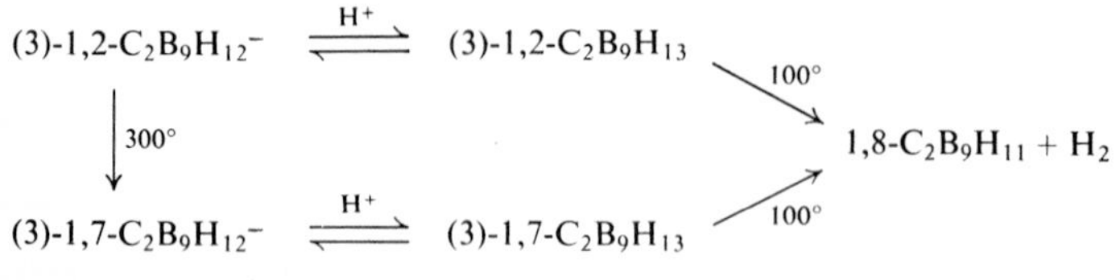

Since the carbon atoms in $1,8\text{-}C_2B_9H_{11}$ are nonadjacent (Fig. 2-1k) (336), the formation of this compound from $(3)\text{-}1,2\text{-}C_2B_9H_{13}$ clearly involves a skeletal rearrangement. A related observation is that both isomeric $C_2B_9H_{12}^-$

ions undergo oxidative degradation of the cage structure by chromic acid at 100° to give the same organic products, strongly suggesting *ortho–meta* rearrangements in the reaction intermediates (*392, 413*).

$C_2B_9H_{11}^{2-}$ (Dicarbollide) Ions

The "extra," or nonterminal, hydrogen atom in (3)-1,2- and (3)-1,7-$C_2B_9H_{12}^-$ is easily removed by reaction with sodium metal (*139*), sodium hydride (*139, 434*), or butyllithium (*250*) in ethereal solvents, forming the respective isomeric $C_2B_9H_{11}^{2-}$ ions (Fig. 9-2).

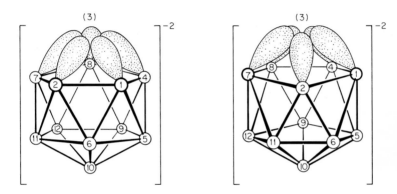

FIG. 9-2. Structures of the (3)-1,2- and (3)-1,7-$C_2B_9H_{11}^{2-}$ ions, showing approximately sp^3 orbitals directed at the vacant (3) vertex of the icosahedron (*137*).

$$C_2B_9H_{12}^- + NaH \rightarrow Na^+C_2B_9H_{11}^{2-} + H_2$$

The same species are produced in low concentration in hot aqueous NaOH solution (*137, 345*); the reversible equilibrium shown has been confirmed by

$$C_2B_9H_{12}^- + OH^- \rightleftharpoons C_2B_9H_{11}^{2-} + H_2O$$

treatment of (3)-1,2-$C_2B_9H_{11}^{2-}$ with water to form (3)-1,2-$C_2B_9H_{12}^-$ and hydroxide ion.

In analogous fashion, the reaction of the $C_2B_9H_{11}L$ compounds (discussed above) with hydride ion removes the "extra" hydrogen atom to yield Lewis base-substituted dicarbollide derivatives (*357*).

$$(3)\text{-}1,2\text{-}C_2B_9H_{11}L + NaH \rightarrow Na^+(3)\text{-}1,2\text{-}C_2B_9H_{10}L^- + H_2$$

9-2. DICARBOLLIDE–BORON INSERTION REACTIONS

The $C_2B_9H_{11}^{2-}$ ions are icosahedra less one boron atom, and it is not surprising that these species react with monoboron reagents such as alkyl- or arylboron dihalides to restore the complete o-carborane icosahedron (*134, 135*). An interesting reaction sequence has been demonstrated by Hawthorne and

$$(3)\text{-}1,2\text{-}C_2B_9H_{11}{}^{2-} + RBCl_2 \xrightarrow[-40°]{\text{THF}} \overset{HC\text{---}CH}{\underset{3\text{-}RB_{10}H_9}{\diagdown O \diagup}} + 2Cl^-$$

$$R = C_6H_5, C_2H_5$$

Wegner (*135*), who prepared 3-substituted o-carborane derivatives from the (3)-1,2-dicarbollide ion as shown, treated the product with ethoxide ion to form the (6)-1,2-$C_2B_9H_{11}R^-$ species (substituted at B(3)), converted the latter ion to the substituted dicarbollide species, and finally carried out a second insertion to yield 3,6-disubstituted o-carboranes.

$$C_2B_9H_{11}{}^{2-} \xrightarrow{RBCl_2} \overset{HC\text{---}CH}{\underset{3\text{-}RB_{10}H_9}{\diagdown O \diagup}} \xrightarrow{OC_2H_5{}^-} C_2B_9H_{11}R^- \xrightarrow{H^-} C_2B_9H_{10}R^{2-}$$

$$\overset{HC\text{---}CH}{\underset{3,6\text{-}RR'B_{10}H_8}{\diagdown O \diagup}} \xleftarrow{R'BCl_2}$$

$$R = C_6H_5, C_2H_5;$$
$$R' = C_6H_5, C_2H_5$$

Attempts to repeat the cycle by degradation of the 3,6-disubstituted o-carboranes have failed (*135*), indicating that only borons 3 and 6 are susceptible to removal by base attack.

The utility of boron-insertion reactions in preparing specific B-substituted o- and m-carborane derivatives from the (3)-1,2- and (3)-1,7-dicarbollide ions is illustrated by the synthesis of 3- and 6-substituted fluoro- (*250*), bromo- (*250*), diethylamino- (*240*), diphenylamino- (*250*), ethoxy- (*240*), n-propyl- (*209*) and n-butyl- (*209*) -o-carboranes, and 3-substituted fluoro- and diphenyl-amino-m-carboranes (*250*), the attacking reagent in each case being the appropriate RBX_2 species. Thus, reaction of the 1-phenyl-(3)-1,2-dicarbollide ion with p-tolyldichloroborane yields optically active 1-phenyl-3-p-tolyl-o-carborane ($[\alpha]_D^{20} + 18.8°$ [C_2H_5OH]) (*434*). Carrying the basic idea a step further, Russian workers have prepared 1,4-bis(3-o-carboranyl)butane (*240*).

$$2\ (3)\text{-}1,2\text{-}C_2B_9H_{11}{}^{2-} + Cl_2B(CH_2)_4BCl_2 \longrightarrow$$

$$\underset{B_{10}H_9}{\overset{HC\text{----}CH}{\diagdown O \diagup}}\text{----}(CH_2)_4\text{----}\underset{B_{10}H_9}{\overset{HC\text{----}CH}{\diagdown O \diagup}}$$

Closely related reactions involving the insertion of heteroatoms are discussed in Section 9-4.

9-3. THE MONOCARBON CARBORANE ANIONS: $CB_{10}H_{13}^-$, $CB_{10}H_{11}^-$, $CB_9H_{10}^-$, $CB_{11}H_{12}^-$

Formal replacement of C by B^- in $nido\text{-}C_2B_9H_{13}$ and $closo\text{-}C_2B_9H_{11}$ produces $CB_{10}H_{13}^-$ and $CB_{10}H_{11}^-$, respectively. The actual synthesis of these monocarbon species and their derivatives, however, does not involve o- or m-carborane at all, instead utilizing decaborane(14) as the starting material.

$CB_{10}H_{13}^-$

Todd *et al.* (*161, 178*), have found that the reaction of decaborane(14) with alkyl isocyanides gives C-amino derivatives of $CB_{10}H_{13}^-$. Treatment of these products with sodium hydride followed by dimethyl sulfate yields the corresponding alkyldimethylamino derivative.

$$B_{10}H_{14} + RN{\equiv}C\colon \longrightarrow RH_2NCB_{10}H_{12} \xrightarrow[\text{2. } (CH_3)_2SO_4]{\textit{1. } H^-} R(CH_3)_2NCB_{10}H_{12}$$

$$R = CH_3,\ C_2H_5,\ n\text{-}C_3H_7,\ t\text{-}C_4H_9$$

Knoth (*175, 178*) has obtained the identical trimethylamino derivative by a different route in which $Na_2B_{10}H_{13}CN$ is protonated on an ion exchange column to give $H_3NCB_{10}H_{12}$, which on treatment with dimethyl sulfate in basic solution yields $(CH_3)_3NCB_{10}H_{12}$. Recently it has been found (*257*) that $H_3NCB_{10}H_{12}$ is also produced by the action of trimethylsilyl- or trimethyltin chloride on the $B_{10}H_{13}CN^{2-}$ ion (the silicon and tin derivatives initially formed are hydrolyzed under basic conditions). Still another route to monocarbon carboranes involves the reaction of $B_{10}H_{13}CN^{2-}$ salts with alkyl iodides, the products in this case being trialkylamine derivatives [the ethyl iodide reaction, however, produced $(C_2H_5)_2HNCB_{10}H_{12}$ (*257*)].

$$B_{10}H_{13}CN^{2-} + (CH_3)_3MCl \xrightarrow{H_2O} H_3NCB_{10}H_{12}$$

$$M = Si, Sn$$

$$CsB_{10}H_{13}CN + 3CH_3I \xrightarrow[C_4H_8O]{H_2O} (CH_3)_3NCB_{10}H_{12} + 2CsI + C_4H_8OHI$$

The amine derivatives may be deaminated by sodium hydride or sodium metal to yield the $CB_{10}H_{13}^-$ ion itself (*162, 175*). Alternatively, treatment of $(CH_3)_3NCB_{10}H_{12}$ with sodium in refluxing tetrahydrofuran gives the adduct $Na_3CB_{10}H_{11}(C_4H_8O)_2$, which on hydrolysis produces $CB_{10}H_{13}^-$ (*162*).

The fact that $CB_{10}H_{13}^-$ is isoelectronic with $C_2B_9H_{13}$ and $B_{11}H_{13}^{2-}$ suggests a possible isostructural relationship. This is supported by NMR and other spectroscopic data for $CB_{10}H_{13}^-$ and its amine derivatives, which are consistent with an icosahedral fragment having two bridge hydrogen atoms on the open face (*161, 162*).

The reaction of $CB_{10}H_{13}^-$ with Br_2 yields B-substituted mono- and dibromo derivatives (the position of substitution has not been established), but Cl_2 at $0°$ decomposes the ion (*162*). A B-dichloro derivative has been obtained, however, from the ultraviolet-catalyzed treatment of $CB_{10}H_{13}^-$ with N-chlorosuccinimide. Compared to the $CB_{10}H_{13}^-$ ion, the $R_3NCB_{10}H_{12}$ compounds are less reactive toward halogens, requiring $AlCl_3$ or ultraviolet light for monobromination; direct chlorination with Cl_2 occurs without the need for a catalyst. Neither $CB_{10}H_{13}^-$ nor its trialkylamine derivatives react with iodine, even under catalytic conditions. B-bromo-C-trialkylamine derivatives of $CB_{10}H_{13}^-$ which are specifically halogenated at B(11)* have been prepared (*162*) from 2-bromodecaborane(14) by reaction with methyl isocyanide and subsequent N-alkylation as described above; the $(CH_3)_3NCB_{10}H_{11}Br$ obtained in this manner is identical with that formed in the monobromination of $(CH_3)_3NCB_{10}H_{12}$.

$CB_{10}H_{11}^-$

Knoth (*175*) has obtained both $NaCB_{10}H_{13}$ (the major product) and small quantities of $NaCB_{10}H_{11}$ from the reaction of sodium with $(CH_3)_3NCB_{10}H_{12}$ in refluxing tetrahydrofuran. High yields of the $CB_{10}H_{11}^-$ ion have been ob-

---

* Numbering based on placement of the carbon atom at position 1 and the icosahedral "hole" at position 3, consistent with that used for the isostructural $C_2B_9H_{12}^-$ and related species (*329*). In the original paper (*162*) the numbering system is based on $B_{10}H_{14}$, placing the bromine atom on B(6).

tained by oxidation of the previously mentioned THF adduct with iodine (*162*). The corresponding trimethylamine derivative, $(CH_3)_3NCB_{10}H_{10}$, is

$$Na_3CB_{10}H_{11}(C_4H_8O)_2 + I_2 \rightarrow NaCB_{10}H_{11} + 2NaI + 2C_4H_8O$$

formed by deprotonation of $(CH_3)_3NCB_{10}H_{12}$ with sodium hydride and subsequent oxidation with iodine (*162*). The ^{11}B NMR spectra of $CB_{10}H_{11}^-$ and $(CH_3)_3NCB_{10}H_{12}$ are nearly identical and are consistent with a closed polyhedral structure similar to that which has been established (*336*) for the isoelectronic *closo*-carborane $C_2B_9H_{11}$ (Fig. 2-1k).

$CB_9H_{10}^-$ and $CB_{11}H_{12}^-$

The synthesis of monocarbon carborane anions isoelectronic with the *closo*-carboranes $C_2B_8H_{10}$ and $C_2B_{10}H_{12}$ has been reported by Knoth (*175*), who has obtained these species by several routes outlined below.

$$2Cs^+CB_{10}H_{13}^- \xrightarrow{300-320°} Cs^+CB_9H_{10}^- + Cs^+CB_{11}H_{12}^- + 2H_2$$

$$\xrightarrow[180°]{(C_2H_5)_3NBH_3} Cs^+CB_{11}H_{12}^-$$

$$NaB_{10}H_{12}CN \cdot S(CH_3)_2 \xrightarrow[\text{ion exchange}]{H_3O^+} B_{10}H_{11}(OH)CNH_3 \xrightarrow[NaOH]{(CH_3)_2SO_4}$$

$$B_{10}H_{11}(OH)CN(CH_3)_3 \xrightarrow[H_2O, THF]{NaOH}$$

$$B_9H_{11}CN(CH_3)_3 \xrightarrow[\text{2. } (CH_3)_4N^+, H_2O]{\text{1. Na, THF; reflux}} (CH_3)_4N^+CB_9H_{10}^- + H_2$$

The ^{11}B NMR spectra (*175*) of $CB_9H_{10}^-$ and $CB_{11}H_{12}^-$ support, but do not prove, Lipscomb's prediction (*195*) that the former is isostructural with $C_2B_8H_{10}$ and $B_{10}H_{10}^{2-}$ (bicapped square antiprism), whereas the latter ion is isostructural with $C_2B_{10}H_{12}$ and $B_{12}H_{12}^{2-}$ (icosahedral). In $CB_9H_{10}^-$, the carbon appears to occupy an apex position. Both carborane anions react with *n*-butyllithium, producing 1-$LiCB_9H_9^-$ and $LiCB_{11}H_{11}^-$, respectively. The reaction of 1-$LiCB_9H_9^-$ with $(CH_3)_3SiCl$ forms $(CH_3)_3SiCB_9H_9^-$, suggesting that a derivative chemistry similar to that of the neutral carboranes may exist for the polyhedral carborane anions (*175*).

9-4. CARBORANES CONTAINING MAIN-GROUP CAGE HETEROATOMS

The discovery of polyhedral boranes containing atoms other than boron and carbon in the cage originally involved only transition metals (see following section), but has recently been extended to include several nontransition elements as well. Future books on the carboranes will no doubt require entire chapters on the heteroatom carborane cage systems, so great is the potential in this area, but for the present we choose to discuss these compounds in the context of the carborane anions and *nido*-carboranes from which most of them have been prepared. Essentially, the following syntheses involve insertion of heteroatoms into open-cage carborane species such as $C_2B_9H_{13}$, $C_2B_9H_{11}^{2-}$, or their monocarbon analogs.

Group II Heteroatoms

$C_2BeB_9H_{12}^-$

Dimethylberyllium attacks (3)-$1,2$-$C_2B_9H_{13}$ in ether to form a polyhedral cage structure containing beryllium (presumably in the originally vacant (3) position) coordinated to an ether molecule (*239*).

$$C_2B_9H_{13} + Be(CH_3)_2 \cdot [O(C_2H_5)_2]_2 \rightarrow C_2BeB_9H_{11} \cdot O(C_2H_5)_2 + (C_2H_5)_2O + 2CH_4$$

The diethyl ether complex is extremely air sensitive, but replacement of the ether with trimethylamine yields the much more stable $C_2BeB_9H_{11} \cdot N(CH_3)_3$ (Fig. 9-3). Both complexes may be regarded as base-substituted derivatives of the $C_2BeB_9H_{12}^-$ ion (as yet uncharacterized) which is isoelectronic with $CB_{11}H_{12}^-$ and $C_2B_{10}H_{12}$. Treatment of the trimethylamine complex with ethanolic KOH produces the (3)-$1,2$-$C_2B_9H_{12}^-$ ion in high yield (*239*).

Group III Heteroatoms

$(C_2H_5)AlC_2B_9H_{11}$

A compound which may be 3-ethyl-1,2-dicarba-3-alana-*closo*-dodecaborane(12) has been prepared from the reaction of $1,2$-$C_2B_9H_{11}^{2-}$ ion with $C_2H_5AlCl_2$, but no structural characterization data have been presented

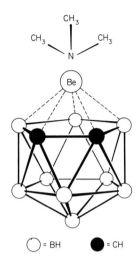

FIG. 9-3. Proposed structure of $BeC_2B_9H_{11}N(CH_3)_3$ (*239*).

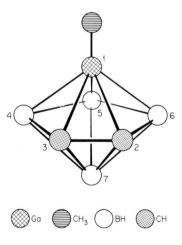

FIG. 9-4. Proposed structure of $CH_3GaC_2B_4H_6$ (*109*).

(*210*). The product crystallizes with two molecules of tetrahydrofuran and decomposes at 120°.

$CH_3GaC_2B_4H_6$

The reaction of $Ga(CH_3)_3$ with *nido*-2,3-$C_2B_4H_8$ (Chapter 3) in the gas phase at 180–215° produces in 20% yield a solid, volatile galladicarba-*closo*-heptaborane(7) species containing a methyl group on gallium (*109*). The NMR and other data virtually establish the closed polyhedral structure but do not distinguish between the adjacent-carbon arrangement, shown in Fig. 9-4, and an alternative structure in which the carbons are equatorial but nonadjacent, as in the isoelectronic 2,4-$C_2B_5H_7$ system. However, the adjacent-carbon arrangement is supported by the rather mild conditions of synthesis as well as by the fact that thermal decomposition of the compound at 285° yields *nido*-2,3-$C_2B_4H_8$ (*110*).

Group IV Heteroatoms

$CH_3GeCB_{10}H_{11}$ and $GeCB_{10}H_{11}^-$

The reaction of $Na_3CB_{10}H_{11}(C_4H_8O)_2$ (see above) with CH_3GeCl_3 in refluxing tetrahydrofuran yields a sublimable solid which has been characterized (*333*) as a *closo*-germacarborane in which the cage germanium atom contains an attached methyl group, analogous to 1-$CH_3C_2B_{10}H_{11}$. Refluxing piperidine apparently removes the methyl group but leaves intact the closed cage structure; subsequent treatment with methyl iodide produces the original methylated compound.

$$Na_3CB_{10}H_{11}(C_4H_8O)_2 \xrightarrow[\text{THF}]{CH_3GeCl_3} CH_3GeCB_{10}H_{11} \underset{CH_3I}{\overset{\text{piperidine}}{\rightleftharpoons}} GeCB_{10}H_{11}^-$$

Both germacarborane species rapidly decompose at 450°.

$C_2SnB_9H_{11}$

The reaction of $C_2B_9H_{11}^{2-}$ ion with $SnCl_2$ in refluxing benzene apparently completes the icosahedron by introducing a tin atom into the cage (*343*). The proposed structure (Fig. 9-5) contains no substituent on tin, and is a homolog of the unknown $C_3B_9H_{11}$ (formally generated from $C_2B_{10}H_{12}$ by replacement of a BH unit with a carbon atom). The lead and germanium

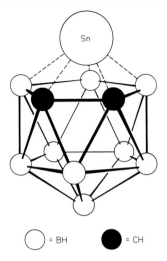

FIG. 9-5. Proposed structure of $SnC_2B_9H_{11}$ (*343*).

analogs of this compound have also been reported (*343*).

Group V Heteroatoms

$CPB_{10}H_{11}$ and $CPB_9H_{11}^-$

The insertion of phosphorus into an icosahedral carborane framework has been accomplished by the addition of phosphorus trichloride to $Na_3CB_{10}H_{11}(C_4H_8O)_2$ in heptane (*199, 334*). The proposed structure contains phosphorus in an *ortho* position relative to carbon, and hence is designated as $1,2\text{-}CPB_{10}H_{11}$. This compound rearranges at 485° to a second isomer (Fig. 9-6) which is probably $1,7\text{-}CPB_{10}H_{11}$, an analog of *m*-carborane [an isostruc-

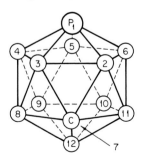

FIG. 9-6. Proposed structure of $1,7\text{-}CPB_{10}H_{11}$ showing numbering of cage atoms (*334*).

tural species, $C_6H_5PB_{11}H_{11}$, is formed in the reaction of phenyldichlorophosphine with the $B_{11}H_{13}^{2-}$ ion (199)].

As expected, 1,2- and 1,7-$CPB_{10}H_{11}$ are chemically similar to their $C_2B_{10}H_{12}$ analogs; thus, the 1,7 isomer may be metallated at carbon and subsequently converted to the C-methyl derivative, while the 1,2 isomer reacts with excess Br_2 over $AlCl_3$ to form 1,2-$CPB_{10}Br_3H_8$ (substituted at boron only) $(199, 334)$. The 1,7 isomer, however, can only be dibrominated (334), indicating a lower polarity than the 1,2 system.

Both isomers of $CPB_{10}H_{11}$ are degraded in refluxing piperidine forming, respectively, the (3)-1,2- and (3)-1,7-$CPB_9H_{11}^-$ ions which are analogs of the $C_2B_9H_{12}^-$ species described earlier, and presumably are open-faced icosahedral species, with the position of the nonterminal hydrogen atom not known. (The 1,7-$CPB_9H_{11}^-$ ion is also formed in the reaction of 1,7-$CPB_{10}H_{11}$ with sodium ethoxide in ethanol, but the latter reagent removes the phosphorus atom from 1,2-$CPB_{10}H_{11}$, giving the $CB_{10}H_{13}^-$ ion (Section 9-3) in low yield). Treatment of the 1,2- and 1,7-$CPB_9H_{11}^-$ ions with methyl iodide yields the p-methyl derivatives (334).

Both $CPB_9H_{11}^-$ ions react with sodium hydride to generate the respective $CPB_9H_{10}^{2-}$ (carbaphosphollide) ions (335). The (3)-1,2- and (3)-1,7-carbaphosphollide ions are isoelectronic and isostructural with the (3)-1,2- and (3)-1,7-$C_2B_9H_{11}^{2-}$ (dicarbollide) anions, with the carbon and phosphorus atoms located on the open face of an icosahedral fragment. This has been established from X-ray studies of transition metal-carbaphosphollyl π-complexes (335) of the type described in the following section. The $CPB_9H_{10}^{2-}$ ions apparently have not been characterized in uncomplexed form.

$CAsB_{10}H_{11}$ AND $CSbB_{10}H_{11}$

The tetrahydrofuran adduct of $Na_3CB_{10}H_{11}$ reacts with $AsCl_3$ and with SbI_3 to yield, respectively, *closo*-1,2-arsacarborane and *closo*-1,2-stibacarborane (333). At temperatures near 500° these molecules rearrange to the corresponding 1,7-*closo* species, although the stibacarborane is extensively decomposed. At 575° 1,7-*closo*-arsacarborane rearranges to 1,12-*closo*-arsacarborane, thus completing the analogy with the icosahedral $C_2B_{10}H_{12}$ system.

$$Na_3CB_{10}H_{11}(C_4H_8O)_2 \begin{array}{l} \xrightarrow[\text{THF, }25°]{AsCl_3} \text{1,2-CAsB}_{10}\text{H}_{11} \xrightarrow{505°} \text{1,7-CAsB}_{10}\text{H}_{11} \xrightarrow{575°} \text{1,12-CAsB}_{10}\text{H}_{11} \\ \quad\quad\quad\quad\quad (25\%) \\ \\ \xrightarrow[\text{THF, }0°]{SbI_3} \text{1,2-CSbB}_{10}\text{H}_{11} \xrightarrow{450°} \text{1,7-CSbB}_{10}\text{H}_{11} \\ \quad\quad\quad\quad\quad (30\%) \quad\quad\quad\quad\quad (20\%) \end{array}$$

Like the analogous *closo*-phosphacarboranes, 1,2- and 1,7-$CAsB_{10}H_{11}$ and 1,2-$CSbB_{10}H_{11}$ are degraded by refluxing piperidine to the respective $CAsB_9H_{11}^-$ and $CSbB_9H_{11}^-$ ions. Treatment with a very large excess of piperidine results in abstraction of the antimony atom from 1,2-$CSbB_{10}H_{11}$, forming the $CB_{10}H_{13}^-$ ion. The synthesis of transition metal complexes of the 1,2- and 1,7-$CAsB_9H_{10}^{2-}$ ions has been reported without details (*333*); these dinegative ions have not been described, but are formal analogs of the known $CPB_9H_{10}^{2-}$ and $C_2B_9H_{11}^{2-}$ species.

9-5. CARBORANE–TRANSITION METAL π-COMPLEXES

General Considerations

The 1,2- and 1,7-dicarbollide ions (Fig. 9-2) each contain an open pentagonal face incorporating two carbon and three boron atoms, with the carbons adjacent in the 1,2 isomer and nonadjacent in the 1,7 isomer. A simple theoretical description (*212*) of the bonding in these anions assumes the formation of three bonding and two antibonding molecular orbitals from the five atomic orbitals which are directed toward the empty apical position centered over the pentagonal face. Filling the bonding orbitals with the six available electrons creates a situation approximating that in cyclopentadienide ion, $C_5H_5^-$, despite the lower symmetry in the dicarbollide species. Recognizing this, Hawthorne and co-workers (*139*) examined the intriguing possibility that the dicarbollide ions might coordinate to transition metals containing partially filled *d*-shells to form sandwich complexes analogous to ferrocene and the other metallocenes. In a remarkable discovery which has enormously broadened the scope of boron chemistry, these workers prepared the ferrocene analogs $(C_2B_9H_{11})_2Fe^{2-}$ and $(C_2B_9H_{11})Fe(C_5H_5)$, followed in short order by a variety of similar complexes involving many transition metals (Table 9-1). The general "sandwich" structure, involving π-bonding between the carborane unit and the metal ion, has been established by X-ray studies on several key compounds as well as by ^{11}B and ^{1}H NMR spectra, and magnetic, electronic, and electrochemical data [the early work on these complexes has been reviewed by Hawthorne (*121*)].

The fusion of the already large areas of boron and transition metal chemistry has generated an entirely new field of research, whose considerable variety and versatility should be clear from the following summary.

TABLE 9-1

TRANSITION METAL–DICARBOLLYL π-COMPLEXES[a]

	Color	mp (°C)	Other data[b]	References
Chromium Complexes				
Cs[(C₂B₉H₁₁)₂Cr]	Violet		IR, E, MAG	251
(CH₃)₄N][(C₂B₉H₁₀(CH₃))₂Cr	Violet		IR, E	251
[(CH₃)₄N][(C₂B₉H₉(CH₃)₂)₂Cr	Blue-violet		IR, E	251
[(CH₃)₄N][(C₂B₉H₁₀(C₆H₅))₂Cr	Violet		IR, E	251
(C₅H₅)Cr(C₂B₉H₁₁)	Dark red	248–249	IR, E	251
(C₅H₅)Cr(C₂B₉H₁₀(CH₃))	Dark red	219–220	IR, E	251
(C₅H₅)Cr(C₂B₉H₉(CH₃)₂)	Dark red	261–262	IR, E	251
(C₅H₅)Cr(C₂B₉H₁₀(C₆H₅))	Dark red	208–209	IR, E	251
[(CH₃)₄N]₂(C₂B₉H₁₁)Cr(CO)₃	Yellow		IR, B	133
Cs[(1,7-C₂B₉H₁₁)₂Cr]	Brown		IR, E	251
(C₅H₅)Cr(1,7-C₂B₉H₁₁)	Dark red	217–218	IR, E	251
Molybdenum and Tungsten Complexes				
[(CH₃)₄N]₂(C₂B₉H₁₁)Mo(CO)₃	Gray		IR	137
[(CH₃)₄N]₂(C₂B₉H₉(CH₃)₂)Mo(CO)₃	Pale yellow		IR	137
[(CH₃)₄N][C₂B₉H₁₁)Mo(CO)₃CH₃	Tan		IR, H	137
[(CH₃)₄N][C₂B₉H₁₁)Mo(CO)₃H	Red		IR	137
[(CH₃)₄N]₂(C₂B₉H₁₁)W(CO)₃	Yellow		IR	137
[(CH₃)₄N][C₂B₉H₁₁)W(CO)₃CH₃	Pale green		IR, H	137
[(CH₃)₄N]₂(C₂B₉H₉(CH₃)₂)Mo₂(CO)₈	Yellow		IR	137

Compound	Color	M.p.	Methods	Ref.
$[(CH_3)_4N]_2(C_2B_9H_{11})W_2(CO)_8$	—		IR	137
$[(CH_3)_4N]_2(C_2B_9H_{11})W(CO)_3Mo(CO)_5$	—		IR	137
$[(CH_3)_4N]_2(C_2B_9H_{11})Mo(CO)_3W(CO)_5$	—		IR	137
Manganese and Rhenium Complexes				
$[(CH_3)_4N](C_2B_9H_{11})Mn(CO)_3$	Pale yellow		IR, H, E	137
$Cs[(C_2B_9H_{11})Mn(CO)_3]$	—		IR, E	137
$[(CH_3)_4N](C_2B_9H_{11})Re(CO)_3$	Pale yellow		IR, H, E	137
$Cs[(C_2B_9H_{11})Re(CO)_3]$	—		IR, E, X	137, 469
Iron(I) Complexes				
$[(CH_3)_4N]_2(C_2B_9H_{11})_2Fe_2(CO)_4$	Dark red		IR, B	133
Iron(II) Complexes				
$[(CH_3)_4N]_2(C_2B_9H_{11})_2Fe$	Lavender		H, E	137
$[(CH_3)_4N]_2(C_2B_9H_9(CH_3)_2)_2Fe$	Blue		H, E	137
$[(CH_3)_4N]_2(C_2B_9H_{10}(C_6H_5))_2Fe$	Orange	158	H, E	137
$[(CH_3)_4N][(C_5H_5)Fe(C_2B_9H_{11})]$			IR, H, E	137
Iron(III) Complexes				
$[(CH_3)_4N]_2(C_2B_9H_{11})_2Fe$	Red	>300	IR, B, E, MAG, PO MB, PR	137 140, 204
$Cs[(C_2B_9H_{11})_2Fe]$	Black			137
$[(CH_3)_4N]_2(C_2B_9H_9(CH_3)_2)_2Fe$	Red	247–249	IR, B, E, MAG, PO PR	137 204
$[(CH_3)_4N](C_2B_9H_{10}(C_6H_5))_2Fe$	Red	>300	IR, B, E, MAG, PO PR	137 204
$(C_5H_5)Fe(C_2B_9H_{11})$	Purple	181–182	IR, B, E, PO MB, PR, X	137 140, 204, 471

continued

TABLE 9-1—*Continued*

	Color	mp (°C)	Other data[b]	References
Cobalt Complexes				
Cs[(C₂B₉H₁₁)₂Co]	Yellow	>300	IR, H, E, PO	137
			NQR, X	115, 470
[(CH₃)₄N][(C₂B₉H₉(CH₃)₂)₂Co]	Red	273–275	IR, H, E, PO	137
[(CH₃)₄N][(C₂B₉H₁₀(C₆H₅))₂Co]	Red	290–293	IR, H, E, PO	137
[(CH₃)₄N][6-C₆H₅C₂B₉H₁₀)₂Co]	Yellow	275–277	H	135
[(CH₃)₄N][(C₂B₉H₈Br₃)₂Co]	Orange		IR, H, E, PO	137
			X	42
(C₅H₅)Co(C₂B₉H₁₁)	Yellow	246–248	IR, H, E, PO	137
[(CH₃)₄N][(C₂B₉H₁₁)Co(CO)₂]	Yellow		IR, B	133
(C₂B₉H₁₀)₂CoS₂CH	Yellow		H, X	39
Cs₂[(C₂B₉H₁₁)Co(C₂B₈H₁₀)Co(C₂B₉H₁₁)]	Red		IR, B, H, E, PO	77
Cs[(1,7-C₂B₉H₁₁)₂Co]	Tan	>350	IR, H, E, PO	137
Nickel(II) Complexes				
[(C₂H₅)₄N]₂(C₂B₉H₁₁)₂Ni]	Brown		IR, MAG	137
Nickel(III) Complexes				
[(CH₃)₄N][(C₂B₉H₁₁)₂Ni]	Yellow-brown	>300	IR, E, MAG	137
Rb[(C₂B₉H₁₁)₂Ni]	Greenish black			137
(C₅H₅)Ni(C₂B₉H₁₁)	Greenish brown		E, MAG, PO	352
[(CH₃)₄N][(1,7-C₂B₉H₁₁)₂Ni]	Olive green	>300	IR, E, MAG	137

Nickel(IV) Complexes				
$(C_2B_9H_{11})_2Ni$	Orange	dec. 265	IR, E, PO	137
$(1,7\text{-}C_2B_9H_{11})_2Ni$	Orange		IR, E, PO	137
Palladium Complexes				
$[(C_2H_5)_4N]_2(C_2B_9H_{11})_2Pd$	Rust		IR, B, E	346
$[(C_2H_5)_4N](C_2B_9H_{11})_2Pd$	Brown		IR, E	346
$[(C_6H_5)_4C_4]Pd(C_2B_9H_{11})$	Red	dec. 310	IR, E	137, 347
$[(C_6H_5)_4C_4]Pd(C_2B_9H_9(CH_3)_2)$	Red		IR, E	137, 347
$(C_2B_9H_{11})_2Pd$	Yellow		IR, E	346
Copper Complexes				
$[(C_2H_5)_4N]_2(C_2B_9H_{11})_2Cu$	Deep blue		IR, E, PO	137, 346
			X	353
$[(CH_3)_4N](C_2B_9H_{11})_2Cu$	—		E	346
$[(C_6H_5)_3PCH_3](C_2B_9H_{11})_2Cu$	Deep red		IR, B, PO	346
			X	354
Gold Complexes				
$[(C_2H_5)_4N]_2(C_2B_9H_{11})_2Au$	Blue		IR, E, PO	346
$[(C_2H_5)_4N](C_2B_9H_{11})_2Au$	Red		IR, B, E	346
$[(C_6H_5)_3PCH_3](C_2B_9H_{11})_2Au$	Red		PO	346

[a] Complexes of the $1,2\text{-}C_2B_9H_{11}^{2-}$ ion or its derivatives, unless otherwise indicated.

[b] IR = infrared spectrum or band positions; B = ^{11}B NMR data; H = proton NMR data; E = electronic spectral data; MAG = magnetic susceptibility data; PR = paramagnetic resonance; PO = polarographic data; MB = Mössbauer data; NQR = nuclear quadropole resonance; X = X-ray crystallographic data.

Dicarbollyl Complexes of Cr, Mo, and W

Reaction of the 1,2- and 1,7-$C_2B_9H_{11}^{2-}$ ions with $CrCl_3$ in nonaqueous solvents under nitrogen yields salts of the bis(π-dicarbollyl)chromium(III)ion (Table 9-1), all of which are stable below 280° and paramagnetic with three unpaired electrons (*251*).

$$2C_2B_9H_{11}{}^{2-} + CrCl_3 \rightarrow (C_2B_9H_{11})_2Cr^- + 3Cl^-$$

An X-ray study (*251*, footnote 4) of the $[1,2\text{-}C_2B_9H_9(CH_3)_2]_2Cr^-$ ion has confirmed the sandwich structure shown in Fig. 9-7. In contrast to the easily

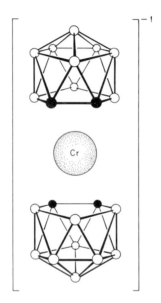

FIG. 9-7. Structure of the $[(3)\text{-}1,2\text{-}(CH_3)_2C_2B_9H_9]_2Cr^-$ ion (*251*). Solid circles represent C—CH$_3$; open circles, B—H.

hydrolyzed chromicinium salts, the bis(π-carbollyl) chromium compounds are stable to air and hot concentrated sulfuric acid (*251*). Attempts to reduce the Cr(III) species to Cr(II) complexes with sodium amalgam have been unsuccessful.

The reaction of cyclopentadienide ion with dicarbollide ion and $CrCl_3$ produces mixed cyclopentadienyldicarbollyl chromium derivatives (*251*) (Table 9-1).

$$C_5H_5{}^- + C_2B_9H_{11}{}^{2-} + CrCl_3 \rightarrow (\pi\text{-}C_5H_5)Cr(C_2B_9H_{11}) + 3Cl^-$$

The hexacarbonyls of chromium, molybdenum, and tungsten react with the

1,2- and 1,7-$C_2B_9H_{11}^{2-}$ ions under ultraviolet light, forming air-sensitive dicarbollyl metal tricarbonyl anions (*133, 137*).

$$C_2B_9H_{11}{}^{2-} + M(CO)_6 \xrightarrow{h\nu} (C_2B_9H_{11})M(CO)_3{}^{2-} + 3CO$$

$$M = Cr, Mo, W$$

The molybdenum and tungsten complexes may be protonated and alkylated in the same manner as the analogous $(\pi\text{-}C_5H_5)M(CO)_3{}^-$ species (*137*).

$$(C_2B_9H_{11})Mo(CO)_3{}^{2-} \xrightarrow{\text{anhydrous HCl}} (C_2B_9H_{11})Mo(CO)_3H^-$$

$$\xrightarrow{CH_3I} (C_2B_9H_{11})Mo(CO)_3CH_3{}^-$$

Reactions of the 1,2-$C_2B_9H_{11}^{2-}$ ion with two moles of $W(CO)_6$ or $Mo(CO_6)$ yield binuclear complexes containing metal–metal bonds; the same products are obtained when the $(C_2B_9H_{11})M(CO)_3^{2-}$ ions are treated with excess $W(CO)_6$ or $Mo(CO)_6$. The latter route may be used to prepare mixed binuclear complexes (*137*). An X-ray investigation (*137*, footnote 34) of one such compound, $(1,2\text{-}C_2B_9H_{11})Mo(CO)_3W(CO)_3^{2-}$, has given the structure shown in Fig. 9-8.

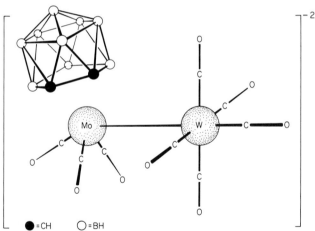

●=CH   ○=BH

FIG. 9-8. Structure of the $[\pi\text{-}(3)\text{-}1,2\text{-}C_2B_9H_{11}]Mo(CO)_3W(CO)_3^{2-}$ ion (*137*).

$$C_2B_9H_{11}{}^{2-} + 2M(CO)_6 \rightarrow (C_2B_9H_{11})M_2(CO)_8{}^{2-} + 4CO$$

$$(C_2B_9H_{11})M(CO)_3{}^{2-} + M'(CO)_6 \rightarrow (C_2B_9H_{11})MM'(CO)_8{}^{2-} + CO$$

$$M = Mo \text{ or } W;$$
$$M' = Mo \text{ or } W$$

Dicarbollyl Complexes of Mn and Re

Bis(dicarbollyl) complexes of manganese and rhenium have not been reported at this writing, but the dicarbollyl metal tricarbonyl anions have been prepared by reaction of $BrMn(CO)_5$ or $BrRe(CO)_5$ with the 1,2-dicarbollide ion (*125, 137*). In both cases the initial reaction at room temperature apparently forms a σ-bonded species, which on refluxing in tetrahydrofuran is converted to the π-complex with loss of CO. The rhenium π-complex has been structurally characterized by X-ray diffraction (Fig. 9-9) (*469*).

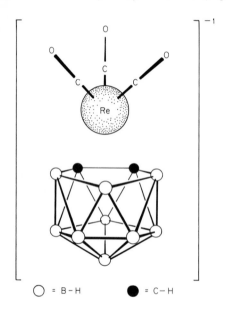

○ = B–H   ● = C–H

FIG. 9-9. Structure of the $[\pi\text{-}(3)\text{-}1,2\text{-}C_2B_9H_{11}]Re(CO)_3^-$ ion (*137*).

$$C_2B_9H_{11}{}^{2-} + BrMn(CO)_5 \xrightarrow{\text{fast}} (\sigma\text{-}C_2B_9H_{11})Mn(CO)_5{}^- + Br^-$$

$$2CO + (C_2B_9H_{11})Mn(CO)_3{}^- \xleftarrow{\substack{\text{slow} \\ \text{reflux}}}$$

Dicarbollyl Complexes of Fe

Under reaction conditions identical to those used in the synthesis of ferrocene, $FeCl_2$ reacts with 1,2-$C_2B_9H_{11}^{2-}$ to form a diamagnetic bis(dicarbollyl) complex of Fe(II) which is readily air-oxidized to the paramagnetic Fe(III) species (Fig. 9-10) (*137, 139*). Reduction of the latter compound with sodium

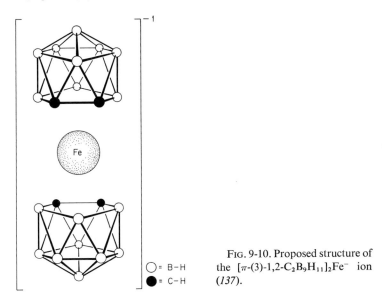

○ = B–H
● = C–H

FIG. 9-10. Proposed structure of the $[\pi\text{-}(3)\text{-}1,2\text{-}C_2B_9H_{11}]_2Fe^-$ ion (*137*).

amalgam affords the original Fe(II) complex (the Fe(III) complex apparently cannot be obtained directly from $FeCl_3$).

$$2C_2B_9H_{11}{}^{2-} + FeCl_2 \longrightarrow (C_2B_9H_{11})_2Fe^{2-} \underset{Na}{\overset{O_2}{\rightleftarrows}} (C_2B_9H_{11})_2Fe^-$$

C-substituted derivatives of $C_2B_9H_{11}^{2-}$ undergo similar reactions to yield the corresponding Fe(II) and Fe(III) complexes.

Reaction of $1,2\text{-}C_2B_9H_{11}^{2-}$ with $FeCl_2$ in the presence of $C_5H_5^-$ ion yields, among other products, a paramagnetic mixed complex of Fe(III) which is easily reduced to the diamagnetic Fe(II) compound (*130, 137*).

$$C_5H_5^- + C_2B_9H_{11}{}^{2-} + FeCl_2 \xrightarrow[\text{reflux}]{\text{THF}} (C_5H_5)Fe(C_2B_9H_{11}) \xrightarrow{Na/Hg} (C_5H_5)Fe(C_2B_9H_{11})^-$$

The structure of the Fe(III) species (Fig. 9-11) has been established by the X-ray method (*471*), and paramagnetic resonance measurements on the iron dicarbollyl complexes are in good agreement with theoretical calculations based on a set of ferrocene-like molecular orbitals in which the unpaired electron occupies an $e_{2g}^{\pm}$ molecular orbital (*204*). The electronic and structural

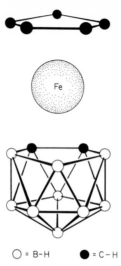

FIG. 9-11. Structure of $(\pi\text{-}(3)\text{-}1,\text{-}2\text{-}C_2B_9H_{11})Fe(\pi\text{-}C_5H_5)$ (*137*).

○ = B–H   ● = C–H

similarity between the dicarbollyl complexes and the metallocenes therefore seems clearly established.

The conversion of the $C_2B_9H_{12}^-$ ion to $C_2B_9H_{11}^{2-}$ in aqueous base, described above (Section 9-1), has been utilized in the preparation of dicarbollyl complexes of iron, cobalt, and nickel by direct reaction of $C_2B_9H_{12}^-$ salts with the metal dihalide in hot concentrated NaOH (*345*).

$$FeCl_2 + C_2B_9H_{12}^- \xrightarrow{OH^-} (C_2B_9H_{11})_2Fe^{2-} \xrightarrow{air} (C_2B_9H_{11})Fe^-$$

Salts of the dicarbollyl complex are stable in aqueous base, but the bis(1,2-dimethyldicarbollyl) and bis(1,2-diphenyldicarbollyl) Fe(III) complexes are quantitatively degraded by bases to the C,C'-disubstituted $RC_2B_9H_{10}^-$ ion and $Fe(OH)_3$ (*137*). Both the substituted and unsubstituted complexes are stable toward strong acids.

Polarographic reduction (*137*) of $(C_2B_9H_{11})_2Fe^-$, $[(CH_3)_2C_2B_9H_9]_2Fe^-$, and $[(C_6H_5)_2C_2B_9H_9]_2Fe^-$ indicates a one-electron change in each case, and all of the resulting Fe(II) complexes (Table 9-1) are diamagnetic, as expected for a d^6 configuration. Mössbauer studies (*140*) of $(C_2B_9H_{11})_2Fe^-$ and $(C_5H_5)Fe(C_2B_9H_{11})$ indicate that the electron vacancy (relative to d^6) is localized on the metal atom, and that all five atoms in the C_2B_3 ring are about equally involved in electron donation to the metal. Calculations of the effective magnetic moments of several iron carbollyl complexes have been reported (*88*).

Reaction of iron pentacarbonyl with the 1,2-dicarbollide ion in refluxing tetrahydrofuran yields a novel bridged ion (*133*) having the composition

$(C_2B_9H_{11})_2Fe_2(CO)_4^{2-}$, whose structure has been determined in an X-ray study (Fig. 9-12) *(100)*. On reaction with sodium amalgam this ion is reduced to a yellow species which may be the $(\pi\text{-}1,2\text{-}C_2B_9H_{11})Fe(CO)_2^{2-}$ ion, containing zero-valent iron *(133)*.

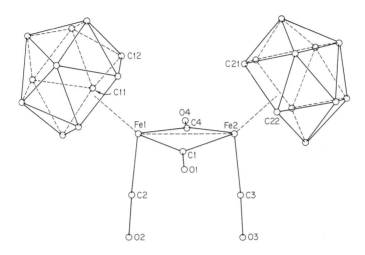

FIG. 9-12. Structure of the $[\pi\text{-}(3)\text{-}1,2\text{-}C_2B_9H_{11}]_2Fe_2(CO)_4^{2-}$ ion *(100)*.

Dicarbollyl Complexes of Co

Diamagnetic bis(1,2-) and bis(1,7-dicarbollyl) cobalt(III) complexes (Table 9-1) have been prepared, like the corresponding Fe(III) species, via both the nonaqueous *(124, 135, 137)* and aqueous *(137, 345)* routes.

$$\tfrac{3}{2}CoCl_2 + 2C_2B_9H_{11}^{2-} \longrightarrow (C_2B_9H_{11})_2Co^- + 3Cl^- + \tfrac{1}{2}Co^0$$

$$CoCl_2 + 2C_2B_9H_{12}^- \xrightarrow[\text{NaOH}]{\text{hot aq}} (C_2B_9H_{11})_2Co^{2-} \xrightarrow{\text{air}} (C_2B_9H_{11})_2Co^- + Co^0$$

Although in both cases the Co(II) complex is believed to form initially and may also be obtained in the electrochemical reduction *(137)* of Co(III) species, no dicarbollyls of Co(II) appear to have been isolated.

Sandwich bonding has been established for several Co(III) complexes by X-ray studies *(39, 42, 470)*, and ^{59}Co nuclear quadropole resonance measurements *(115)* have shown the carbon–metal and boron–metal bonding in

$(1,2\text{-}C_2B_9H_{11})_2Co^-$ to be essentially identical, in agreement with the previously mentioned Mössbauer studies of Fe(III) dicarbollyl compounds.

Bromination of $(1,2\text{-}C_2B_9H_{11})_2Co^-$ in glacial acetic acid yields the bis(tribromo) derivative (*137*), whose structure, determined by the X-ray method (*42*), is depicted in Fig. 9-13. It will be noted that the bromine atoms are located

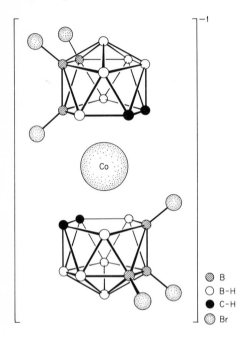

⊘	B
○	B-H
●	C-H
⊛	Br

FIG. 9-13. Structure of the $[\pi\text{-}(3)\text{-}1,2\text{-}C_2B_9H_8Br_3]_2Co^-$ ion (*137*).

as far as possible from carbon, as is also the case in the products of the electrophilic halogenation of *o*- and *m*-carborane (Chapters 6 and 7).

The π-cyclopentadienyl-1,2-dicarbollylcobalt(III) complex is obtained (*124, 137*) by direct reaction of $C_5H_5^-$, $1,2\text{-}C_2B_9H_{11}^{2-}$, and $CoCl_2$, in analogous fashion to the preparation of the corresponding Fe(III) species. As expected for a low-spin d^6 complex, $(1,2\text{-}C_2B_9H_{11})Co(C_5H_5)$ is diamagnetic. The yellow, air-sensitive $(C_2B_9H_{11})Co(CO)_2^-$ ion is formed from $1,2\text{-}C_2B_9H_{11}^{2-}$ and $Co_2(CO)_8$ in refluxing tetrahydrofuran (*133*).

A compound obtained in very small yield (1 %) from the reaction of $CoCl_2$ with $1,2\text{-}C_2B_9H_{11}^{2-}$ in hot aqueous base (the major product of which is $(C_2B_9H_{11})_2Co^-$, as described above) has been characterized (*77*) as a salt of $(C_2B_9H_{11})Co(C_2B_8H_{10})Co(C_2B_9H_{11})^{2-}$, whose X-ray-determined structure (*251a*) is presented in Fig. 9-14. This same complex is formed in larger yield by the treatment of $(C_2B_9H_{11})_2Co^-$ salts with $CoCl_2$ and aqueous NaOH at 100°. The central carborane cage, $C_2B_8H_{10}^{4-}$ (Fig. 9-15), has been designated the

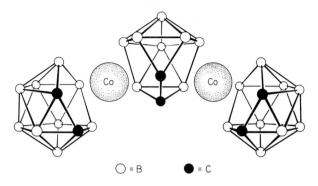

○ = B ● = C

FIG. 9-14. Structure of the $(C_2B_9H_{11})Co(C_2B_8H_{10})Co(C_2B_9H_{11})^{2-}$ ion (77).

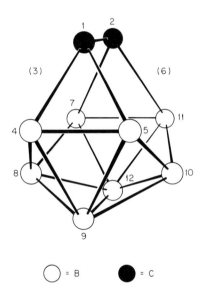

○ = B ● = C

FIG. 9-15. Structure of the (3,-6)-1,2-$C_2B_8H_{10}^{4-}$ ion (77).

(3,6)-1,2-dicarbacanastide(4–) ion by Hawthorne (from the Spanish noun for basket).

Another somewhat exotic complex has been isolated (39) from the room-temperature reaction of the $(C_2B_9H_{11})_2Co^-$ ion with $AlCl_3$ in CS_2 solvent; addition of HCl increases the reaction rate and affords yields of 70%. This compound, formulated as $(C_2B_9H_{10})_2CoS_2CH$, has been structurally character-ized by X-ray diffraction (39) (Fig. 9-16).

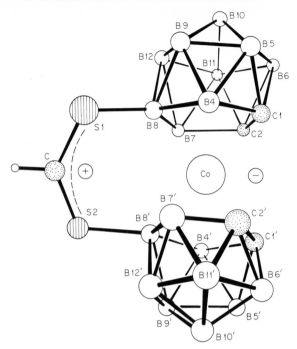

FIG. 9-16. Structure of $(C_2B_9H_{10})_2CoS_2CH$ (*39*).

Dicarbollyl Complexes of Ni and Pd

Treatment of nickel(II) salts with the 1,2- or $1,7\text{-}C_2B_9H_{11}^{2-}$ ion followed by air oxidation produces bis(1,2-) or bis(1,7-dicarbollyl)nickel(III) species (*137*), $(C_2B_9H_{11})_2Ni^-$. Both of these paramagnetic complexes are easily oxidized with aqueous ferric ion, yielding the corresponding $(C_2B_9H_{11})_2Ni$, compounds which are diamagnetic and contain nickel in the unusual +4 oxidation state (*137, 345*). These neutral complexes are highly stable and strongly acidic in the Lewis sense, forming crystalline adducts with Lewis bases (*346*). The $(1,2\text{-}C_2B_9H_{11})_2Ni^-$ ion can be reduced to the extremely air-sensitive $(1,2\text{-}C_2B_9H_{11})_2Ni^{2-}$, which is paramagnetic with two unpaired electrons. Paralleling the reactions of $FeCl_2$ and $CoCl_2$, hot aqueous basic solutions of $1,2\text{-}C_2B_9H_{12}^-$ react with $NiCl_2$, forming $(C_2B_9H_{11})_2Ni^-$ on exposure to air (*137, 345*).

The Ni(III) ion, $(1,2\text{-}C_2B_9H_{11})_2Ni^-$, has been shown in X-ray studies (*346*, footnote 12, *354*) to have a symmetrical sandwich structure analogous to that of the dicarbollyl Fe(III) complex. The mixed complex $(C_5H_5)Ni(1,2\text{-}$

$C_2B_9H_{11}$) has been prepared and found to be paramagnetic with one unpaired electron (352).   The $(1,2\text{-}C_2B_9H_{11})_2Ni^-$ ion may be oxidized by cyclic voltammetry to the previously mentioned $Ni(IV)$ compound, $(1,2\text{-}C_2B_9H_{11})_2Ni$, and in this respect contrasts with the nickelocene analog $(C_5H_5)_2Ni$, which may be electrolytically oxidized to $(C_5H_5)_2Ni^{2+}$ only in very cold ($-40°$) solutions. Polarographic and chemical redox reactions indicate that all of the above interconversions of oxidation states in the nickel dicarbollyl species are reversible (137, 352).

The preparation of dicarbollyl palladium complexes by direct reaction of $Pd(II)$ salts with the $1,2\text{-}C_2B_9H_{11}^{2-}$ ion has been accomplished only when the latter reactant is in very large excess (346). The resulting diamagnetic $(C_2B_9H_{11})_2Pd^{2-}$ ion is oxidized by iodine to the air-stable, diamagnetic $(C_2B_9H_{11})_2Pd$. An intermediate ion containing $Pd(III)$ is formed when the $Pd(II)$ and $Pd(IV)$ species are mixed in equimolar amounts; as expected for a d^7 configuration, the product contains one unpaired electron (346).

$$(C_2B_9H_{11})_2Pd^{2-} + (C_2B_9H_{11})_2Pd \rightarrow 2(C_2B_9H_{11})_2Pd^-$$

The π-tetraphenylcyclobutadienyl-π-1,2-dicarbollylpalladium(II) complex has been prepared, as has its 1,2-dimethyl derivative (137, 347) (Fig. 9-17), in a

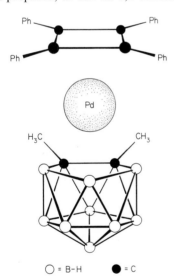

FIG. 9-17. Structure of $[\pi\text{-}(3)\text{-}1,\text{-}2 - (CH_3)_2C_2B_9H_9]Pd[C_4(C_6H_5)_4]$ (137).

$\bigcirc$ = B–H $\bullet$ = C

reaction analogous to that used to synthesize the corresponding π-tetraphenylcyclobutadienyl-π-cyclopentadienylpalladium(II) cation (137).

$$[(C_6H_5)_4C_4PdCl_2]_2 + 2(1,2\text{-}C_2B_9H_{11})^{2-} \rightarrow 2[(C_6H_5)_4C_4]Pd(C_2B_9H_{11}) + 4Cl^-$$
$$(10\%)$$

Dicarbollyl Complexes of Cu and Au

Copper(II) salts interact with the $1,2\text{-}C_2B_9H_{12}^{2-}$ ion in cold, 40% NaOH solution (*137*) to yield the $(1,2\text{-}C_2B_9H_{11})_2Cu^{2-}$ ion, whose structure (*353*) (Fig. 9-18) contains a curious and significant feature: the dicarbollide ligands

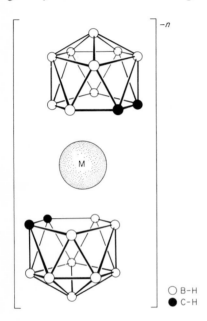

Fig. 9-18. Structure of the dicarbollyl complexes of d^8 and d^9 metal atoms. M = Ni(II), Cu(II), Cu(III), and Au(III). Proposed structure of M = Pd(II) and Au(II) (*346*).

○ B-H
● C-H

are situated so that the boron-metal atom distances are shorter than the carbon–metal distances, and the two dicarbollide cages are shifted laterally by 0.6 Å from the more symmetrical conformation found in the dicarbollyls discussed previously. Similar "slipped" structures have been found for the bis(dicarbollyl) complexes of Ni(II), Cu(II), Cu(III), and Au(III), each of which contains eight or more d electrons. Indeed it appears, so far without exception, that all dicarbollyl sandwich complexes of metals having d^8 or d^9 configurations have "slipped" structures, while all those with fewer than eight d electrons have the more symmetrical structure [bis(dicarbollyls) of d^3, d^5, d^6, d^7, d^8, and d^9 metals have been characterized]. Wing (*353, 354*) has described the "slipped" structures in terms of π-allylic bonding arising from the electron-richness of the d^8 and d^9 complexes and from the heterocyclic nature of the coordinating C_2B_3 rings. Warren and Hawthorne, on the other hand, suggest (*346*) that the slipped conformations arise from requirements of orbital symmetry—specifically, that in d^8 and d^9 systems the metal $nd_{z^2} - (n+1)$ s hybrid acceptor orbital lacks the proper symmetry to interact with the filled dicarbollide molecular orbitals unless distortion from pseudo-C_5 symmetry occurs. Such an effect is believed to exist in the benzene complexes of Ag(I) and Cu(I), both of which contain d^{10} metal configurations. This explanation, although qualitative, does account for the sharp structural differences between the d^8 and d^9 dicarbollyl complexes as opposed to those of metals having fewer d electrons. However, the experimental evidence for this phenomenon rests on X-ray structural investigations of only a few dicarbollyl–metal complexes. If the slip-distortions in fact arise from more complex factors than have so far been suggested, "slipped" complexes of metals having fewer than eight d electrons may yet be found. Detailed molecular orbital treatments of this problem are clearly required.

The deep blue $(1,2\text{-}C_2B_9H_{11})_2Cu^{2-}$ ion contains one unpaired electron and is stable in cold aqueous basic solution (346). In acetonitrile, acetone, or diethyl ether solution, this ion is air-oxidized to the red diamagnetic $(C_2B_9H_{11})_2Cu^-$ ion, an example of the very few known Cu(III) species. Significantly, this complex represents the most stable oxidation state of all bis(dicarbollyl) copper complexes, as shown by cyclic voltammetry studies (the existence of a corresponding Cu(I) complex is also suggested by these data) (346). Both the Cu(II) and Cu(III) bis(dicarbollyls) have the "slipped" configuration described above, as disclosed from X-ray diffraction studies (353, 354).

The bis(dicarbollyl) gold(II) ion is formed in the reaction of $1,2\text{-}C_2B_9H_{11}^{2-}$ salts with gold(III) chloride in nonaqueous solvents (346). Oxidation of the blue-green paramagnetic (d^9) $(1,2\text{-}C_2B_9H_{11})_2Au^{2-}$ ion yields the red, diamagnetic Au(III) complex, which may be reduced to the Au(II) species with sodium amalgam. As with the copper system, cyclic voltammetry also suggests the existence of a Au(I) bis(dicarbollyl) complex.

$$AuCl_3 + 2C_2B_9H_{11}{}^{2-} \longrightarrow (C_2B_9H_{11})_2Au^{2-} \underset{Na/Hg}{\overset{H_2O_2,\ H^+}{\rightleftarrows}} (C_2B_9H_{11})_2Au^-$$

The bis(dicarbollyl) complexes of copper, gold, and palladium are particularly interesting and significant in that the analogous π-cyclopentadienyl compounds (e.g., the hypothetical cupricene, auricene, and palladocene) are unknown at present. Thus, the 1,2- and 1,7-dicarbollide ions seem capable of forming an even greater number and variety of π-complexes than is the case with cyclopentadienide ion and other organic ring systems. Dicarbollyls of silver and platinum have not been characterized despite several attempts (346) to prepare them via the aqueous and nonaqueous routes described above (a σ-bonded o-carboranyl platinum derivative has been prepared (Section 6-14)).

The reaction of $AgNO_3$ with $C_2B_9H_{12}^-$ salts in basic solution has been reported (26) to yield a bis(dicarbollyl) sandwich complex of Ag(I), but few details are given and the characterization appears doubtful at present.

Monocarbollyl—Transition Metal π-Complexes

The synthesis of the tetrahydrofuran adduct of $Na_3CB_{10}H_{11}$ has been described in Section 9-3. The $CB_{10}H_{11}^{3-}$ (monocarbollide) ion is isoelectronic and isostructural with the 1,2- and 1,7-$C_2B_9H_{11}^{2-}$ dicarbollide ions and, as expected, forms similar π-bonded complexes with transition metal ions (160, 176, 179). A number of such species have been prepared, employing as ligands the $CB_{10}H_{11}^{3-}$ ion and certain of its amine-substituted derivatives (Table 9-2);

TABLE 9-2

TRANSITION METAL–MONOCARBOLLYL π-COMPLEXES

Compound	Color	Refs.
$Cs_3[(CB_{10}H_{11})_2Cr]\cdot H_2O$		176
$[(CH_3)_3NH]_2(CB_{10}H_{11})_2Mn$		176
$[(CH_3)_4N]_3(CB_{10}H_{11})_2Fe$	Red	160
$[(CH_3)_4N](NH_3CB_{10}H_{10})_2Fe$		176
$[(CH_3)_4N]_3(CB_{10}H_{11})_2Co$	Yellow	160, 176
$Cs_3[(CB_{10}H_{11})_2Co]\cdot H_2O$		176
$[(CH_3)_4N]_2(CB_{10}H_{11})_2Co$	Black	176
$[(CH_3)_4N](CB_{10}H_{10}NH_2C_2H_5)_2Co$	Orange	160
$[(CH_3)_4N]_2(NH_2CB_{10}H_{10})Co(NH_3CB_{10}H_{10})$		176
$Cs_2[(CB_{10}H_{11})_2Ni]$	Yellow	160, 176
$(C_3H_7NH_2CB_{10}H_{10})_2Ni$	Orange	160
$(C_3H_7(CH_3)NHCB_{10}H_{10})_2Ni$	Orange	160
$((CH_3)_2NHCB_{10}H_{10})_2Ni$		176
$[(CH_3)_4N]_2(HOCB_{10}H_{10})_2Ni$		176

the syntheses involve both aqueous and nonaqueous routes, corresponding generally to the reactions used in the preparation of the metal dicarbollyl complexes (see above). Thus, the treatment of the previously described $CB_{10}H_{13}^-$ and $NH_3CB_{10}H_{12}$ (both isoelectronic with $C_2B_9H_{12}^-$) with aqueous base yields monocarbollyl complexes of iron, cobalt, and nickel, while reactions in tetrahydrofuran produce monocarbollyls of these metals as well as of manganese and chromium (Table 9-2).

$$C_3H_7NH_2CB_{10}H_{12} + NiCl_2 \xrightarrow{OH^-} (C_3H_7NH_2CB_{10}H_{10})_2Ni$$

$$[(CH_3)(C_3H_7)NHCB_{10}H_{10}]_2Ni \xleftarrow[CH_3I]{NaHCO_3}$$

$$Na_3CB_{10}H_{11} \cdot 2THF \xrightarrow[OH^-]{NiCl_2} (CB_{10}H_{11})_2Ni^{2-}$$

$$\xrightarrow[THF]{CoCl_2} (CB_{10}H_{11})_2Co^{3-}$$

The preparation of $(CB_{10}H_{11})Mn(CO)_3^{2-}$ and $(CB_{10}H_{11})_2Cu^{3-}$ has been reported without details (*160*). At this writing, none of the monocarbollyl transition metal complexes has been examined by X-ray diffraction, but π-bonded sandwich structures seem certain for these compounds in view of their close similarity to the dicarbollyl complexes in electronic, magnetic, and chemical properties.

Carbaphosphollyl–Transition Metal Complexes

The 1,2- and 1,7-$CPB_9H_{10}^{2-}$ (carbaphosphollide) ions (Section 9-4), like their isoelectronic analogs $C_2B_9H_{11}^{2-}$ and $CB_{10}H_{11}^{3-}$, readily form π-complexes with transition metals (*335*) (Table 9-3). Treatment of the carbaphosphollyl

TABLE 9-3

TRANSITION METAL–CARBAPHOSPHOLLYL π-COMPLEXES

Compound	Color	mp (°C)	Other data	Refs.
(1,7-$CH_3CPB_9H_{10}$)Mn(CO)$_3$				*335*
(1,2-$CH_3CPB_9H_{10}$)$_2$Fe				*335*
[(CH$_3$)$_4$N]$_2$(1,7-CPB_9H_{10})$_2$Fe				*335*
(1,7-$CH_3CPB_9H_{10}$)$_2$Fe (isomer 1)	Brown	240	X-ray	*335*
(1,7-$CH_3CPB_9H_{10}$)$_2$Fe (isomer 2)	Red	233–234	X-ray	*335*
(C_5H_5)Fe(1,7-$CH_3CPB_9H_{10}$)				*335*
[(CH$_3$)$_4$N](1,7-CPB_9H_{10})$_2$Co				*335*
(1,7-$CH_3CPB_9H_{10}$)$_2$Co (isomer 1)		278–279		*335*
(1,7-$CH_3CPB_9H_{10}$)$_2$Co (isomer 2)		231–233		*335*
(1,7-$CH_3CPB_9H_{10}$)$_2$Ni				*335*

complexes of iron, cobalt, or nickel with methyl iodide results in methylation of the previously unsubstituted phosphorus atom. In the case of the iron complex, two stereoisomers are obtained; an X-ray study (*335*) of a cocrystallite containing both forms has established the structure shown in Fig. 9-19.

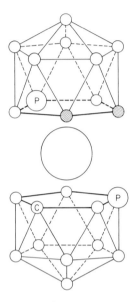

FIG. 9-19. Structure of the $(CPB_9H_{10})_2Fe^{2-}$ ion (*335*). Established carbon position in lower cage is shown; possible carbon sites in upper cage are dotted.

Analogous complexes of the 1,2- and $1,7-CAsB_9H_{10}^{2-}$ ions have been prepared (*333*) although no details have been published.

$$1,7\text{-}CPB_9H_{11}^- \xrightarrow[\text{2. FeCl}_2]{\text{1. NaH}} (1,7\text{-}CPB_9H_{10})_2Fe^{2-} \xrightarrow{\text{CH}_3\text{I}} (CH_3PCB_9H_{10})_2Fe$$

(probably *d,d-l,l*-racemate + *d,l meso* form)

Nonicosahedral—Transition Metal π-Complexes

Although all of the π-complexes described above apparently involve structures in which the metal atom completes an icosahedral cage system, several transition metal complexes of smaller carborane systems have recently been characterized (Table 9-4). It seems highly probable, in fact, that a very large family of carborane–metal complexes exists which is by no means restricted to icosahedral geometry.

TABLE 9-4

NONICOSAHEDRAL TRANSITION METAL π-COMPLEXES

Compound	Color	mp (°C)	Other data[a]	Refs.
[(CH$_3$)$_4$N](1,6-C$_2$B$_7$H$_9$)$_2$Co	Brown	240	IR, B, H, E	87, 128
[(CH$_3$)$_4$N](6,7-C$_2$B$_7$H$_9$)$_2$Co	Red	235	IR, H	87
[(CH$_3$)$_4$N](1,10-C$_2$B$_7$H$_9$)$_2$Co		192	B, H	87
Cs[(1,10-C$_2$B$_7$H$_9$)$_2$Co]	Orange	210	IR, E	86, 87
(C$_5$H$_5$)Co(1,6-C$_2$B$_7$H$_9$)	Red	158–159	IR, B, H, MS, E	87, 128
(C$_5$H$_5$)Co(1,10-C$_2$B$_7$H$_9$)	Orange	113	IR, B, H, MS, E	86, 87
(C$_5$H$_5$)Co(8-COCH$_3$-1,6-C$_2$B$_7$H$_8$)	Red	162	IR, B, H	95
(C$_5$H$_5$)Co(1,6-C$_2$B$_7$H$_7$Br$_2$)	Orange	129–130	H	95
[(CH$_3$)$_4$N](C$_2$B$_6$H$_8$)Mn(CO)$_3$	Amber		IR, B, H, E	85, 132
[(CH$_3$)$_4$N]((C$_6$H$_5$)C$_2$B$_6$H$_7$)Mn(CO)$_3$	Amber		B, H	85
[(CH$_3$)$_4$N]((CH$_3$)$_2$C$_2$B$_6$H$_6$)Mn(CO)$_3$	Orange		B, H	85
(2-CH$_3$-2,3,4-C$_3$B$_3$H$_5$)Mn(CO)$_3$	Yellow-orange		IR, B, H, MS	156

[a] IR = infrared spectrum or band positions; B = ^{11}B NMR data; H = proton NMR data; MS = mass spectral data; E = electronic spectral data.

COMPLEXES OF C$_2$B$_7$H$_9^{2-}$

The reaction of the open-cage carborane 1,3-C$_2$B$_7$H$_{13}$ (Section 5-1) with sodium hydride generates a new ion, C$_2$B$_7$H$_{11}^{2-}$ (Fig. 9-20) (128). The C$_2$B$_7$H$_{11}^{2-}$

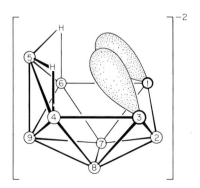

FIG. 9-20. Proposed structure of the 1,3-C$_2$B$_7$H$_{11}^{2-}$ ion (87).

ion reacts with anhydrous $CoCl_2$ in ether at 25°, forming cobalt metal and a red Co(III) complex characterized as $(C_2B_7H_9)_2Co^-$ (87, 128). When the reaction

$$2C_2B_7H_{11}^{2-} + \tfrac{3}{2}Co^{2+} \rightarrow \tfrac{1}{2}Co^0 + 2H_2 + (C_2B_7H_9)_2Co^-$$

is carried out at 70°, a brown Co(III) complex is obtained which is isomeric with the product formed at 25°. From NMR, electronic spectra, and preliminary X-ray data, the structure indicated in Fig. 9-21 has been postulated

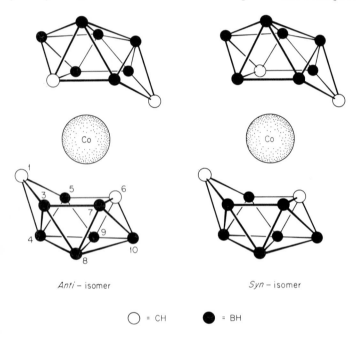

FIG. 9-21. Structures of the isomers of the $(\pi\text{-}1,6\text{-}C_2B_7H_9)_2Co^-$ ion (87).

for the brown complex (*syn* and *anti* forms are possible although no more than one form has been observed), and the structure in Fig. 9-22 has been proposed for the red complex (87). Hawthorne (87) has designated the $C_2B_7H_9^{2-}$ unit (which has not been isolated in uncomplexed form) as the dicarbazapide ion, from the Spanish zapato (shoe).

Both the red (6,7) and brown (1,6) isomers rearrange at 250–315° to a third isomer (86, 87) which has been assigned the structure presented in Fig. 9-23. The thermal isomerization is analogous to the rearrangement of 1,6- to 1,10-$C_2B_8H_{10}$ (Section 5-2), which also involves a ten-membered polyhedral cage. However, all evidence indicates that the 6,7 and 1,6 isomers of the $(C_2B_7H_9)_2Co^-$ ion are *not* thermally interconvertible (87).

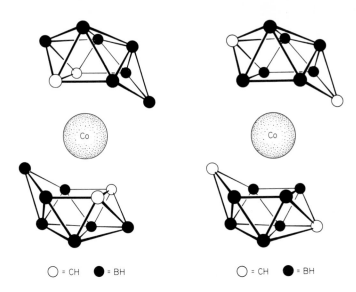

○ = CH   ● = BH   ○ = CH   ● = BH

FIG. 9-22. (Left). Proposed structure of the $(\pi\text{-}6,7\text{-}C_2B_7H_9)_2Co^-$ ion (87).
FIG. 9-23. (Right). Proposed structure of the $(\pi\text{-}1,10\text{-}C_2B_7H_9)_2Co^-$ ion (87).

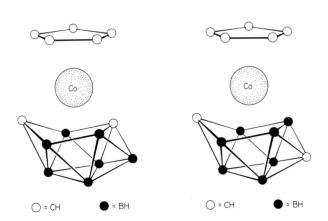

○ = CH   ● = BH   ○ = CH   ● = BH

FIG. 9-24. (Left). Proposed structure of $(\pi\text{-}C_5H_5)Co(\pi\text{-}1,6\text{-}C_2B_7H_9)$. (87).
FIG. 9-25. (Right). Proposed structure of $(\pi\text{-}C_5H_5)Co(\pi\text{-}1,10\text{-}C_2B_7H_9)$ (87).

The sequence of reactions involving these species is indicated below.

$$1,3\text{-}C_2B_7H_{13} \xrightarrow[\text{(C}_2\text{H}_5)_2\text{O}]{\text{NaH}} C_2B_7H_{11}^{2-} \xrightarrow{25°, \text{CoCl}_2} [\pi\text{-}1,6\text{-}C_2B_7H_9]_2\text{Co}^-$$
(brown)

$$\xrightarrow{70°, \text{CoCl}_2} [\pi\text{-}6,7\text{-}C_2B_7H_9]_2\text{Co}^- \quad 315°$$
(red)

315°

$$[\pi\text{-}1,10\text{-}C_2B_7H_9]_2\text{Co}^- \quad \text{(orange)}$$

The reaction of $C_2B_7H_{11}^{2-}$ with $C_5H_5^-$ and anhydrous $CoCl_2$ yields a mixed complex, $(\pi\text{-}C_5H_5)Co(1,6\text{-}C_2B_7H_9)$ (87, 128), which in turn rearranges at 315° to $(\pi\text{-}C_5H_5)Co(1,10\text{-}C_2B_7H_9)$ (86, 87). Figs. 9-24 and 9-25 depict the proposed structures of these species.

A few chemical reactions of $(\pi\text{-}C_5H_5)Co(\pi\text{-}1,6\text{-}C_2B_7H_9)$ have been studied (95). Treatment with excess bromine in CCl_4 yields a B-dibromo derivative with traces of a tribromo species, and concentrated HNO_3 attacks the complex to yield an extremely shock sensitive nitro derivative. The complex itself can be acylated by acetyl chloride, but only at a boron cage atom; significantly, no substitution on the cyclopentadienyl ring occurs (in contrast, ferrocene is easily acylated under these conditions).

$$(C_5H_5)Co(1,6\text{-}C_2B_7H_9) \xrightarrow[\text{AlCl}_3]{\text{CH}_3\text{COCl}} (C_5H_5)Co(C_2B_7H_8COCH_3)$$

The acetyl derivative is soluble in HCl (unlike the parent complex, which is insoluble in water and acid). Attempts to effect a thermal cage rearrangement of the acetyl derivative resulted in decomposition at 250°.

The isomeric complex $(C_5H_5)Co(1,10\text{-}C_2B_7H_9)$ fails to acylate under the conditions employed for the acylation of the 1,6 isomer (95).

COMPLEXES OF $C_2B_6H_8^{2-}$

Treatment of the $C_2B_7H_{11}^{2-}$ ion with $BrMn(CO)_5$ in refluxing tetrahydrofuran (85, 132) leads to partial decomposition of the boron substrate, yielding H_2, $Mn_2(CO)_{10}$, and unidentified solids as well as a new ion characterized as $(C_2B_6H_8)Mn(CO)_3^-$. Electronic and NMR spectra of this species suggest the structure shown in Fig. 9-26. The same complex is produced in the reaction of $C_2B_7H_{11}^{2-}$ with $Mn_2(CO)_{10}$. The C-phenyl and C,C'-dimethyl derivatives have been obtained from the corresponding $C_2B_7H_{11}^{2-}$ substituted derivatives (85).

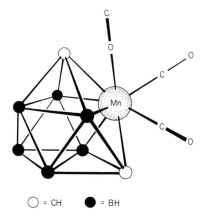

FIG. 9-26. Proposed structure of the $(\pi\text{-}C_2B_6H_8)Mn(CO)_3{}^-$ ion (132).

COMPLEXES OF $C_3B_3H_6{}^-$

Methyl derivatives of the the *nido*-carborane $2,3,4\text{-}C_3B_3H_7$ contain one hydrogen bridge, which may be abstracted with sodium hydride to yield the corresponding $C_3B_3H_6{}^-$ derivative (Section 3-3). This species has a nearly planar C_3B_2 basal ring which is capable of coordinating to transition metal atoms to form π-bonded complexes. Such complexes have been obtained in

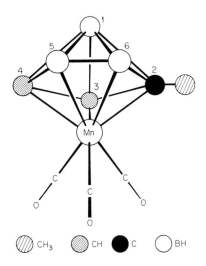

FIG. 9-27. Proposed structure of $(\pi\text{-}CH_3C_3B_3H_5)Mn(CO)_3$ (156).

reactions of methyl derivatives of $C_3B_3H_7$ with $Mn_2\,(CO)_{10}$(*156*). The proposed structure of the 2-monomethyl derivative, an isoelectronic analog of the known $(\pi\text{-}C_5H_5)Mn(CO)_3$, is presented in Fig. 9-27.

$$2\text{-}CH_3C_3B_3H_6 + \tfrac{1}{2}Mn_2(CO)_{10} \xrightarrow[\text{sealed tube}]{175\text{--}200^\circ} (\pi\text{-}2\text{-}CH_3C_3B_3H_5)Mn(CO)_3 + \tfrac{1}{2}H_2 + 2CO$$

The same complex has been obtained from the reaction of $2\text{-}CH_3C_3B_3H_5^-$ ion with $BrMn(CO)_5$ followed by refluxing in THF; a σ-bonded complex probably forms in the reaction at room temperature, which proceeds without evolution of CO (*157*).

$$CH_3C_3B_3H_6 + NaH \longrightarrow CH_3C_3B_3H_5^- \xrightarrow[25^\circ]{BrMn(CO)_5} \sigma\text{-}CH_3C_3B_3H_5Mn(CO)_5 \xrightarrow{\Delta}$$
$$\text{wine red}$$

$$(\pi - CH_3C_3B_3H_5)Mn(CO)_3 + 2CO$$
$$\text{yellow}$$

Supplementary Sources of Information

BORANE AND CARBORANE CHEMISTRY

The classic work on the synthesis of the binary boron hydrides, still useful to experimentalists after nearly four decades, is that of Stock (*325*). Lipscomb's book (*195*) is heavily oriented toward structure and bonding theory, while the more recent monograph of Muetterties and Knoth (*215*) is a concise presentation of the chemistry and structures of polyhedral borane species. A detailed summary of research in the borane field in the 1950's and 1960's, including much previously unpublished data, has been prepared by Hughes, Smith, and Lawless (*158*) (edited by Holzmann). Hawthorne (*122, 123*) has twice recently reviewed the general area of boron hydride and carborane chemistry. A volume edited by Adams (*2*) provides coverage of the boron hydride literature up to 1964, although at that time little carborane work had been publicly reported.

Review articles dealing with the carboranes as a class have been authored by Williams (*349*), Niedenzu (*215a*), Koster and Grassberger (*185*), Issleib *et al.* (in German) (*163*), Muetterties and Knoth (*213*), Onak (*220*), and Bobinski (*19*). The derivative chemistry of *o*-, *m*-, and *p*-carborane has recently been reviewed by Bregadze and Okhlobystin (*30*). Some aspects of carborane polymer chemistry are discussed in an article on inorganic polymers by Gerrard (*89*). Hawthorne (*121*) has summarized the work on transition metal–carborane π-complexes through part of 1968. A very recent review by Todd (*332a*) on the same subject has appeared.

NUCLEAR MAGNETIC RESONANCE SPECTROSCOPY

The proton and boron-11 NMR spectra of boranes and carboranes have been discussed in comprehensive detail by Eaton and Lipscomb (61); the same subject was reviewed earlier by Schaeffer (254). Although no generally applicable theory of boron-11 chemical shifts exists at present, Boer et al. (20), have attempted to account for the [11]B chemical shifts of the $C_2B_{10}H_{12}$ icosahedral isomers in terms of the paramagnetic shielding of boron nuclei. Onak (221) has carried out [11]B decoupling experiments on nido-2,3-$C_2B_4H_8$, and the same author in cooperation with several associates has presented a correlation of [11]B chemical shifts of some boron cage systems with values calculated from a ring-current model (227).

MASS SPECTROSCOPY

A useful discussion of mass spectroscopy as applied specifically to boranes and carboranes has been given by Ditter, Gerhart, and Williams (46).

INFRARED AND RAMAN SPECTROSCOPY

Infrared investigations of icosahedral carborane derivatives have been reported by Leites et al. (193), and by Bregadze and co-workers (27). Bukalov et al. (37) have obtained the Raman spectra of some o- and m-carboranyl compounds using a helium–neon laser.

STRUCTURE AND THEORY

The papers of Lipscomb and his associates dealing with molecular orbital calculations on carborane frameworks prior to 1963 are summarized in

Lipscomb's monograph (*195*) (references to later work are given in Chapter 2 of this book). Waddington (*344*) has described the bonding in the icosahedral carboranes in terms of a free-electron model. Boer, Potenza, and Lipscomb (*21*) have calculated the charge distribution in $C_2B_8H_{10}$ isomers, and Haas (*111*) has determined the numbers of possible geometrical and optical substitution isomers in $B_{10}H_{10}^{2-}$, $B_{12}H_{12}^{2-}$, and the three $C_2B_{10}H_{12}$ icosahedral carboranes. Scott (*265a*) has presented a ligand field model for several types of transition metal dicarbollyl π-complexes.

RELATED BORON CAGE SYSTEMS

Only a few examples are known of carbon-free boron cages with skeletal heteroatoms. Species having cage nitrogen and sulfur atoms have been reported (*141*) [related borane derivatives with external B—S and B—N bonds have also been described (*177, 214*)]. Complexes of various borane cage structures with transition metals are known, but those prepared by Klanberg and co-workers (*171, 172*), involving cage systems with 9 to 11 boron atoms, seem most closely related to the carborane–metal complexes. A molecular framework incorporating four boron and eight nitrogen atoms has been reported by Miller and Johnson (*211*).

REFERENCES

1. Adams, R. M., *Inorg. Chem.* **2**, 1087 (1963).
2. Adams, R. M., ed., "Boron, Metallo-Boron Compounds, and Boranes." Wiley (Interscience), New York, 1964.
3. Akimov, B. A., Bekasova, N. I., Zhigach, A. F., Zamyatina, V. A., Korshak, V. V., Sarishvili, I. G., and Sobolevskii, M. V., *Plast. Massy* **11**, 16 (1965).
4. Akimov, B. A., Zhigach, A. F., Korshak, V. V., Sarishvili, I. G., and Sobolevskii, M. V., *Plast. Massy* **7**, 21 (1966).
5. Akimov, B. A., Zhigach, A. F., Korshak, V. V., Sarishvili, I. G., and Sobolevskii, M. V., *Plast. Massy* **12**, 22 (1966).
6. Akovbyan, E. M., Chaikina, E. A., Gal'braikh, L. S., Bregadze, V. I., Okhlobystin, O. Yu., and Rogovin, Z. A. *Vysokomol. Soedin., Ser. B* **10**, 428 (1968).
7. Alexander, R. P., and Schroeder, H. A., *Inorg. Chem.* **2**, 1107 (1963).
8. Alexander, R. P., and Schroeder, H. A., *Inorg. Chem.* **5**, 493 (1966).
9. Alexander, R. P., and Schroeder, H. A., U.S. Pat. 3,373,193 (1968).
10. Andrianov, V. G., Stanko, V. I., Struchkov, Yu. T., and Klimova, A. I., *Zh. Strukt. Khim.* **8**, 707 (1967).
11. Beall, H. A., and Lipscomb, W. N., *Inorg. Chem.* **6**, 874 (1967).
12. Beaudet, R. A., and Poynter, R. L., *J. Amer. Chem. Soc.* **86**, 1258 (1964).
13. Beaudet, R. A., and Poynter, R. L., *J. Chem. Phys.* **43**, 2166 (1965).
14. Berry, T. E., Tebbe, F. N., and Hawthorne, M. F., *Tetrahedron Lett.* p. 715 (1965).
15. Bilevich, K. A., Zakharkin, L. I., and Okhlobystin, O. Yu., *Izv. Akad. Nauk SSSR, Ser. Khim.* p. 1914 (1965).
16. Bilevich, K. A., Zakharkin, L. I., and Okhlobystin, O. Yu., *Izv. Akad. Nauk SSSR, Ser. Khim.* p. 448 (1967).
17. Binger, P., *Tetrahedron Lett.* p. 2675 (1966).
18. Blay, N. J., Williams, J., and Williams, R. L., *J. Chem. Soc., London* p. 424 (1960).
19. Bobinski, J., *J. Chem. Educ.* **41**, 500 (1964).
20. Boer, F. P., Hegstrom, R. A., Newton, M. D., Potenza, J. A., and Lipscomb, W. N., *J. Amer. Chem. Soc.* **88**, 5340 (1966).
21. Boer, F. P., Potenza, J. A., and Lipscomb, W. N., *Inorg. Chem.* **5**, 1301 (1966).
22. Boer, F. P., Streib, W. E., and Lipscomb, W. N., *Inorg. Chem.* **3**, 1666 (1964).
23. Boone, J. L., Brotherton, R. J., and Petterson, L. L., *Inorg. Chem.* **4**, 910 (1965).
24. Bramlett, C. L., and Grimes, R. N., *J. Amer. Chem. Soc.* **88**, 4269 (1966).
25. Brattsev, V. A., and Stanko, V. I., *Zh. Obshch. Khim.* **38**, 1657 (1968).
26. Brattsev, V. A., and Stanko, V. I., *Zh. Obshch. Khim.* **38**, 2820 (1968).
27. Bregadze, V. I., Chumaevskii, N. A., and Shkirtil, E. B., *Dokl. Akad. Nauk SSSR* **181**, 910 (1968).
28. Bregadze, V. I., and Okhlobystin, O. Yu., *Dokl. Akad. Nauk SSSR* **177**, 347 (1967).
29. Bregadze, V. I., and Okhlobystin, O. Yu., *Izv. Akad. Nauk SSSR, Ser. Khim.* p. 2084 (1967).
30. Bregadze, V. I., and Okhlobystin, O. Y., *Organometal. Chem. Rev. Sect. A* **4**, 345 (1969).
31. Bresadola, S., Rigo, P., and Turco, A., *Chem. Commun.* **20**, 1205 (1968).
32. Bresadola, S., Rossetto, F., and Tagliavini, G., *Ann. Chim. (Rome)* **58**, 597 (1968).
33. Bresadola, S., Rossetto, F., and Tagliavini, G., *Chem. Commun.* p. 623 (1966).
34. Bresadola, S., Rossetto, F., and Tagliavini, G., *Chim. Ind. (Milan)* **50**, 452 (1968).

35. Bresadola, S., Rossetto, F., and Tagliavini, G., *Eur. Polym. J.* **4**, 75 (1968).
36. Bryukhova, E. V., Stanko, V. I., Klimova, A. I., Titova, N. S., and Semin, G. K., *Zh. Strukt. Khim.* **9**, 39 (1968).
37. Bukalov, S. S., Leites, L. A., and Aleksanyan, V. T., *Izv. Akad. Nauk SSSR, Ser. Khim.* p. 929 (1968).
38. Burg, A. B., and Kratzer, R., *Inorg. Chem.* **1**, 725 (1962).
39. Churchill, M. R., Gold, K., Francis, J. N., and Hawthorne, M. F., *J. Amer. Chem. Soc.* **91**, 1222 (1969).
39a. Clark, S. L., and Mangold, D. J., U.S. Pat. 3,092,664 (1963).
40. Council of the American Chemical Society, *Inorg. Chem.* **7**, 1945 (1968).
41. Davis, M. A., and Soloway, A. H., *J. Med. Chem.* **10**, 730 (1967).
42. DeBoer, B. G., Zalkin, A., and Templeton, D. H., *Inorg. Chem.* **7**, 2288 (1968).
43. Delman, A. D., Kelly, J. J., Stein, A. A., and Simms, B. B., *J. Polym. Sci., Part A-1* **5**, 2119 (1967).
44. Delman, A. D., Stein, A. A., Kelly, J. J., and Simms, B. B., *J. Appl. Polym. Sci.* **11**, 1979 (1967).
45. Ditter, J. F., *Inorg. Chem.* **7**, 1748 (1968).
46. Ditter, J. F., Gerhart, F. J., and Williams, R. E., *Advan. Chem. Ser.* **72**, 191 (1968).
47. Ditter, J. F., Klusmann, E. B., Oakes, J. D., and Williams, R. E., *Inorg. Chem.* **9**, 889 (1970).
48. Ditter, J. F., Spielman, J. R., and Williams, R. E., *Inorg. Chem.* **5**, 118 (1966).
49. Dobrott, R. D., and Lipscomb, W. N., *J. Chem. Phys.* **37**, 1779 (1962).
50. Dobson, J., Gaines, D. F., and Schaeffer, R., *J. Amer. Chem. Soc.* **87**, 4072 (1965).
51. Dobson, J., Keller, P. C., and Schaeffer, R., *Inorg. Chem.* **7**, 399 (1968).
52. Dobson, J., Keller, P. C., and Schaeffer, R., *J. Amer. Chem. Soc.* **87**, 3522 (1965).
53. Dobson, J., and Schaeffer, R., *Inorg. Chem.* **7**, 402 (1968).
54. Dunks, G. B., and Hawthorne, M. F., *Inorg. Chem.* **7**, 1038 (1968).
54a. Dunks, G. B., and Hawthorne, M. F., *Inorg. Chem.*, **8**, 2667 (1969).
54b. Dunks, G. B., and Hawthorne, M. F., *Inorg. Chem.*, **9**, 893 (1970).
55. Dunks, G. B., and Hawthorne, M. F., *J. Amer. Chem. Soc.* **90**, 7355 (1968).
56. Dupont, J. A., and Hawthorne, M. F., *J. Amer. Chem. Soc.* **86**, 1643 (1964).
57. Dupont, J. A., and Hawthorne, M. F., U.S. Pat. 3,228,986 (1966).
58. Dupont, J. A., and Hawthorne, M. F., U.S. Pat. 3,228,987 (1966).
59. Dupont, J. A., and Hawthorne, M. F., U.S. Pat. 3,254,117 (1966).
60. Dupont, J. A., and Schaeffer, R., *J. Inorg. Nucl. Chem.* **15**, 310 (1960).
61. Eaton, G. R., and Lipscomb, W. N., "NMR Studies of Boron Hydrides and Related Compounds." Benjamin, New York, 1969.
62. Echeistova, A. I., Syrkin, Ya. K., Stanko, V. I., and Klimova, A. I., *Zh. Strukt. Khim.* **8**, 933 (1967).
63. Edwards, L. J., and Makhlouf, J. M., *J. Amer. Chem. Soc.* **88**, 4728 (1966).
64. Enrione, R. E., Boer, F. P., and Lipscomb, W. N., *Inorg. Chem.* **3**, 1659 (1964).
65. Enrione, R. E., Boer, F. P., and Lipscomb, W. N., *J. Amer. Chem. Soc.* **86**, 1451 (1964).
66. Fein, M. M., Bobinski, J., Mayes, N., Schwartz, N. N., and Cohen, M. S., *Inorg. Chem.* **2**, 1111 (1963).
67. Fein, M. M., and Cohen, M. S., U.S. Pat. 3,376,347 (1968).
68. Fein, M. M., Cohen, M. S., and Nebel, C. W., U.S. Pat. 3,256,326 (1966).
69. Fein, M. M., Grafstein, D., Paustian, J. E., Bobinski, J., Lichstein, B. M., Mayes, N., Schwartz, N. N., and Cohen, M. S., *Inorg. Chem.* **2**, 1115 (1963).
70. Fein, M. M., Green, J., and Mayes, N., U.S. Pat. 3,354,193 (1967).
71. Fein, M. M., Green, J., and O'Brien, E. L., U.S. Pat. 3,355,478 (1967).

72. Fein, M. M., and Paustian, J. E., *Ind. Eng. Chem., Process Des. Develop.* **4**, 129 (1965).
73. Fein, M. M., and Paustian, J. E., U.S. Pat. 3,217,031 (1965).
74. Fein, M. M., and Schwartz, N. N., U.S. Pat. 3,290,357 (1966).
75. Fein, M. M., Schwartz, N. N., and Karlan, S. I., French Pat. 1,436,574 (1966).
76. Fetter, N. R., *Can. J. Chem.* **44**, 1463 (1966).
77. Francis, J. N., and Hawthorne, M. F., *J. Amer. Chem. Soc.* **90**, 1663 (1968).
78. Franz, D. A., and Grimes, R. N., *Abstr. Pap., 158th Nat. Meeting, Amer. Chem. Soc.* No. INOR-100 (1969).
78a. Franz, D. A., and Grimes, R. N., *J. Amer. Chem. Soc.* **92**, 1438 (1970).
79. Franz, D. A., Howard, J. W., and Grimes, R. N., *J. Amer. Chem. Soc.* **91**, 4010 (1969).
80. Fritchie, C. J., *Inorg. Chem.* **6**, 1199 (1967).
81. Garrett, P. M., George, T. A., and Hawthorne, M. F., *Inorg. Chem.* **8**, 2008 (1969).
82. Garrett, P. M., Smart, J. C., Ditta, G. S., and Hawthorne, M. F., *Inorg. Chem.* **8**, 1907 (1969).
83. Garrett, P. M., Smart, J. C., and Hawthorne, M. F., *J. Amer. Chem. Soc.* **91**, 4707 (1969).
84. Garrett, P. M., Tebbe, F. N., and Hawthorne, M. F., *J. Amer. Chem. Soc.* **86**, 5016 (1964).
85. George, T. A., and Hawthorne, M. F., *Inorg. Chem.* **8**, 1801 (1969).
86. George, T. A., and Hawthorne, M. F., *J. Amer. Chem. Soc.* **90**, 1661 (1968).
87. George, T. A., and Hawthorne, M. F., *J. Amer. Chem. Soc.* **91**, 5475 (1969).
88. German, E. D., Dyatkina, M. E., *Zh. Strukt. Khim.* **7**, 866 (1966).
89. Gerrard, W., *Plast. Inst., Trans. J.* **35**, 509 (1967).
90. Grafstein, D., Bobinski, J., Dvorak, J., Paustian, J. E., Smith, H. F., Karlan, S. I., Vogel, C., and Fein, M. M., *Inorg. Chem.* **2**, 1125 (1963).
91. Grafstein, D., Bobinski, J., Dvorak, J., Smith, H. F., Schwartz, N. N., Cohen, M. S., and Fein, M. M., *Inorg. Chem.* **2**, 1120 (1963).
92. Grafstein, D., and Dvorak, J., *Inorg Chem.* **2**, 1128 (1963).
93. Grafstein, D., and Dvorak, J. J., U.S. Pat. 3,226,429 (1965).
94. Grassberger, M. A., Hoffmann, E. G., Schomburg, G., and Köster, R., *J. Amer. Chem. Soc.* **90**, 56 (1968).
95. Graybill, B. M., and Hawthorne, M. F., *Inorg. Chem.* **8**, 1799 (1969).
96. Green, J., and Mayes, N., *J. Macromol. Sci., Chem.* **1**, 135 (1967).
97. Green, J., Mayes, N., and Cohen, M. S., *J. Polym. Sci., Part A* **2**, 3113 (1964).
98. Green, J., Mayes, N., Kotolby, A. P., and Cohen, M. S., *J. Polym. Sci., Part A* **2**, 3135 (1964).
99. Green, J., Mayes, N., Kotloby, A., Fein, M. M., O'Brien, E. L., and Cohen, M. S., *J. Polym. Sci., Part B* **2**, 109 (1964).
100. Green, P. T., and Bryan, R. F., *Inorg. Chem.* **9**, 1464 (1970).
101. Gregor, V., Hermanek, S., and Plesek, J., *Collect. Czech. Chem. Commun.* **33**, 980 (1968).
102. Gregor, V., Plesek, J., and Hermanek, S., *Int. Union. Pure Appl. Chem., Int. Symp. Macromol. Chem. 1965*; see *J. Polym. Sci., Part C* p. 4623 (1965).
103. Grimes, R. N,, *J. Amer. Chem. Soc.* **88**, 1070 (1966).
104. Grimes, R. N., *J. Amer. Chem. Soc.* **88**, 1895 (1966).
105. Grimes, R. N., *J. Organometal. Chem.* **8**, 45 (1967).
106. Grimes, R. N., and Bramlett, C. L., *J. Amer. Chem. Soc.* **89**, 2557 (1967).
107. Grimes, R. N., Bramlett, C. L., and Vance, R. L., *Inorg. Chem.* **7**, 1066 (1968).
108. Grimes, R. N., Bramlett, C. L., and Vance, R. L., *Inorg. Chem.* **8**, 55, (1969).
109. Grimes, R. N., and Rademaker, W. J., *J. Amer. Chem. Soc.* **91**, 6498 (1969).
110. Grimes, R. N., and Rademaker, W. J., unpublished results (1969).

111. Haas, T. E., *Inorg. Chem.* **3**, 1053 (1964).

112. Harmon, K. M., and Harmon, A. B., *J. Amer. Chem. Soc.* **88**, 4093 (1966).

113. Harmon, K. M., Harmon, A. B., and MacDonald, A. A., *J. Amer. Chem. Soc.* **86**, 5036 (1964).

114. Harmon, K. M., Harmon, A. B., and Thompson B., C., *J. Amer. Chem. Soc.* **89**, 5309 (1967).

115. Harris, C. B., *Inorg. Chem.* **7**, 1517 (1969).

116. Harrison, B. C., Solomon, I. J., Hites, R. D., and Klein, M. J., *J. Inorg. Nucl. Chem.* **14**, 195 (1960).

117. Hart, H., and Lipscomb, W. N., *Inorg. Chem.* **7**, 1070 (1968).

118. Hart, H., and Lipscomb, W. N., *J. Amer. Chem. Soc.* **89**, 4220 (1967).

119. Hart, H., and Lipscomb, W. N., *J. Amer. Chem. Soc.* **91**, 771 (1969).

120. Haslinger, F., and Soloway, A. H., *J. Med. Chem.* **9**, 792 (1966).

121. Hawthorne, M. F., *Accounts Chem. Res.* **1**, 281 (1968).

122. Hawthorne, M. F., *in* "The Chemistry of Boron and its Compounds" (E. L. Muetterties, ed.), Chapter 5, p. 223. Wiley, New York, 1967.

123. Hawthorne, M. F., *Endeavour* **25**, 146 (1966).

124. Hawthorne, M. F., and Andrews, T. D., *Chem. Commun.* **19**, 443 (1965).

125. Hawthorne, M. F., and Andrews, T. D., *J. Amer. Chem. Soc.* **87**, 2496 (1965).

126. Hawthorne, M. F., Andrews, T. D., Garrett, P. M., Olsen, F. P., Reintjes, M., Tebbe, F. N., Warren, L. F., Wegner, P. A., and Young, D. C., *Inorg. Syn.* **10**, 91–118 (1967).

127. Hawthorne, M. F., Berry, T. E., and Wegner, P. A., *J. Amer. Chem. Soc.* **87**, 4746 (1965).

128. Hawthorne, M. F., and George, T. A., *J. Amer. Chem. Soc.* **89**, 7114 (1967).

129. Hawthorne, M. F., and Owen, D. A., *J. Amer. Chem. Soc.* **90**, 5912 (1968).

130. Hawthorne, M. F., and Pilling, R. L., *J. Amer. Chem. Soc.* **87**, 3987 (1965).

131. Hawthorne, M. F., and Pitochelli, A. R., *J. Amer. Chem. Soc.* **81**, 5519 (1959).

132. Hawthorne, M. F., and Pitts, A. D., *J. Amer. Chem. Soc.* **89**, 7115 (1967).

133. Hawthorne, M. F., and Ruhle, H. W., *Inorg. Chem.* **8**, 176 (1969).

134. Hawthorne, M. F., and Wegner, P. A., *J. Amer. Chem. Soc.* **87**, 4392 (1965).

135. Hawthorne, M. F., and Wegner, P. A., *J. Amer. Chem. Soc.* **90**, 896 (1968).

136. Hawthorne, M. F., Wegner, P. A., and Stafford, R. C., *Inorg. Chem.* **4**, 1675 (1965).

137. Hawthorne, M. F., Young, D. C., Andrews, T. D., Howe, D. V., Pilling, R. L., Pitts, A. D., Reintjes, M., Warren, L. F., and Wegner, P. A., *J. Amer. Chem. Soc.* **90**, 879 (1968).

138. Hawthorne, M. F., Young, D. C., Garrett, P. M., Owen, D. A., Schwerin, S. G., Tebbe, F. N., and Wegner, P. A., *J. Amer. Chem. Soc.* **90**, 862 (1968).

139. Hawthorne, M. F., Young, D. C., and Wegner, P. A., *J. Amer. Chem. Soc.* **87**, 1818 (1965).

140. Herber, R. H., *Inorg. Chem.* **8**, 174 (1969).

141. Hertler, W. R., Klanberg, F., and Muetterties, E. L., *Inorg. Chem.* **6**, 1696 (1967).

142. Heying, T. L., Ager, J. W., Clark, S. L., Alexander, R. P., Papetti, S., Reid, J. A., and Trotz, S. I., *Inorg. Chem.* **2**, 1097 (1963).

143. Heying, T. L., Ager, J. W., Clark, S. L., Mangold, D. J., Goldstein, H. L., Hillman, M., Polak, R. J., and Szymanski, J. W., *Inorg. Chem.* **2**, 1089 (1963).

144. Heying, T. L., Papetti, S., and Schaffling, O. G., French Pat. 1,484,253 (1967).

145. Heying, T. L., Papetti, S., and Schaffling, O. G., U.S. Pat. 3,388,090 (1968).

146. Heying, T. L., Papetti, S., and Schaffling, O. G., U.S. Pat. 3,388,091 (1968).

147. Heying, T. L., Papetti, S., and Schaffling, O. G., U.S. Pat. 3,388,092 (1968).

148. Hill, W. E., *Inorg. Chem.* **7**, 222 (1968).

240   REFERENCES

149. Hoard, J. L., and Hughes, R. E., *in* "The Chemistry of Boron and its Compounds" (E. L. Muetterties, ed.), Chapter 2, p. 25. Wiley, New York, 1967.

150. Hoffmann, R., and Lipscomb, W. N., *Inorg. Chem.* **2**, 231 (1963).

151. Hoffmann, R., and Lipscomb, W. N, *J. Chem. Phys.* **36**, 2179 (1962).

152. Hoffmann, R., and Lipscomb, W. N., *J. Chem. Phys.* **36**, 3489 (1962).

153. Hoffmann, R., and Lipscomb, W. N., *J. Chem. Phys.* **37**, 520 (1962).

154. Hoffmann, R., and Lipscomb, W. N,, *J. Chem. Phys.* **37**, 2872 (1962).

155. Hota, N. K., and Matteson, D. S., *J. Amer. Chem. Soc.* **90**, 3570 (1968).

156. Howard, J. W., and Grimes, R. N., *J. Amer. Chem. Soc.* **91**, 6499 (1969).

157. Howard, J. W., and Grimes, R. N., to be published.

158. Hughes, R. L., Smith, I. C., and Lawless, E. W., *in* "Production of the Boranes and Related Research" (R. T. Holzmann, ed.). Academic Press, New York, 1967.

159. Hurd, D. T., *J. Amer. Chem. Soc.* **70**, 2053 (1948).

160. Hyatt, D. E., Little, J. L., Moran, J. T., Scholer, F. R., and Todd, L. J., *J. Amer. Chem. Soc.* **89**, 3342 (1967).

161. Hyatt, D. E., Owen, D. A., and Todd, L. J., *Inorg. Chem.* **5**, 1749 (1966).

162. Hyatt, D. E., Scholer, F. R., Todd, L. J., and Warner, J. L., *Inorg. Chem.* **6**, 2229 (1967).

163. Issleib, K., Lindner, R., and Tzschach, A., *Z. Chem.* **1**, 1 (1966).

164. Jacobson, R. A., and Lipscomb, W. N., *J. Chem. Phys.* **31**, 605 (1959).

165. Joy, F., Lappert, M. F., and Prokai, B., *J. Organometal. Chem.* **5**, 506 (1966).

166. Kaczmarczyk, A., Dobrott, R. D., and Lipscomb, W. N., *Proc. Nat. Acad. Sci. U.S.* **48**, 729 (1962).

167. Kaesz, H. D., Bau, R., Beall, H. A., and Lipscomb, W. N., *J. Amer. Chem. Soc.* **89**, 4218 (1967).

168. Kalinin, V. N., and Zakharkin, L. I., *Zh. Obshch. Khim.* **37**, 2136 (1967).

169. Kauffman, J. M., Green, J., Cohen, M. S., Fein, M. M., and Cottrill, E. L., *J. Amer. Chem. Soc.* **86**, 4210 (1964).

170. Klanberg, F., Eaton, D. R., Guggenberger, L. J., and Muetterties, E. L., *Inorg. Chem.* **6**, 1271 (1967).

171. Klanberg, F., Muetterties, E. L., and Guggenberger, L. J., *Inorg. Chem.* **7**, 2272 (1968).

172. Klanberg, F., Wegner, P. A., Parshall, G. W., and Muetterties, E. L., *Inorg. Chem.* **7**, 2072 (1968).

173. Klebanskii, A. L., Gridina, V. F., Dorofeenko, L. P., Krupnova, L. E., Zakharova, G. E., and Shkambarnaya, N. I., *Khim. Geterotsikl. Soedin.* p. 570 (1967).

174. Klebanskii, A. L., Gridina, V. F., Dorofeenko, L. P., Zhigach, A. F., Kozlova, N. V. Krupnova, L. E., Zakharova, G. E., and Shkambarnaya, N. I., *Khim. Geterotsikl. Soedin.* p. 976 (1968).

175. Knoth, W. H., *J. Amer. Chem. Soc.* **89**, 1274 (1967).

176. Knoth, W. H., *J. Amer. Chem. Soc.* **89**, 3342 (1967).

177. Knoth, W. H., Hertler, W. R., and Muetterties, E. L., *Inorg. Chem.* **4**, 280 (1965).

178. Knoth, W. H., Little, J. L., Lawrence, J. R., Scholer, F. R., and Todd, L. J., *Inorg. Syn.* **11**, 33 (1968).

179. Knoth, W. H., Little, J. L., and Todd, L. J., *Inorg. Syn.* **11**, 41 (1968).

180. Knoth, W. H., Miller, H. C., England, D. C., Parshall, G. W., and Muetterties, E. L., *J. Amer. Chem. Soc.* **84**, 1056 (1962).

181. Koetzle, T. F., Scarbrough, F. E., and Lipscomb, W. N., *Inorg. Chem.* **7**, 1076 (1968).

182. Kongpricha, S., and Schroeder, H., *Inorg. Chem.* **8**, 2449 (1969).

183. Korshak, V. V., Sobolevskii, M. V., Zhigach, A. F., Sarishvili, I. G., Frolova, Z. M., Gol'din, G. S., and Baturina, L. S., *Vysokomol. Soedin., Ser. B* **10**, 584 (1968).

184. Köster, R., and Grassberger, M. A., *Angew. Chem., Int. Ed. Engl.* **5**, 580 (1966).

185. Köster, R., and Grassberger, M. A., *Angew. Chem. Int. Ed. Engl.* **6**, 218 (1967).
186. Köster, R., Grassberger, M. A., Hoffmann, E. G., and Rotermund, G. W., *Tetrahedron Lett.* p. 905 (1966).
187. Köster, R., Horstschaefer, H. J., and Binger, P., *Angew. Chem., Int. Ed. Engl.* **5**, 730 (1966).
188. Köster, R., and Rotermund, G. W., *Tetrahedron Lett.* p. 1667 (1964).
189. Köster, R., and Rotermund, G. W., *Tetrahedron Lett.* p. 777 (1965).
190. Laubengayer, A. W., and Rysz, W. R., *Inorg. Chem.* **4**, 1513 (1965).
191. Ledoux, W. A., and Grimes, R. N., *Abstr. Pap., 158th Nat. Meeting, Amer. Chem. Soc.* No. INOR-101 (1969).
192. Leites, L. A., Ogorodnikova, N. A., and Zakharkin, L. I., *J. Organometal. Chem.* **15**, 287 (1968).
193. Leites, L. A., Vinogradova, L. E., Kalinin, V. N., and Zakharkin, L. I., *Izv. Akad. Nauk SSSR, Ser. Khim.* p. 1016 (1968).
194. Lindner, H. H., and Onak, T. P., *J. Amer. Chem. Soc.* **88**, 1886 (1966).
195. Lipscomb, W. N., "Boron Hydrides." Benjamin, New York, 1963.
196. Lipscomb, W. N., *Proc. Nat. Acad. Sci. U.S.* **47**, 1791 (1961).
197. Lipscomb, W. N., *Science* **153**, 373 (1966).
198. Lipscomb, W. N., Pitochelli, A. R., and Hawthorne, M. F., *J. Amer. Chem. Soc.* **81**, 5833 (1959).
199. Little, J. L., Moran, J. T., and Todd, L. J., *J. Amer. Chem. Soc.* **89**, 5495 (1967).
200. Longuet-Higgins, H. C., and Roberts, M. de V., *Proc. Roy. Soc. Ser. A* **230**, 110 (1955).
201. Lutz, C. A., and Ritter, D. M., *Can. J. Chem.* **41**, 1344 (1963).
202. L'vov, A. I., and Zakharkin, L. I., *Izv. Akad. Nauk SSSR, Ser. Khim.* p. 2653 (1967).
203. Maguire, R. G., Solomon, I. J., and Klein, M. J., *Inorg. Chem.* **2**, 1133 (1963).
204. Maki, A. H., and Berry, T. E., *J. Amer. Chem. Soc.* **87**, 4437 (1965).
205. Maruca, R., Schroeder, H. A., and Laubengayer, A. W., *Inorg. Chem.* **6**, 572 (1967).
206. Mastryukov, V. S., Vilkov, L. V., Zhigach, A. F., and Siryatskaya, V. N., *Zh. Strukt. Khim.* **10**, 136 (1969).
207. Mayes, N., and Green, J., *Inorg. Chem.* **4**, 1082 (1965).
208. Mayes, N., Green, J., and Cohen, M. S., *J. Polym. Sci., Part A-1* **5**, 365 (1967).
209. Mikhailov, B. M., and Potapova, T. V., *Izv. Akad. Nauk SSSR, Ser. Khim.* p. 1629 (1967).
210. Mikhailov, B. M., and Potapova, T. V., *Izv. Akad. Nauk SSSR, Ser. Khim.* p. 1153 (1968).
211. Miller, J. J., and Johnson, F. A., *J. Amer. Chem. Soc.* **90**, 218 (1968).
212. Moore, E. B., Lohr, L. L., and Lipscomb, W. N., *J. Chem. Phys.* **35**, 1329 (1961).
213. Muetterties, E. L., and Knoth, W. H., *Chem. & Eng. News* **44**, 88 (1965).
214. Muetterties, E. L., and Knoth, W. H., *Inorg. Chem.* **4**, 1498 (1965).
215. Muetterties, E. L., and Knoth, W. H., "Polyhedral Boranes." Marcel Dekker, New York, 1968.
215a. Niedenzu, K., *Naturwissenschaften* **56**, 305 (1969).
216. Obenland, C. O., and Papetti, S., *J. Org. Chem.* **31**, 3868 (1966).
217. Okhlobystin, O. Y., and Zakharkin, L. I., *J. Organometal. Chem.* **3**, 257 (1965).
218. Olsen, F. P., and Hawthorne, M. F., *Inorg. Chem.* **4**, 1839 (1965).
219. Olsen, R. R., and Grimes, R. N., *J. Amer. Chem. Soc.* **92**, 5072 (1970).
220. Onak, T. P., *Advan. Organometal. Chem.* **3**, 263–363 (1965).
221. Onak, T. P., *Inorg. Chem.* **7**, 1043 (1968).
222. Onak, T. P., Drake, R. P., and Dunks, G. B., *Inorg. Chem.* **3**, 1686 (1964).
223. Onak, T. P., Drake, R., and Dunks, G. B., *J. Amer. Chem. Soc.* **87**, 2505 (1965).

224. Onak, T. P., and Dunks, G. B., *Inorg. Chem.* **5**, 439 (1966).

225. Onak, T. P., Dunks, G. B., Beaudet, R. A., and Poynter, R. L., *J. Amer. Chem. Soc.* **88**, 4622 (1966).

226. Onak, T. P., Dunks, G. B., Spielman, F. R., Gerhart, F. J., and Williams, R. E., *J. Amer. Chem. Soc.* **88**, 2061 (1966).

227. Onak, T. P., Marynick, D., Mattschei, P., and Dunks, G., *Inorg. Chem.* **7**, 1754 (1968).

228. Onak, T. P., Mattschei, P., and Groszek, E., *J. Chem. Soc., A* p. 1990 (1969).

229. Onak, T. P., Gerhart, F. J., and Williams, R. E., *J. Amer. Chem. Soc.* **85**, 3378 (1965).

230. Onak, T. P., Williams, R. E., and Weiss, H. G., *J. Amer. Chem. Soc.* **84**, 2830 (1962).

230a. Owen, D. A., and Hawthorne, M. F., *J. Amer. Chem. Soc.* **91**, 6002 (1969).

231. Papetti, S., and Heying, T. L., *Inorg. Chem.* **2**, 1105 (1963).

232. Papetti, S., and Heying, T. L., *Inorg. Chem.* **3**, 1448 (1964).

233. Papetti, S., and Heying, T. L., *J. Amer. Chem. Soc.* **86**, 2295 (1964).

234. Papetti, S., Obenland, C. O., and Heying, T. L., *Ind. Eng. Chem., Prod. Res. Develop.* **5**, 334 (1966).

235. Papetti, S., Schaeffer, B. B., Grey, A. P., and Heying, T. L., *J. Polym. Sci., Part A-1* **4**, 1623 (1966).

236. Papetti, S., Schaeffer, B. B., Troscianiec, H. J., and Heying, T. L., *Inorg. Chem.* **3**, 1444 (1964).

237. Pilling, R. L., Tebbe, F. N., and Hawthorne, M. F., *Proc. Chem. Soc., London* p. 402 (1964).

238. Pitochelli, A. R., and Hawthorne, M. F., *J. Amer. Chem. Soc.* **82**, 3228 (1960).

239. Popp, G., and Hawthorne, M. F., *J. Amer. Chem. Soc.* **90**, 6553 (1968).

240. Potapova, T. V., and Mikhailov, B. M., *Izv. Nauk SSSR, Ser. Khim.* p. 2367 (1967).

241. Potenza, J. A., and Lipscomb, W. N., *Inorg. Chem.* **3**, 1673 (1964).

242. Potenza, J. A., and Lipscomb, W. N., *Inorg. Chem.* **5**, 1471 (1966).

243. Potenza, J. A., and Lipscomb, W. N., *Inorg. Chem.* **5**, 1478 (1966).

244. Potenza, J. A., and Lipscomb, W. N., *Inorg. Chem.* **5**, 1483 (1966).

245. Potenza, J. A., and Lipscomb, W. N., *J. Amer. Chem. Soc.* **86**, 1874 (1964).

246. Potenza, J. A., and Lipscomb, W. N., *Proc. Nat. Acad. Sci. U.S.* **56**, 1917 (1966).

247. Potenza, J. A., Lipscomb, W. N., Vickers, G. D., and Schroeder, H. A., *J. Amer. Chem. Soc.* **88**, 628 (1966).

248. Prince, S. R., and Schaeffer, R., *Chem. Commun.* p. 451 (1968).

249. Reiner, J. R., Alexander, R. P., and Schroeder, H. A., *Inorg. Chem.* **5**, 1460 (1966).

250. Roscoe, J. S., Kongpricha, S., and Papetti, S., *Inorg. Chem.* **9**, 1561 (1970).

251. Ruhle, H. W., and Hawthorne, M. F., *Inorg. Chem.* **7**, 2279 (1968).

251a. St. Clair, D., Zalkin, A., and Templeton, D. H., *Inorg. Chem.* **8**, 2080 (1969).

252. Salinger, R. M., and Frye, C. L., *Inorg. Chem.* **4**, 1815 (1965).

253. Sarishvili, I. G., Zhigach, A. F., Sobolevskii, M. V., Akimov, B. A., and Kozyreva E. M., *Plast. Massy* **10**, 37 (1967).

254. Schaeffer, R., *in* "Progress in Boron Chemistry" (H. Steinberg and A. L. McCloskey, eds.), Vol. 1, Chapter 10, p. 417. MacMillan, New York, 1964.

255. Schaeffer, R., Johnson, Q., and Smith, G., *Inorg. Chem.* **4**, 917 (1965).

256. Schoenfelder, C. W., and Fein, M. M., U.S. Pat. 3,355,496 (1967).

257. Scholer, F. R., and Todd, L. J., *J. Organometal. Chem.* **14**, 261 (1968).

257a. Schroeder, H., private communication (1969).

258. Schroeder, H., *Inorg. Macromol. Rev.* **1**, 45 (1970).

259. Schroeder, H. A., *Rubber Age (New York)* p. 58 (1969).

260. Schroeder, H., Heying, T. L., and Reiner, J. R., *Inorg. Chem.* **2**, 1092 (1963).

261. Schroeder, H., Papetti, S., Alexander, R. P., Sieckhaus, J. F., and Heying, T. L., *Inorg. Chem.* **8**, 2444 (1969).

262. Schroeder, H., Reiner, J. R., Alexander, R. P., and Heying, T. L., *Inorg. Chem.* **3**, 1464 (1964).

263. Schroeder, H., Schaffling, O. G., Larchar, T. B., Frulla, F. F., and Heying, T. L., *Rubber Chem. Technol.* **39**, 1184 (1966).

264. Schroeder, H., and Vickers, G. D., *Inorg. Chem.* **2**, 1317 (1963).

265. Schwartz, N. N., O'Brien, E. I., Karlan, S. L., and Fein, M. M., *Inorg. Chem.* **4**, 661 (1965).

265a. Scott, D. R., *J. Organometal. Chem.* **6**, 429 (1966).

266. Scott, J. E., and Spielman, J. R., *U.S. Clearinghouse Fed. Sci. Tech. Inform.* AD 666314 (1968) (Avail. CFSTI).

267. Seklemian, H. V., and Williams, R. E., *Inorg. Nucl. Chem. Lett.* **3**, 289 (1967).

268. Semenuk, N. S., Papetti, S., and Schroeder, H. A., *Angew. Chem., Int. Ed. Engl.* **6**, 997 (1967).

269. Semenuk, N. S., Papetti, S., and Schroeder, H., *Inorg. Chem.* **8**, 2441 (1969).

270. Seyferth, D., and Prokai, B., *J. Organometal. Chem.* **8**, 366 (1967).

271. Shapatin, A. S., Golubtsov, S. A., Solov'ev, A. A., Zhigach, A. F., and Siryatskaya, V. N., *Plast. Massy* **12**, 19 (1965).

272. Shapatin, A. S., Krasovskaya, T. A., Golubtsov, S. A., Vishnevskii, F. N., and Smazhok, M. P., *Kremniiorg. Soedin., Tr. Soveshch.* **3**, 162 (1967).

273. Shapiro, I., Good, C. D., and Williams, R. E., *J. Amer. Chem. Soc.* **84**, 3837 (1962).

274. Shapiro, I., Keilin, B., Williams, R. E., and Good, C. D., *J. Amer. Chem. Soc.* **85**, 3167 (1963).

275. Shapiro, I., and Weiss, H. G., U.S. Pat. 3,086,996 (1963).

275a. Sieckhaus, J. F., private communication (1969).

276. Sieckhaus, J. F., Semenuk, N. S., Knowles, T. A., and Schroeder, H., *Inorg. Chem.* **8**, 2452 (1969).

276a. Sieckhaus, J. F., Semenuk, N. S., Knowles, T. A., and Schroeder, H., *U.S. Clearinghouse Fed. Sci. Tech. Inform.* AD 665306 (1968) (Avail. CFSTI).

277. Smart, J. C., Garrett, P. M., and Hawthorne, M. F., *J. Amer. Chem. Soc.* **91**, 1031 (1969).

278. Smith, H. D., *Inorg. Chem.* **8**, 676 (1969).

279. Smith, H. D., *J. Amer. Chem. Soc.* **87**, 1817 (1965).

280. Smith, H. D., and Hohnstedt, L. F., *Inorg. Chem.* **7**, 1061 (1968).

281. Smith, H. D., Knowles, T. A., and Schroeder, H., *Inorg. Chem.* **4**, 107 (1965).

282. Smith, H. D., Obenland, C. O., and Papetti, S., *Inorg. Chem.* **5**, 1013 (1966).

283. Smith, H. D., Robinson, M. A., and Papetti, S., *Inorg. Chem.* **6**, 1014 (1967).

284. Sobolevskii, M. V., Zhigach, A. F., Grinevich, K. P., Sarishvili, I. G., Siryatskaya, V. N., and Kozyreva, E. M., *Plast. Massy* **1**, 21 (1966).

285. Sobolevskii, M. V., Zhigach, A. F., Sarishvili, I. G., Grinevich, K. P., and Beyul, S. S. *Plast. Massy* **4**, 19 (1966).

286. Soloway, A. H., and Butler, D. N., *J. Med. Chem.* **9**, 411 (1966).

287. Sperling, L. H., Cooper, S. L., and Tobolsky, A. V., *J. Appl. Polym. Sci.* **10**, 1725 (1966).

288. Sperling, L. H., Zaganiaris, E. J., and Tobolsky, A. V., *U.S. Clearinghouse Fed. Sci. Tech. Inform.* AD 656049 (1967) (Avail. CFSTI).

289. Spielman, J. R., Dunks, G. B., and Warren, R., *Inorg. Chem.* **8**, 2172 (1969).

290. Spielman, J. R., and Scott, J. E., *J. Amer. Chem. Soc.* **87**, 3512 (1965).

291. Spielman, J. R., Warren, R., Dunks, G. B., Scott, J. E., and Onak, T., *Inorg. Chem.* **7**, 216 (1968).

292. Stanko, V. I., *Zh. Obshch. Khim.* **35**, 1139 (1965).

293. Stanko, V. I., and Anorova, G. A., *Zh. Obshch. Khim.* **36**, 946 (1966).
294. Stanko, V. I., and Anorova, G. A., *Zh. Obshch. Khim.* **37**, 507 (1967).
295. Stanko, V. I., and Anorova, G. A., *Zh. Obshch. Khim.* **37**, 2359 (1967).
296. Stanko, V. I., and Anorova, G. A., *Zh. Obshch. Khim.* **38**, 1404 (1968).
297. Stanko, V. I., and Anorova, G. A., *Zh. Obshch. Khim.* **38**, 2818 (1968).
298. Stanko, V. I., Anorova, G. A., and Klimova, T. P., *Zh. Obshch. Khim.* **36**, 1774 (1966).
299. Stanko, V. I., Anorova, G. A., and Klimova, T. P., *Zh. Obshch. Khim.* **38**, 1193 (1968).
299a. Stanko, V. I., Anorova, G. A., and Klimova, T. P., *Zh. Obshch. Khim.* **39**, 1073 (1969).
300. Stanko, V. I., and Bobrov, A. V., *Zh. Obshch. Khim.* **35**, 2003 (1965).
301. Stanko, V. I., and Brattsev, V. A., *Zh. Obshch. Khim.* **37**, 515 (1967).
302. Stanko, V. I., and Brattsev, V. A., *Zh. Obshch. Khim.* **38**, 662 (1968).
303. Stanko, V. I., Brattsev, V. A., Al'perovich, N. E., and Titova, N. S., *Zh. Obshch. Khim.* **36**, 1862 (1966).
304. Stanko, V. I., Brattsev, V. A., Al'perovich, N. E., and Titova, N. S., *Zh. Obshch. Khim.* **38**, 1056 (1968).
305. Stanko, V. I., Brattsev, V. A., Vostrikova, T. N., and Danilova, G. N., *Zh. Obshch. Khim.* **38**, 1348 (1968).
306. Stanko, V. I., Bregadze, V. I., Klimova, A. I., Okhlobystin, O. Yu., Kashin, A. N., Butin, K. P., and Beletskaya, I. P., *Izv. Akad. Nauk SSSR, Ser. Khim.* p. 421 (1968).
307. Stanko, V. I., Echeistova, A. I., Astakhova, I. S., Klimova, A. I., Struchkov, Yu. T., and Syrkin, Ya. K., *Zh. Strukt. Khim.* **8**, 928 (1967).
307a. Stanko, V. I., and Gol'tyapin, Yu. V., *Zh. Obshch. Khim.* **39**, 711 (1969).
308. Stanko, V. I., Gol'tyapin, Yu. V., and Brattsev, V. A., *Zh. Obshch. Khim.* **37**, 2360 (1967).
309. Stanko, V. I., Gol'tyapin, Yu. V., and Volkov, A. F., *Zh. Obshch. Khim.* **37**, 514 (1967).
310. Stanko, V. I., and Klimova, A. I., *Zh. Obshch. Khim.* **35**, 753 (1965).
311. Stanko, V. I., and Klimova, A. I., *Zh. Obshch. Khim.* **35**, 1141 (1965).
312. Stanko, V. I., and Klimova, A. I., *Zh. Obshch. Khim.* **35**, 1503 (1965).
313. Stanko, V. I., and Klimova, A. I., *Zh. Obshch. Khim.* **36**, 159 (1966).
314. Stanko, V. I., and Klimova, A. I., *Zh. Obshch. Khim.* **36**, 432 (1966).
315. Stanko, V. I., and Klimova, A. I., *Zh. Obshch. Khim.* **36**, 2219 (1966).
316. Stanko, V. I., and Klimova, A. I., *Zh. Obshch. Khim.* **38**, 1194 (1968).
317. Stanko, V. I., Klimova, A. I., Chapovskii, Yu. A., and Klimova, T. P., *Zh. Obshch. Khim.* **36**, 1779 (1966).
318. Stanko, V. I., Klimova, A. I., and Klimova, T. P., *Zh. Obshch. Khim.* **37**, 2236 (1967).
319. Stanko, V. I., Klimova, A. I., and Titova, N. S., *Zh. Obshch. Khim.* **38**, 2817 (1968).
320. Stanko, V. I., Kopylov, V. V., and Klimova, A. I., *Zh. Obshch. Khim.* **35**, 1433 (1965).
321. Stanko, V. I., and Struchkov, Yu. T., *Zh. Obshch. Khim.* **35**, 930 (1965).
322. Stanko, V. I., Struchkov, Yu. T., and Klimova, A. I., *Zh. Strukt. Khim.* **7**, 629 (1966).
323. Stanko, V. I., Struchkov, Yu. T., Klimova, A. I., Bryukhova, L. V., and Semin, G. K., *Zh. Obshch. Khim.* **36**, 1707 (1966).
324. Steck, S. J., Pressley, G. A., Stafford, F. E., Dobson, J., and Schaeffer, R., *Inorg. Chem.* **8**, 830 (1969).
325. Stock, A., "Hydrides of Boron and Silicon." Cornell Univ. Press, Ithaca, New York, 1933.
326. Streib, W. E., Boer, F. P., and Lipscomb, W. N., *J. Amer. Chem. Soc.* **85**, 2331 (1963).
327. Tebbe, F. N., Garrett, P. M., and Hawthorne, M. F., *J. Amer. Chem. Soc.* **86**, 4222 (1964).
328. Tebbe, F. N., Garrett, P. M., and Hawthorne, M. F., *J. Amer. Chem. Soc.* **88**, 607 (1966).
329. Tebbe, F. N., Garrett, P. M., and Hawthorne, M. F., *J. Amer. Chem. Soc.* **90**, 869 (1968).

330. Tebbe, F. N., Garrett, P. M., Young, D. C., and Hawthorne, M. F., *J. Amer. Chem. Soc.* **88**, 609 (1966).

331. Thiokol Chem. Corp., *U.S. Clearinghouse Fed. Sci. Tech. Inform.*, AD 624735 (1967) (Avail. CFSTI).

332. Thiokol Chem. Corp., Brit. Pat. 1,079,523 (1967).

332a. Todd, L. J., *Adv. Organometal. Chem.* **8**, 87 (1970).

333. Todd, L. J., Burke, A. R., Silverstein, H. T., Little, J. L., and Wikholm, G. S., *J. Amer. Chem. Soc.* **91**, 3376 (1969).

334. Todd, L. J., Little, J. L., and Silverstein, H. T., *Inorg. Chem.* **8**, 1698 (1969).

335. Todd, L. J., Paul, I. C., Little, J. L., Welcker, P. S., and Peterson, C. R., *J. Amer. Chem. Soc.* **90**, 4489 (1968).

336. Tsai, C., and Streib, W. E., *J. Amer. Chem. Soc.* **88**, 4513 (1966).

337. Vickers, G. D., Agahigian, H., Pier, E. A., and Schroeder, H., *Inorg. Chem.* **5**, 693 (1966).

338. Vilkov, L. V., Khaikin, L. S., Zhigach, A. F., and Siryatskaya, V. N., *Zh. Strukt. Khim.* **9**, 889 (1968).

339. Vilkov, L. V., Mastryukov, V. S., Akishin, P. A., and Zhigach, A. F., *Zh. Strukt. Khim.* **6**, 447 (1965).

340. Vilkov, L. V., Mastryukov, V. S., Zhigach, A. F., and Siryatskaya, V. N., *Zh. Strukt. Khim.* **8**, 3 (1967).

341. Voet, D., and Lipscomb, W. N., *Inorg. Chem.* **3**, 1679 (1964).

342. Voet, D., and Lipscomb, W. N., *Inorg. Chem.* **6**, 113 (1967).

343. Voorhees, R. L., and Rudolph, R. W., *J. Amer. Chem. Soc.* **91**, 2173 (1969).

344. Waddington, T. C., *Trans. Faraday Soc.* **63**, 42 (1967).

345. Warren, L. F., and Hawthorne, M. F., *J. Amer. Chem. Soc.* **89**, 470 (1967).

346. Warren, L. F., and Hawthorne, M. F., *J. Amer. Chem. Soc.* **90**, 4823 (1968).

347. Wegner, P. A., and Hawthorne, M. F., *Chem. Commun.* p. 861 (1966).

348. Wiesboeck, R. A., and Hawthorne, M. F., *J. Amer. Chem. Soc.* **86**, 1642 (1964).

349. Williams, R. E., *in* "Progress in Boron Chemistry," Vol. 2, Chapter 2, p. 37. Pergamon Press, Oxford, 1970.

350. Williams, R. E., and Gerhart, F. J., *J. Amer. Chem. Soc.* **87**, 3513 (1965).

351. Williams, R. E., and Onak, T. P., *J. Amer. Chem. Soc.* **86**, 3159 (1964).

352. Wilson, R. J., Warren, L. F., and Hawthorne, M. F., *J. Amer. Chem. Soc.* **91**, 758 (1969).

353. Wing, R. M., *J. Amer. Chem. Soc.* **89**, 5599 (1967).

354. Wing, R. M., *J. Amer. Chem. Soc.* **90**, 4828 (1968).

355. Wunderlich, J. A., and Lipscomb, W. N., *J. Amer. Chem. Soc.* **82**, 4427 (1960).

356. Yoshizaki, T., Shiro, M., Nakagawa, Y., and Watanabe, H., *Inorg. Chem.* **8**, 698 (1969).

357. Young, D. C., Howe, D. V., and Hawthorne, M. F., *J. Amer. Chem. Soc.* **91**, 859 (1969).

358. Zaborowski, R., and Cohn, K., *Inorg. Chem.* **8**, 678 (1969).

359. Zaganiaris, E. J., Sperling, L. H., and Tobolsky, A. V., *J. Macromol. Sci., Chem.* **1**, 1111 (1967).

360. Zakharkin, L. I., *Izv. Akad. Nauk SSSR, Ser. Khim.* p. 158 (1965).

361. Zakharkin, L. I., *Izv. Akad. Nauk SSSR, Ser. Khim.* p. 1114 (1965).

362. Zakharkin, L. I., *Dokl. Akad. Nauk SSSR* **162**, 817 (1965).

363. Zakharkin, L. I., *Tetrahedron Lett.* p. 2255 (1964).

364. Zakharkin, L. I., Brattsev, V. A., and Chapovskii, Yu. A., *Zh. Obshch. Khim.* **35**, 2160 (1965).

365. Zakharkin, L. I., Brattsev, V. A., and Stanko, V. I., *Zh. Obshch. Khim.* **36**, 886 (1966).

366. Zakharkin, L. I., Bregadze, V. I., and Okhlobystin, O. Y., *Izv. Akad. Nauk SSSR, Ser. Khim.* p. 1539 (1964).

367. Zakharkin, L. I., Bregadze, V. I., and Okhlobystin, O. Yu., *J. Organometal. Chem.* **4**, 211 (1965).

368. Zakharkin, L. I., Bregadze, V. I., and Okhlobystin, O. Y., *J. Organometal. Chem.* **6**, 228 (1966).

369. Zakharkin, L. I., Bregadze, V. I., and Okhlobystin, O. Yu., *Zh. Obshch. Khim.* **36**, 761 (1966).

370. Zakharkin, L. I., and Chapovskii, Yu. A., *Izv. Akad. Nauk SSSR, Ser. Khim.* p. 772 (1964).

371. Zakharkin, L. I., and Chapovskii, Yu. A., *Tetrahedron Lett.* p. 1147 (1964).

372. Zakharkin, L. I., Chapovskii, Yu. A., Brattsev, V. A., and Stanko, V. I., *Zh. Obshch. Khim.* **36**, 878 (1966).

373. Zakharkin, L. I., Chapovskii, Y. A., and Stanko, V. I., *Izv. Akad. Nauk SSSR, Ser. Khim.* p. 2208 (1964).

374. Zakharkin, L. I., and Grebennikov, A. V., *Izv. Akad. Nauk SSSR, Ser. Khim.* p. 2019 (1966).

375. Zakharkin, L. I., and Grebennikov, A. V., *Izv. Akad. Nauk SSSR, Ser. Khim.* p. 696 (1967).

376. Zakharkin, L. I., and Grebennikov, A. V., *Izv. Akad. Nauk SSSR Ser. Khim.* p. 1376 (1967).

377. Zakharkin, L. I., Grebennikov, A. V., and Kazantsev, A. V., *Izv. Akad. Nauk SSSR, Ser. Khim.* p. 2077 (1967).

378. Zakharkin, L. I., Grebennikov, A. V., Kazantsev, A. V., and L'vov, A. I., *Izv. Akad. Nauk SSSR, Ser. Khim.* p. 2079 (1967).

379. Zakharkin, L. I., Grebennikov, A. V., and Savina, L. A., *Izv. Akad. Nauk SSSR, Ser. Khim.* p. 1130 (1968).

380. Zakharkin, L. I., Grebennikov, A. V., Vinogradova, L. E., and Leites, L. A., *Zh. Obshch. Khim.* **38**, 1048 (1968).

381. Zakharkin, L. I., and Kalinin, V. N., *Dokl. Akad. Nauk SSSR* **163**, 110 (1965).

382. Zakharkin, L. I., and Kalinin, V. N., *Dokl. Akad. Nauk SSSR* **164**, 577 (1965).

383. Zakharkin, L. I., and Kalinin, V. N., *Dokl. Akad. Nauk SSSR* **169**, 590 (1966).

384. Zakharkin, L. I., and Kalinin, V. N., *Dokl. Akad. Nauk SSSR* **170**, 92 (1966).

385. Zakharkin, L. I., and Kalinin, V. N., *Dokl. Akad. Nauk SSSR* **173**, 1091 (1967).

386. Zakharkin, L. I., and Kalinin, V. N., *Izv. Akad. Nauk SSSR, Ser. Khim.* p. 579 (1965).

387. Zakharkin, L. I., and Kalinin, V. N., *Izv. Nauk SSSR, Ser. Khim.* p. 1311 (1965).

388. Zakharkin, L. I., and Kalinin, V. N., *Izv. Akad. Nauk SSSR, Ser. Khim.* p. 2206 (1965).

389. Zakharkin, L. I., and Kalinin, V. N., *Izv. Akad. Nauk SSSR, Ser. Khim.* p. 575 (1966).

390. Zakharkin, L. I., and Kalinin, V. N., *Izv. Akad. Nauk SSSR, Ser. Khim.* p. 586 (1966).

391. Zakharkin, L. I., and Kalinin, V. N., *Izv. Akad. Nauk SSSR, Ser. Khim.* p. 2014 (1966).

392. Zakharkin, L. I., and Kalinin, V. N., *Izv. Akad. Nauk SSSR, Ser. Khim.* p. 462 (1967).

393. Zakharkin, L. I., and Kalinin, V. N., *Izv. Akad. Nauk SSSR, Ser. Khim.* p. 473 (1967).

394. Zakharkin, L. I., and Kalinin, V. N., *Izv. Akad. Nauk SSSR, Ser. Khim.* p. 937 (1967).

395. Zakharkin, L. I., and Kalinin, V. N., *Izv. Akad. Nauk SSSR, Ser. Khim.* p. 2577 (1967).

396. Zakharkin, L. I., and Kalinin, V. N., *Izv. Akad. Nauk SSSR, Ser. Khim.* p. 2585 (1967).

397. Zakharkin, L. I., and Kalinin, V. N., *Izv. Akad. Nauk SSSR, Ser. Khim.* p. 685 (1968).

398. Zakharkin, L. I., and Kalinin, V. N., *Izv. Akad. Nauk SSSR, Ser. Khim.* p. 1423 (1968).

399. Zakharkin, L. I., and Kalinin, V. N., *Izv. Akad. Nauk SSSR, Ser. Khim.* p. 194 (1969).

399a. Zakharkin, L. I., and Kalinin, V. N., *Izv. Akad. Nauk SSSR, Ser. Khim.* p. 607 (1969).

400. Zakharkin, L. I., and Kalinin, V. N., *Tetrahedron Lett* p. 407 (1965).

401. Zakharkin, L. I., and Kalinin, V. N., *Zh. Obshch. Khim.* **35**, 1691 (1965).
402. Zakharkin, L. I., and Kalinin, V. N., *Zh. Obshch. Khim.* **35**, 1882 (1965).
403. Zakharkin, L. I., and Kalinin, V. N., *Zh. Obshch. Khim.* **36**, 362 (1966).
404. Zakharkin, L. I., and Kalinin, V. N., *Zh. Obshch. Khim.* **36**, 2218 (1966).
405. Zakharkin, L. I., and Kalinin, V. N., *Zh. Obshch. Khim.* **37**, 281 (1967).
406. Zakharkin, L. I., and Kalinin, V. N., *Zh. Obshch. Khim.* **37**, 964 (1967).
407. Zakharkin, L. I., Kalinin, V. N., and Gedymin, V. V., *J. Organometal. Chem.* **16**, 371 (1969).
408. Zakharkin, L. I., Kalinin, V. N., Kvasov, B. A., and Fedin, E. I., *Izv. Akad. Nauk SSSR, Ser. Khim.* p. 2415 (1968).
409. Zakharkin, L. I., Kalinin, V. N., and Lozovskaya, V. S., *Izv. Akad. Nauk SSSR, Ser. Khim.* p. 1780 (1968).
410. Zakharkin, L. I., Kalinin, V. N., and L'vov, A. I., *Izv. Akad. Nauk SSSR, Ser. Khim.* p. 1091 (1966).
411. Zakharkin, L. I., Kalinin, V. N., and Podvisotskaya, L. S., *Izv. Akad. Nauk SSSR, Ser. Khim.* p. 1713 (1965).
412. Zakharkin, L. I., Kalinin, V. N., and Podvisotskaya, L. S., *Izv. Akad. Nauk SSSR, Ser. Khim.* p. 1495 (1966).
413. Zakharkin, L. I., Kalinin, V. N., and Podvisotskaya, L. S., *Izv. Akad. Nauk SSSR, Ser. Khim.* p. 1867 (1966).
414. Zakharkin, L. I., Kalinin, V. N., and Podvisotskaya, L. S., *Izv. Akad. Nauk SSSR, Ser. Khim.* p. 2310 (1967).
415. Zakharkin, L. I., Kalinin, V. N., and Podvisotskaya, L. S., *Izv. Akad. Nauk SSSR, Ser. Khim.* p. 679 (1968).
416. Zakharkin, L. I., Kalinin, V. N., and Podvisotskaya, L. S., *Izv. Akad. Nauk SSSR, Ser. Khim.* p. 688 (1968).
417. Zakharkin, L. I., Kalinin, V. N., and Podvisotskaya, L. S., *Izv. Akad. Nauk SSSR, Ser. Khim.* p. 2661 (1968).
418. Zakharkin, L. I., Kalinin, V. N., and Podvisotskaya, L. S., *Zh. Obshch. Khim.* **36**, 1786 (1966).
419. Zakharkin, L. I., Kalinin, V. N., and Shepilov, I. P., *Dokl. Akad. Nauk SSSR* **174**, 606 (1967).
420. Zakharkin, L. I., Kalinin, V. N., and Shepilov, I. P., *Izv. Akad. Nauk SSSR, Ser. Khim.* p. 1286 (1966).
421. Zakharkin, L. I., Kalinin, V. N., and Snyakin, A. P., *Izv. Akad. Nauk SSSR, Ser. Khim.* p. 1878 (1967).
422. Zakharkin, L. I., Kalinin, V. N., and Snyakin, A. P., *Izv. Akad. Nauk SSSR, Ser. Khim.* p. 197 (1968).
422a. Zakharkin, L. I., Kalinin, V. N., Snyakin, A. P., and Kvasov, B. A., *J. Organometal. Chem.* **18**, 19 (1969).
423. Zakharkin, L. I., Kalinin, V. N., and Zhigareva, G. G., *Izv. Akad. Nauk SSSR, Ser. Khim.* p. 2081 (1967).
424. Zakharkin, L. I., Kalinin, V. N., Zhigareva, G. G., and Grebennikov, A. V., *Izv. Akad. Nauk SSSR, Ser. Khim.* p. 1862 (1968).
425. Zakharkin, L. I., and Kazantsev, A. V., *Izv. Akad. Nauk SSSR, Ser. Khim.* p. 2190 (1965).
426. Zakharkin, L. I., and Kazantsev, A. V., *Izv. Akad. Nauk SSSR, Ser. Khim.* p. 568 (1966).
427. Zakharkin, L. I., and Kazantsev, A. V., *Izv. Akad. Nauk SSSR, Ser. Khim.* p. 1840 (1966).

428. Zakharkin, L. I., and Kazantsev, A. V., *Zh. Obshch. Khim.* **35**, 1123 (1965).

429. Zakharkin, L. I., and Kazantsev, A. V., *Zh. Obshch. Khim.* **36**, 944 (1966).

430. Zakharkin, L. I., and Kazantsev, A. V., *Zh. Obshch. Khim.* **36**, 945 (1966).

431. Zakharkin, L. I., and Kazantsev, A. V., *Zh. Obshch. Khim.* **36**, 1285 (1966).

432. Zakharkin, L. I., and Kazantsev, A. V., *Zh. Obshch. Khim.* **37**, 554 (1967).

433. Zakharkin, L. I., and Kazantsev, A. V., *Zh. Obshch. Khim.* **37**, 1211 (1967).

434. Zakharkin, L. I., Kukulina, E. I., and Podvisotskaya, L. S., *Izv. Akad. Nauk SSSR, Ser. Khim.* p. 1866 (1966).

435. Zakharkin, L. I., and L'vov, A. I., *Izv. Akad. Nauk SSSR, Ser. Khim.* p. 151 (1966).

436. Zakharkin, L. I., and L'vov, A. I., *J. Organometal. Chem.* **5**, 313 (1966).

437. Zakharkin, L. I., and L'vov, A. I., *Zh. Obshch. Khim.* **36**, 761 (1966).

438. Zakharkin, L. I., and L'vov, A. I., *Zh. Obshch. Khim.* **37**, 742 (1967).

439. Zakharkin, L. I., and L'vov, A. I., *Zh. Obshch. Khim.* **37**, 1217 (1967).

440. Zakharkin, L. I., L'vov, A. I., and Grebennikov, A. V., *Izv. Akad. Nauk SSSR, Ser. Khim.* p. 2282 (1968).

441. Zakharkin, L. I., L'vov, A. I., and Podvisotskaya, L. S., *Izv. Akad. Nauk SSSR, Ser. Khim.* p. 1905 (1965).

442. Zakharkin, L. I., L'vov, A. I., Soshka, S. A., and Shepilov, I. P., *Zh. Obshch. Khim.* **38**, 255 (1968).

443. Zakharkin, L. I., and Ogorodnikova, N. A., *Izv. Akad. Nauk SSSR, Ser. Khim.* p. 346 (1966).

444. Zakharkin, L. I., and Ogorodnikova, N. A., *J. Organometal. Chem.* **12**, 13, (1968).

445. Zakharkin, L. I., and Ogorodnikova, N. A., *Zh. Obshch. Khim.* **38**, 2595 (1968).

446. Zakharkin, L. I., Okhlobystin, O. Yu., Semin, G. K., and Babushkina, T. A., *Izv. Akad. Nauk SSSR, Ser. Khim.* p. 1913 (1965).

447. Zakharkin, L. I., and Podvisotskaya, L. S., *Izv. Akad. Nauk SSSR, Ser. Khim.* p. 1464 (1965).

448. Zakharkin, L. I., and Podvisotskaya, L. S., *Izv. Akad. Nauk SSSR, Ser. Khim.* p. 771 (1966).

449. Zakharkin, L. I., and Podvisotskaya, L. S., *Izv. Akad. Nauk SSSR, Ser. Khim.* p. 1369 (1967).

450. Zakharkin, L. I., and Podvisotskaya, L. S., *Izv. Akad. Nauk SSSR, Ser. Khim.* p. 417 (1968).

451. Zakharkin, L. I., and Podvisotskaya, L. S., *Izv. Akad. Nauk SSSR, Ser. Khim.* p. 681 (1968).

452. Zakharkin, L. I., and Podvisotskaya, L. S., *J. Organometal. Chem.* **7**, 385 (1967).

453. Zakharkin, L. I., and Podvisotskaya, L. S., *Zh. Obshch. Khim.* **37**, 506 (1967).

454. Zakharkin, L. I., Ponomarenko, A. A., and Okhlobystin, O. Yu., *Izv. Akad. Nauk SSSR, Ser. Khim.* p. 2210 (1964).

455. Zakharkin, L. I., Stanko, V. I., Brattsev, V. A., Chapovskii, Yu. A., Klimova, A. I., Okhlobystin, O. Yu., and Ponomarenko, A. A., *Dokl. Akad. Nauk SSSR* **155**, 1119 (1964).

456. Zakharkin, L. I., Stanko, V. I., Brattsev, V. A., Chapovskii, Yu. A., and Okhlobystin, O. Yu., *Izv. Akad. Nauk SSSR, Ser. Khim.* p. 2238 (1963).

457. Zakharkin, L. I., Stanko, V. I., Brattsev, V. A., Chapovskii, Yu. A., and Struchkov, Yu. T., *Izv. Akad. Nauk SSSR, Ser. Khim.* p. 2069 (1963).

458. Zakharkin, L. I., Stanko, V. I., and Chapovskii, Yu. A., *Izv. Akad. Nauk SSSR, Ser. Khim.* p. 582 (1964).

459. Zakharkin, L. I., Stanko, V. I., and Chapovskii, Yu. A., *Izv. Akad. Nauk SSSR, Ser. Khim.* p. 1539 (1964).

460. Zakharkin, L. I., Stanko, V. I., and Klimova, A. I., *Izv. Akad. Nauk SSSR, Ser. Khim.* p. 771 (1964).

461. Zakharkin, L. I., Stanko, V. I., and Klimova, A. I., *Izv. Akad. Nauk SSSR, Ser. Khim.* p. 1946 (1966).

462. Zakharkin, L. I., Stanko, V. I., and Klimova, A. I., *Zh. Obshch. Khim.* 35, 394 (1965).

463. Zakharkin, L. I., Stanko, V. I., Klimova, A. I., and Chapovskii, Yu. A., *Izv. Akad. Nauk SSSR, Ser. Khim.* p. 2236 (1963).

464. Zakharkin, L. I., and Zhigareva, G. G., *Izv. Akad. Nauk SSSR, Ser. Khim.* p. 932 (1965).

465. Zakharkin, L. I., and Zhigareva, G. G., *Izv. Akad. Nauk SSSR, Ser. Khim.* p. 1358 (1967).

465a. Zakharkin, L. I., and Zhigareva, G. G., *Izv. Akad. Nauk SSSR, Ser. Khim.* p. 611 (1969).

466. Zakharkin, L. I., and Zhigareva, G. G., *Zh. Obshch. Khim.* 37, 1791 (1967).

467. Zakharkin, L. I., and Zhigareva, G. G., *Zh. Obshch. Khim.* 37, 2781 (1967).

468. Zakharkin, L. I., Zhigareva, G. G., and Kazantsev, A. V., *Zh. Obshch. Khim.* 38, 89 (1968).

469. Zalkin, A., Hopkins, T. E., and Templeton, D. H., *Inorg. Chem.* 5, 1189 (1966).

470. Zalkin, A., Hopkins, T. E., and Templeton, D. H., *Inorg. Chem.* 6, 1911 (1967).

471. Zalkin, A., Templeton, D. H., and Hopkins, T. E., *J. Amer. Chem. Soc.* 87, 3988 (1965).

472. Zhigach, A. F., Sobolevskii, M. V., Sarishvili, I. G., and Akimov, B. A., *Plast. Massy* 5, 20 (1965).

AUTHOR INDEX

Numbers in parentheses are reference numbers and indicate that an author's work is referred to although his name is not cited in the text. Numbers in italics show the page on which the complete reference is listed.

Adams, R. M., 12(1), 233(2), *236*
Agahigian, H., 55(337), 157(337), 177(337), *245*
Ager, J. W., 3(143), 54(143), 56(143), 57(143), 58(142, 143), 59(142, 143), 60(142), 61(142, 143), 62(142), 63(143), 65(142, 66(142), 67(142), 69(142), 70(142, 143), 72(142, 143), 73(142, 143), 75(142), 78(142), 81(142), 82 (142), 83(142), 85(142, 143), 88(142), 90(142), 91(142), 92(142), 93(142), 95 (142), 98(142), 103(142), 109(143), 110 (142), 113(142), 120(142), 140(142, 143), *239*
Akimov, B. A., 182(253, 472), 183(4, 5), 189(3), *236, 242, 249*
Akishin, P. A., 55(339), *245*
Akovbyan, E. M., 95(6), *236*
Aleksanyan, V. T., 55(37), 152(37), 234(37), *237*
Alexander, R. P., 56(7), 58(142), 59(142), 60(142), 61(7, 142), 62(142), 65(142), 66(142), 67(142), 69(142), 70(7, 142), 72(142), 73(143), 75(142), 78(142), 81(142), 82(142), 83(142), 85(142), 88(142), 90(142), 91(142), 93(142), 95(142), 98(142), 100(249), 103(142), 110(142), 113(142), 115(7), 116(7), 117(9), 120(142), 136(262), 139(262), 140(142), 141(262), 153(249), 154(8, 249), 156(262), 166(249), 168(8, 249),

174(262), 189(7), *236, 239, 242, 243*
Al'perovich, N. E., 60(304), 85(304), 99(303, 304), 103(304), 104(303, 304), 115(303), 153(304), 166(303, 304), *244*
Andrews, T. D., 13(137), 46(137), 193(126), 194(137), 197(137), 208(137), 209(137), 210(137), 211(137), 213(137), 214(125, 137), 215(137), 216(137), 217(124, 137), 218(124, 137), 220(137), 221(137), 222(137), *239*
Andrianov, V. G., 23(10), 55(10), *236*
Anorova, G. A., 58(293, 298), 62(296, 299a), 63(299a), 64(293, 299a), 67(292), 68 (292, 299), 69(293, 297, 298), 83(292, 293, 298), 136(294, 295), 140(294, 295, 296, 299a), 145(299), 146(299a), 152 (298), 155(299a), 163(298), 174(299a), *243, 244*
Astakhova, I. S., 62(307), 63(307), 64(307), 135(307), 154(307), 155(307), 156(307), 172(307), 173(307), *244*

Babushkina, T. A., 78(446), 136(446), *248*
Baturina, L. S., 190(183), *240*
Bau, R., 12(167), 158(167), 159(167), *240*
Beall, H. A., 12(167), 158(167), 159(167), 172(11), 173(11), *236, 240*
Beaudet, R. A., 3(12), 41(12, 13, 225), 43(12, 13, 225), *236, 242*

251

Bekasova, N. I., 189(3), *236*

Beletskaya, I. P., 141(306), 150(306), 151(306), 161(306), 177(306), *244*

Berry T. E., 48(14), 50(14), 57(127), 58(127), 75(127), 82(127), 89(127), 108(127), 152(127), 162(127), 196(14), 209(204), 215(204), *236, 239, 241*

Beyul, S. S., 183(285), *243*

Bilevich, K. A., 74(15, 16), *236*

Binger, P., 25(17), 27(17), 29(17), 30(17), 35(187), 38(187), 42(187), *236, 241*

Blay, N. J., 24(18), 36(18), *236*

Bobinski, J., 3(66), 54(66), 56(66, 69, 91), 57(66, 69, 91), 58(66, 69), 59(90), 60(66, 91), 63(91), 65(90, 91), 66(91), 67(91), 69(66, 69, 91), 70(66, 69), 72(91), 73(91), 75(66, 69, 91), 76(66, 69, 91), 77(66), 78(66, 91), 79(91), 80(66, 69), 82(90, 91), 83(91), 85(69, 91), 86(91), 88(90), 89(91), 91(90), 92(69, 90, 91), 93(90, 91), 95(90, 91), 97(90), 98(91), 111(91), 193(111), 195(66), *237, 238*

Boer, F. P., 17(326), 23(22), 46(64, 65), 234 (20), 235(21), *236, 237, 244*

Boone, J. L., 61(23), 110(23), *236*

Borrov, A. V., 57(300), 58(300), 108(300), *244*

Bramlett, C. L., 24(106, 107), 26(107), 27(24, 107), 30(24, 106, 107), 34(106, 108), 35(106), 36(106, 108), 37(106), 40(108), 42(108), 43(106, 108), 45(106, 108), 52(106), *236, 238*

Brattsev, A., 143(308), 175(308), *244*

Brattsev, V. A., 54(456, 457), 55(457), 56(364, 365, 455, 457), 57(364, 365, 455, 457), 58(372), 59(372), 60(304, 372, 455), 61(364, 365), 65(364, 372, 455), 66(455), 67(364, 455), 69(365), 70(365, 455), 72(364, 365), 73(364, 455), 75(455), 82(455), 85(304, 372, 455), 86(364, 372), 87(372, 455), 88(365, 372, 455), 89(372, 455), 90(364), 92(365, 455), 95(372), 96(364), 97(364, 372), 99(303, 304), 103(304, 372, 455), 104(303), 108(365), 109(364, 372, 455), 113(364, 455), 115(303), 135(305), 162(302), 166(303), 172(305), 173(305), 193(301, 302), 195(25), 196(301), 224 (25), *236, 244, 245, 246, 248*

Bregadze, V. I., 61(367), 62(367), 95(6), 115(366, 367), 117(367), 118(367), 120

(366, 367), 127(366, 367), 128(29, 367), 129(29), 141(306), 148(366, 367, 368, 369), 149(28, 368, 369), 150(306, 368), 151(306), 161(306), 233(30), 234(27), *236, 244, 246*

Bresadola, S., 147(31), 170(32, 33, 35), 176(34), 187(35), 189(34), *236, 237*

Brotherton, R. J., 61(23), 110(23), *236*

Bryan, R. F., 217(100), *238*

Bryukhova, E. V., 134(323), 135(36, 323), 172(323), 173(36, 323), *237, 244*

Bukalov, S. S., 55(37), 152(37), 234(37), *237*

Burg, A. B., 46(38), 217(38), *237*

Burke, A. R., 204(333), 206(333), 207(333), 226(333), *245*

Butin, K. P., 141(306), 150(306), 151(306), 161(306), 177(306), *244*

Butler, D. N., 108(286), 110(286), *243*

Chaikina, E. A., 95(6), *236*

Chapovskii, Yu. A., 54(456, 457), 55(457), 56(364, 455, 457), 57(364, 455, 457), 58(371, 372, 373, 463), 59(317, 371, 372, 373, 463), 60(317, 372, 455), 61(317, 364, 371), 65(364, 372, 455), 66(455, 463), 67(364, 455, 458), 70(455), 72(364), 73(364, 455), 75(455), 82(455, 458, 463), 84(373, 463), 85(317, 371, 372, 455), 86(317, 364, 370, 371, 372), 87(372, 455, 459), 88(372, 455), 89(372, 455), 90(317, 364, 373), 92(455), 94(317), 95(372), 96(364), 97(364, 372), 100(317), 103(372, 455, 459), 104(317), 109(364, 370, 371, 372, 455), 113(317, 364, 370, 371, 455), *244, 245, 246, 248, 249*

Chumaevskii, N. A., 234(27), *236*

Churchill, M. R., 210(39), 219(39), 220(39), *237*

Clark, S. L., 3(143), 54(143), 56(143), 57(143), 58(142, 143), 59(142, 143), 60(142), 61(142, 143), 62(142), 63(143), 65(142), 66(142), 67(142), 70(142, 143), 72(39a, 142, 143), 73(142, 143), 75(142), 78(142), 81(142), 82(142), 83(142), 85(142, 143), 88(142), 90(142), 91(142), 92(142), 93(142), 95(143), 98(143), 103(142), 109(143), 110(142), 113(142), 120(142), 140(142, 143), *237, 239*

Cohen, M. S., 3(66), 54(66), 56(66, 69, 91), 57(66, 69, 91), 58(66, 69), 60(66, 91, 169), 63(91), 65(91, 99), 66(91), 67(91), 69(66, 69, 91), 70(66, 69), 72(91), 73(91), 75(66, 69, 91), 76(66, 69, 91), 77(66), 78(66, 91), 79(91), 80(66, 69), 82(91), 83(91), 85(69, 91), 86(91), 89(91), 92(69, 91), 93(91), 95(68, 91), 96(99), 98(91, 97, 98, 99), 109(169), 111(91), 124(208), 126(208), 140(67), 161(67), 172(67), 182(97, 98, 99), 183(208), 185(208), 189(169), 190(99, 208), 191(208), 193(91), 195(66), *237*, *238*, *240*, *241*

Cohn, K., 61(358), 115(358), 117(358), 119(318), *245*

Cooper, S. L., 186(287), *243*

Cottrill, E. L., 60(169), 109(169), 189(169), *240*

Danilova, G. N., 135(305), 172(305), 173(305), *244*

Davis, M. A., 115(41), *237*

DeBoer, B. G., 210(42), 217(42), 218(42), *237*

Delman, A. D., 185(44), 186(44), 192(43), *237*

Ditta, G. S., 47(82), 49(82), 52(82), *238*

Ditter, J. F., 33(47), 34(47), 35(45), 46(48), 234(46), *237*

Dobrott, R. D., 3(49), 48(49), 158(166), *237*, *240*

Dobson, J., 46(50, 51, 52, 53, 324), *237*, *244*

Dorofeenko, L. P., 120(173), 125(174), *240*

Drake, R. P., 26(222), 30(222), 35(222, 223), 39(223), 40(222), 41(222), 42(222), 43(222, 223), 44(223), *241*

Dunks, G. B., 25(54a, 55), 26(222, 226, 227), 27(54a, 55, 289), 28(224, 289, 291), 30(54a, 55, 222, 224, 226, 227, 289), 35(54, 222, 223), 39(223), 40(222), 41(222, 225), 42(222), 43(222, 223, 225), 44(223), 47(54), 48(54b, 55), 52(54b), 234(227), *237*, *240*, *241*, *243*

Dupont, J. A., 55(56, 57, 58), 58(56, 59), 76(56, 57, 58, 59), 81(59), *237*

Dvorak, J., 55(92, 93), 56(91, 92), 57(91, 92), 59(90), 60(91), 63(91), 65(90, 91), 66(91), 67(91), 69(91, 92), 72(91), 73(91), 75(91), 76(91), 78(91), 79(91), 82(90, 91, 92), 83(91), 84(92), 85(91), 86(91), 88(90), 89(91), 91(90), 92(90, 91), 93(90, 91), 95(90, 91), 97(90), 98(91), 111(91), 151(92), 152(92, 93), 153(92), 154(92), 156(92, 93), 157(92), 158(92), 160(92), 161(92), 162(92), 163(92, 93), 164(92, 93), 165(92), 168(92, 93), 193(91), *238*

Dyatkina, M. E., 216(88), *238*

Eaton, D. R., 48(170), *240*

Eaton, G. R., 234(61), *237*

Echeistova, A. I., 56(62), 62(62, 307), 63(62, 307), 64(62, 307), 135(62, 307), 152(62), 154(62, 307), 155(62, 307), 156(62, 307), 157(62), 172(307), 173(62, 307), *237*, *244*

Edwards, L. J., 4(63), 196(63), *237*

England, D. C., 3(180), *240*

Enrione, R. E., 46(64, 65), *237*

Fedin, E. I., 62(408), 112(408), *247*

Feinn, M. M., 3(66), 54(66), 56(66, 69), 57(66, 69, 265), 58(66, 69), 59(90), 60(66, 169), 62(265), 65(90, 99), 66(91), 67(91), 69(66, 69, 91), 70(66, 69), 72(91), 73(91), 75(66, 69, 72, 91), 76(66, 69, 91), 77(66), 78(66, 91), 79(91), 80(66, 69), 82(90, 91), 83(91), 85(69), 86(91), 88(90), 89(91), 91(90), 92(69, 73, 90, 91), 93(90, 91), 95(68, 90, 91), 96(99), 97(90), 98(91, 99), 109(169), 111(91), 120(265), 124(75, 265), 125(265), 126(70, 71, 265), 140(67), 157(256), 161(67), 172(67), 182(97, 99), 189(169), 190(99), 193(91), 195(66), *237*, *238*, *240*, *242*, *243*

Fetter, N. R., 82(76), 91(76), *238*

Francis, J. N., 210(39, 77), 218(77), 219(39, 77), 220(30), *237*, *238*

Franz, D. A., 23(78a), 24(79), 25(78, 78a), 27(78a), 29(79), 30(78a), 31(78a), *238*

Fritchie, C. J., 196(80), *238*

Frolova, Z. M., 190(183), *240*

Frula, F. F., 187(263), *243*

Frye, C. L., 55(252), 62(252), 120(252), 154(252), 157(252), 169(252), *242*

Gaines, D. F., 46(50), *237*

Gal'braikh, L. S., 95(6), *236*

Garrett, P. M., 35(329), 46(81, 84, 327, 328, 329), 47(81, 82, 329, 330), 48(329), 49(82, 83, 277, 329), 50(83, 329), 51(329), 52(82, 83, 327, 329, 330), 146(277), 147(277), 152(84), 157(329), 193(84, 126, 138), 194(138), 195(84, 138), 196(84, 327, 329), 200(329), *238, 239, 243 244, 245*

Gedymin, V. V., 59(407), 61(407), 62(407), 63(407), 109(407), 112(407), *247*

George, T. A., 46(81), 47(81, 128), 227(85, 86, 87, 128), 228(86, 87, 128), 229(87), 230(85, 86, 87), *238, 239*

Gerhart, F. J., 26(226), 30(226, 229), 35(229), 41(229), 42(229), 43(229), 45(350), 48(350), 52(350), 234(46), *237, 242, 245*

German, E. D., 216(88), *238*

Gerrard, W., 233(89), *238*

Gold, K., 210(39), 219(39), 220(39), *237*

Gol'din, G. S., 190(183), *240*

Goldstein, H. L., 3(143), 54(143), 56(143), 57(143), 58(143), 59(143), 61(143), 63(143), 70(143), 72(143), 73(143), 85(143), 109(143), 140(143), *239*

Gol'tyapin, Yu. V., 143(308), 144(309), 175(308), 178(307a), 179(307a), 196(309), *244*

Golubtson, S. A., 126(271), 190(272), 191(272), *243*

Good, C. D., 3(273, 274), 26(275), 33(273, 274), 34(273, 274), 40(273), 41(274), 42(273, 274), 43(274), *243*

Grafstein, D., 55(92, 93), 56(69, 91, 92), 57(69, 91, 92), 58(69), 59(90), 60(91), 63(91), 65(90, 91), 66(91), 67(91), 69(69, 91, 92), 70(69), 72(91), 73(91), 75(69, 91), 76(69, 91), 78(91), 79(91), 80(69), 82(90, 91, 92), 83(91), 84(92), 85(69, 91), 86(91), 88(90), 89(91), 91(90), 92(69, 90, 91), 93(90, 91), 95(90, 91), 97(90), 98(91), 111(91), 151(92), 152(92, 93), 153(92), 154(92), 156(92, 93), 157(92), 158(92), 160(92), 161(92), 162(92), 163(92, 93), 164(92, 93), 165(92), 168(92, 93), 193(91), *237, 238*

Grassberger, M. A., 25(94, 184), 27(94, 184), 28(94, 184), 30(94), 32(185), 35(94, 184, 185, 186), 37(185), 38(184, 185, 186), 39(184, 185, 186), 43(186), 233(185), *238, 240, 241*

Graybill, B. M., 227(95), 230(95), *238*

Grebennikov, A. V., 58(376, 380), 59(379, 380), 66(376, 377), 67(377), 68(380), 69(375, 380), 70(375), 71(375, 377, 380), 72(377), 75(377), 76(379), 78(379), 82(376), 83(375, 378, 379), 93(375, 378), 101(378),102(378, 379), 104(440), 106(440), 107(440), 113(375), 114(378), 117(440), 136(424), 145(424), 146(424), 152(376, 379, 380), 161(377), 163(376, 379), 193(374), 195(374), *246, 247, 248*

Green, J., 60(169), 65(99), 96(99, 207), 98(97, 98, 99), 109(169), 124(208), 126(70, 71, 208), 182(97, 98, 99), 183(208), 185(96, 208), 186(96), 189(169), 190(96, 99, 208), 191(208), 192(96), *237, 238, 240, 241*

Green, P. T., 217(100), *238*

Gregor, V., 95(101), 190(102), *238*

Grey, A. P., 183(235), 184(235), 186(235), *242*

Gridina, V. F., 120(173), 125(174), *240*

Grimes, R. N., 23(78a), 24(79, 106, 107, 191), 25(78, 78a), 26(107), 27(24, 78a, 106, 157, 191), 29(79, 109, 156, 157), 30(24, 78a, 106, 107, 157), 31(78a), 34(103, 104, 105, 106, 108, 121), 35(106, 121), 36(104, 105, 106, 108), 37(106), 40(103, 104, 105, 108), 41(104), 42(103, 104, 105, 108), 43(104, 106, 108), 44(219), 45(106, 108), 52(106), 203(109), 204(109, 110), 227(156), 231(156), 232(156, 157), *236, 238, 240, 241*

Grinevich, K. P., 191(284), 183(285), *243*

Groszek, E., 26(228), 30(228), 35(228), 39(228), 43(228), *242*

Guggenberger, L. J., 48(170), 235(171), *240*

Harmon, A. B., 75(112, 113), *239*

Harmon, K. M., 74(114), 75(112, 113), *239*

Harris, C. B., 210(115), 217(115), *239*

Harrison, B. C., 24(116), 36(116), *239*

Hart, H., 12(119), 48(117, 118), 159(119), 172(119), *239*

Haslinger, F., 133(120), *239*

Hass, T. E., 235(111), *239*

Hawthorne, M. F., 3(131, 238), 13(137), 18(139), 25(54a, 55), 27(54a, 55), 30(54a, 55), 35(54, 329), 46(81, 84, 137, 327, 328, 329, 348), 47(54, 81, 82, 128,

329, 330), 48(14, 54b, 55, 329), 49(82, 83, 277, 329), 50(83, 329), 51(129, 230a, 329, 357), 52(54b, 82, 83, 327, 329, 330), 53(129, 230a), 55(56, 57, 58), 56(135), 57(127, 135), 58(56, 59, 127), 75(127), 76(56, 57, 58, 59), 81(59), 82(127), 89(127), 108(127), 135(238), 146(277), 147(277), 152(84, 127), 157(329), 162 (127), 193(84, 126, 136, 138, 348), 194 (135, 136, 138, 348), 195(84, 138, 397), 196(84, 218, 327, 329, 348, 357), 197(137, 139, 345, 357), 198(134, 135), 200(329), 202(239), 203(239), 207(121, 139), 208(133, 137, 251), 209(133, 137), 210(39, 133, 135, 137, 352), 211(137, 346, 347), 212(251), 213(133, 137), 214(125, 137), 215(130, 137, 139), 216(133, 137, 345), 217(124, 133, 135, 137, 345), 218(124, 133, 137, 346), 219 (39), 220(39, 137, 345, 346), 221(137, 346, 347, 352), 222(137, 346,) 223(346), 227(85, 86, 87, 95, 128, 132), 228(86, 87, 128), 229(87), 230(85, 86, 87, 95, 132), 231(132), 233(121, 122, 123), *236, 237, 238, 241, 242, 243, 244, 245*
Hegstrom, R. A., 234(20), *236*
Herber, R. H., 209(140), 216(140), *239*
Hermanek, S., 95(101), 190(102), *238*
Hertler, W. R., 235(141, 177), *239, 240*
Heying, T. L., 3(143), 54(143), 55(233, 234), 56(143, 233, 260), 57(143), 58(142, 143), 59(142, 143), 60(142), 61(142, 143), 62(142, 231), 63(143), 65(142, 231, 236), 66(142), 67(142), 69(142), 70(142, 143), 72(142, 143), 73(142, 143), 75(142), 78(142), 81(142), 82(142), 83(142), 85(142, 143), 88(142), 90(142), 91(142), 92(142), 93(142), 95(145), 98(145), 103(145), 109(143), 110(142), 113(142), 120(142, 231), 121(231, 236), 122(231, 236), 123(236), 127(261), 136(260, 262), 139(262), 140(142, 143), 141(260, 262), 142(260), 152(233), 154(232), 156(234, 262), 169(232, 261), 170(261), 174(262), 177(232, 234), 178(233), 179(261), 183(235), 184(144, 145, 146, 147, 235), 186(235), 187(261, 263), *239, 242, 243*
Hill, W. E., 77(148), 80(148), *239*

Hillman, M., 3(143), 54(143), 56(143), 57(143), 58(143), 59(143), 61(143), 63(143), 70(143), 72(143), 73(143), 85(143), 109(143), 140(143), *239*
Hites, R. D., 24(116), 36(116), *239*
Hoard, J. L., 2(240), *240*
Hoffmann, E. G., 25(94), 27(94), 28(94), 30(94), 35(94, 186), 38(186), 39(186), 43(186), *238, 241*
Hoffmann, R., 3(151, 152, 154), 17(150, 152), 19(151, 142, 153, 154), 20(151), 40(152), 158(150), 159(150), 172(152), *240*
Hohnstedt, L. F., 132(280), 133(280), *243*
Hopkins, T. E., 209(469, 471), 210(470), 214(469), 215(471), *249*
Horstschaefer, H. J., 35(187), 38(187), 42 (187), *241*
Hota, N. K., 71(155), *240*
Howard, J. W., 25(79), 27(157), 29(79, 156, 157), 30(157), 227(156), 231(156), 232(156, 157), *238, 240*
Howe, D. V., 13(137), 46(137), 51(357), 194(137), 195(357), 196(357), 197(137, 357), 208(137), 209(137), 210(137), 211(137), 213(137), 214(137), 215(137), 216(137), 217(137), 218(137), 220(137), 221(137), 222(137), *239, 245*
Hughes, R. E., 2(149), *240*
Hughes, R. L., 233(158), *240*
Hurd, D. T., 24(159), 36(159), *240*
Hyatt, D. E., 199(161), 200(161, 162), 201(162), 224(160), 225(160), *240*

Issleib, K., 233(163), *240*

Jacobson, R. A., 48(164), *240*
Johnson, F. A., 235(211), *241*
Johnson, O., 4(255), *242*
Joy, F., 25(165), *240*

Kaczmarczyk, A., 158(166), *240*
Kaesz, H. D., 12(167), 158(167), 159(167), *240*
Kalinin, V. N., 55(403), 56(193), 57(382, 388, 419, 420), 58(382, 388, 398, 419, 420), 59(397, 407, 410, 411, 418), 60(381, 398, 411, 418), 61(382, 396, 402,

407, 418), 62(193, 389, 390, 391, 397, 407, 408), 63(193, 389, 390, 391, 397, 407), 64(193, 389, 390, 391), 69(423), 70(418), 76(398), 77(394), 78(394), 82(410), 84(398, 419, 420), 85(410, 417, 419), 87(410), 89(398), 90(398, 419, 420, 422a), 99(398), 107(411, 418), 108(382, 388, 420), 109(382, 388, 397, 400, 407), 110(402, 410), 111(402, 410), 112(407, 408), 118(415), 128(415), 134(384, 387, 389, 409), 135(387, 389, 403), 136(385, 394, 406, 424), 137(382, 385, 390, 391, 406), 138(390, 391), 143(393, 395), 144(395, 421, 422), 145 (423, 424), 146(415, 424), 148(415, 416), 149(415), 151(422), 152(193, 382, 388, 398, 419), 153(402), 154(402), 155(193, 383, 389, 423), 156(193, 382, 403), 157 (168), 158(382), 159(382, 399a, 405), 160(412, 414), 161(399, 416), 162(382, 388, 401, 419, 420, 422a), 167(388, 396, 402, 418), 172(382, 389, 406, 409), 173(382, 403, 409), 174(385, 403, 423), 175(393, 421, 423), 176(416), 178(417), 179(399, 417), 193(381, 382, 386, 392, 400, 401), 194(381, 386, 400), 196(395), 197(392, 413), 234(193), *240, 241, 246, 247*

Karlan, S. I., 57(265), 59(90), 62(265), 65(90), 82(90), 88(90), 91(90), 92(90), 93(90), 95(90), 97(90), 120(265), 124 (75, 265), 125(265), 126(265), *238, 243*

Kashin, A. N., 141(306), 150(306), 151(306), 161(306), 177(306), *244*

Kauffman, J. M., 60(169), 109(169), 189 (169), *240*

Kazantsev, A. V., 56(425), 57(425), 58(425), 59(426), 60(426, 430), 65(425), 66(377), 67(377, 425), 71(377), 72(377), 75(377), 83(378), 88(431), 89(427), 93(378, 426, 431), 94(430, 432), 97(427, 431), 101 (378), 102(378, 429, 430, 432), 103(432), 104(432), 105(429, 432), 113(432), 114(378, 426), 115(431), 140(468), 152(428, 433), 153(433), 161(377, 428, 433), 164(433), 165(426, 433), 166(433), 168(433), *246, 247, 248, 249*

Keilin, B., 3(274), 33(274), 34(274), 41(274), *243*

Keller, P. C., 46(51, 52), *237*

Kelley, J. J., 185(44), 186(44), 192(43), *237*

Khaikin, L. S., 173(338), *245*

Klanberg, F., 48(170), 235(141, 171, 172), *239, 240*

Klebanskii, A. L., 120(173), 125(174), *240*

Klein, M. J., 24(116, 203), 36(116, 203), *239, 241*

Klimova, A. I., 23(10), 55(10, 314, 315), 56(62, 320, 455), 57(310, 320, 455), 58(313, 463), 59(312, 317, 463), 60(317, 455, 462), 61(312, 317), 62(62, 307, 318), 63(62, 307, 318), 64(62, 307, 318, 461), 65(455), 66(455, 463), 67(455), 70(320, 455), 72(320, 455), 75(320, 455), 82(455, 463), 84(313, 463), 85(317, 455), 86(317), 87(455), 88(455), 89(455), 90(317), 92(455), 93(311), 94(317), 100(317), 103(455), 104(310, 312, 317), 108(462), 109(455), 113(312, 317, 455), 134(316, 323, 461), 135(36, 62, 307, 318, 322, 323, 461), 136(318, 322, 460, 461), 137(461), 139(460, 461), 141(306, 460, 461), 144(460, 461), 150(306), 151(306, 314), 152(62, 313), 154(62, 307, 318), 155(62, 307, 318, 322), 156 (62, 307, 314, 315, 318), 157(62, 315), 161(306), 163(313), 172(307, 314, 318, 319, 322, 323, 461), 173(36, 62, 307, 322, 323), 174(461), 177(306), *236, 237, 244, 248, 249*

Klimova, T. P., 58(298), 59(317), 60(317), 61(317), 63(299a, 318), 64(299a, 318), 68(299), 69(298), 83(298), 85(317), 86(317), 90(317), 94(317), 100(317), 104(317), 113(317), 135(318), 136(318), 140(299a), 145(299), 146(299a), 152 (298), 154(318), 155(299a, 318), 156 (318), 163(298), 172(318), 174(299a), *244*

Klusmann, E. B., 33(47), 34(47), *237*

Knoth, W. H., 3(180, 215), 11(215), 158 (215), 199(175, 178), 200(175), 201 (175), 224(176, 179), 233(215), 235(177, 214), *240, 241*

Knowles, T. A., 56(281), 70(281), 115(281), 117(281), 135(281), 136(281), 142(281), 155(281), 156(281), 159(276), 161(281), 172(281), 174(281), 177(276), 178(276), 179(276), 180(276, 267a), *243*

Köster, R., 25(94, 184), 27(94, 184), 28(94, 184), 30(94), 32(185), 35(94, 184, 185,

186, 187, 188, 189), 37(185, 188, 189), 38(184, 185, 186, 187), 39(184, 185, 186), 40(188), 42(187, 188), 43(186), 233(185), *238, 240, 241*

Koetzle, T. F., 48(181), *240*

Kongpricha, S., 61(250), 62(250), 63(250), 64(182), 137(182), 138(250), 141(182), 152(182), 153(250), 154(250), 155(182), 156(182), 159(250), 167(250), 172(250), 173(182), 174(182, 250), 175(182), 178(182), 179(182, 250), 180(182), 193(250), 197(250), 198(250), *240, 242*

Kopylov, V. V., 56(320), 57(320), 70(320), 72(320), 75(320), *244*

Korshak, V. V., 183(4, 5), 189(3), 190(183), *236, 240*

Kotolby, A. P., 65(99), 96(99), 98(98, 99), 182(98, 99), 190(99), *238*

Kozlova, N. V., 174(125), *240*

Kozyreva, E. M., 191(284), 182(253), *242, 243*

Krasovskaya, T. A., 190(272), 191(272), *243*

Kratzer, R., 46(38), 217(38), *237*

Krupnova, L. E., 120(173), 125(174), *240*

Kukulina, E. I., 197(434), 198(434), *448*

Kvasov, B. A., 62(408), 90(422a), 112(408), 162(422a), *247*

Lappert, M. F., 25(165), *240*

Larchar, T. B., 187(263), *243*

Laubengayer, A. W., 56(190, 205), 63(205), 135(205), 152(190, 205), 155(205), 156(190), 157(205), 177(190), 178(190), *241*

Lawless, E. W., 233(158), *240*

Lawrence, J. R., 199(178), *240*

Ledoux, W. A., 24(191), 27(191), 34(191), 35(191), *241*

Leites, L. A., 55(37), 56(193), 58(380), 59(380), 62(193), 63(193), 64(192, 193), 68(380), 69(380), 71(380), 141(192), 142(192), 152(37, 192, 380), 155(193), 156(193), 234(37, 193), *237, 241, 246*

Lichstein, B. M., 56(69), 57(69), 58(69), 69(69), 70(69), 75(69), 76(69), 80(69), 85(69), 92(69), *237*

Linder, H. H., 24(194), 36(194), *241*

Lindner, R., 233(163), *240*

Lipscomb, W. N., 1(195, 197), 2(195), 3(49, 151, 152, 154, 196, 198, 212, 355),

5(195), 11(195, 197), 12(119), 13(195), 14(195), 15(195), 16(195), 17(150, 152, 195, 196, 326), 18(195), 19(151, 152, 153, 154), 20(151, 195, 196), 21(195, 196), 23(22), 40(152), 41(197), 46(64, 65, 342), 48(49, 117, 118, 164, 181, 198), 55(241, 245, 341), 63(247), 64(247), 134(242, 247), 135(242, 243, 244, 247), 136(241, 245, 247), 157(246), 158(150, 166, 197), 159(119, 150, 197), 172(11, 119, 152), 173(11),201(195), 207(212), 233(195), 234(20, 61), 235(21, 195), *236, 237, 239, 240, 241, 242, 244, 245*

Little, J. L., 199(178), 204(333), 205(199, 334), 206(199, 333, 334, 335), 207(333), 224(160, 179), 225(160, 335), 226(333, 335), *240, 241, 245*

Lohr, L. L., 3(212), 207(212), *241*

Longuet-Higgins, H. C., 3(200), 18(200), *241*

Lozovskaya, V. S., 134(409), 172(409), 173(409), *247*

Lutz, C. A., 24(201), 36(201), *241*

L'vov, A. I., 57(410), 59(410), 60(202, 435, 436, 437, 438, 439, 442), 61(437), 82(410), 83(378), 85(410), 86(436), 87(410), 93(378), 94(202, 436, 439), 95(202, 439), 97(202, 439), 99(437, 438), 100(435), 101(378, 435), 102(378), 103(202, 438), 104(202, 435, 439, 440), 105(442), 106(202, 439, 440, 442), 107(440), 109(435, 437), 110(410), 111(410), 113(437), 114(378, 437), 115(437, 438, 439), 152(202, 441), 153(202, 439, 441, 442), 154(441), 155(441), 162(220), 163(441), 164(202, 439), 165(439), 166(202, 441), 167(442), 175(441), 217(440), *241, 246, 247, 248*

MacDonald, A. A., 75(113), *239*

Maguire, R. G., 24(203), 36(203), *241*

Makhlouf, J. M., 4(63), 196(63), *237*

Maki, A. H., 209(204), 215(204), *241*

Mangold, D. J., 3(143), 54(143), 56(143), 57(143), 58(143), 59(143), 61(143), 63(143), 70(143), 72(39a, 143), 73(143), 85(143), 109(143), 140(143), *237, 239*

Maruca, R., 56(205), 63(205), 135(205), 152(205), 155(205), 157(205), *241*

Marynick, D., 26(227), 30(227), 234(227), *242*

Mastryukov, V. S., 55(339), 157(206), *241, 245*

Matteson, D. S., 71(155), *240*

Mattschei, P., 26(227, 228), 30(227, 228), 35(228), 39(228), 43(228), 234(227), *242*

Mayes, N., 3(66), 54(66), 56(66, 69), 57(66, 69), 58(66, 69), 60(66), 65(99), 69(66, 69), 70(66, 69), 75(66, 69), 76(66, 69), 77(66), 78(66), 80(66, 69), 85(69), 92(69), 96(99, 207), 98(97, 98, 99), 124(208), 126(70, 208), 182(97, 98, 99), 183(208), 185(96, 208), 186(96), 190(96, 99, 208), 191(208), 192(96), 195(66), *237, 238, 241*

Mikhailov, B. M., 56(209), 198(209, 240), 204(209), *241, 242*

Miller, H. C., 3(180), *240*

Miller, J. J., 235(211), *241*

Moore, E. B., 3(212), 207(212), *241*

Moran, J. T., 205(199), 206(199), 224(160), 225(160), *240, 241*

Muetterties, E. L., 3(180, 215), 11(215), 48(170), 158(215), 233(215), 235(171, 172, 177, 214), *239, 240, 241*

Nakagawa, Y., 193(356), 195(356), *245*

Nebel, C. W., 95(68), *237*

Newton, M. D., 234(20), *236*

Niedenzu, K., 233(215a), *241*

Oakes, J. D., 33(47), 34(47), *237*

Obenland, C. O., 55(234), 62(282), 129(282), 130(282), 131(282), 132(282), 152(216), 153(216), 154(282), 156(234), 162(216), 164(216), 165(216), 167(282), 171(282), 172(216), 177(234), *241, 242, 243*

O'Brien, E. L., 57(265), 62(265), 65(99), 96(99), 98(99), 120(265), 124(265), 125(265), 126(71, 265), 182(99), 190(99), *237, 243*

Ogorodnikova, N. A., 64(192, 443, 444), 75(443), 136(444), 139(444), 141(192, 443, 444, 445), 142(192), 144(443), 156(444), 172(444), 175(443), *241, 248*

Okhlobystin, O. Y., 69(217), 115(366), 120(366), 127(366), 148(366, 368), 149(368), 150(368), *241, 246*

Okhlobystin, O. Yu., 54(456), 56(454, 455), 57(455), 58(454, 455), 60(455), 61(367), 62(367), 65(455), 66(455), 67(455), 70(454, 455), 73(455), 74(15, 16), 75(454, 455), 78(446), 80(454), 82(455), 85(455), 87(455), 88(455), 89(455), 92(455), 95(6), 103(455), 109(455), 113(455), 115(367), 117(367), 118(367), 120(367), 127(367), 128(29, 367), 129(29), 136(446), 141(306), 148(367, 369), 149(28, 369), 150(306), 151(306), 161(306), 177(306), 233(30), *236, 244, 246, 248*

Olsen, F. P., 193(126), 196(218), *239, 241*

Olsen, R. R., 44(219), *241*

Onak, T. P., 17(230), 24(194), 26(222, 226, 227, 228, 230), 28(224, 291), 30(221, 222, 224, 226, 227, 228, 229, 230, 351), 35(222, 223, 228, 229), 36(194), 39(223, 228), 40(222), 41(222, 225, 229), 42(222, 229), 43(222, 223, 225, 228, 229), 44(223), 233(220), 234(221, 227), *240, 241, 243, 245*

Owen, D. A., 51(129, 230a), 53(129, 230a), 193(138), 194(138), 195(138), 199(161), 200(161), *239, 240, 242*

Papetti, S., 55(233, 234), 56(233), 58(142), 59(142), 60(142), 61(142, 250), 62(142, 231, 250, 269, 282), 63(250), 65(142, 231, 236, 66(142), 67(142), 69(142), 70(142), 72(142), 73(142), 75(142), 78(142), 81(142), 82(142), 83(142), 85(14), 88(142), 90(142), 91(142), 92(142), 93(142), 95(142), 98(142), 103(142), 110(142), 113(142), 120(142, 231), 121(231, 236), 122(231, 236), 123(236), 127(261), 129(282), 130(282), 131(282, 283), 132(282, 283), 138(250), 140(142), 152(216, 233), 153(216, 250), 154(232, 250, 269, 282, 283), 156(234), 159(250), 162(216), 164(216), 165(216), 167(250, 282), 169(232, 261), 170(261), 171(269, 282), 172(216, 250), 174(250), 177(233, 234), 178(233, 269), 179(250, 261), 183(235), 184(144, 145, 146, 147, 235), 186(235), 187(261), 188(268), 193(250), 197(250), 198(250), *239, 241, 242, 243*

Parshall, G. W., 3(180), 235(172), *240*

Paterson, C. R., 206(335), 225(335), 226(335), *245*

Paul, I. C., 206(335), 225(335), 226(335), *245*

Paustian, J. E., 56(69), 57(69), 58(69), 59(90), 65(90), 69(69), 70(69), 75(69, 72), 76(69), 80(69), 82(90), 85(69), 88(90), 91(90), 92(69, 73, 90), 93(90), 95(90), 97(90), *237, 238*

Petterson, L. L., 61(23), 110(23), *236*

Pier, E. A., 55(337), 157(337), 177(337), *245*

Pilling, R. L., 13(137), 46(137), 135(237), 194(137), 197(137), 208(137), 209(137), 210(137), 211(137), 213(137), 214(137), 215(130, 137), 216(137), 217(137), 218(137), 220(137), 221(137), 222(137), *242, 239*

Pitochelli, A. R., 3(131, 198, 238), 48(198), *239, 241, 242*

Pitts, A. D., 13(137), 46(137), 194(137), 197(137), 208(137), 209(137), 210(137), 211(137), 213(137), 214(137), 215(137), 216(137), 217(137), 218(137), 220(137), 221(137), 222(137), 227(132), 230(132), 231(132), *239*

Plesek, J., 95(101), 190(102), *238*

Podvisotskaya, L. S., 59(411, 418), 60(411, 418), 61(418), 62(447), 63(447), 69(453), 70(418), 85(417), 87(450), 94(450), 104(449), 105(449), 107(411, 418), 113(449), 118(415), 128(415), 139(447), 145(447), 146(415, 448), 148(415, 416), 149(415), 150(451), 152(441), 153(441), 154(441), 155(441), 160(412, 414, 453), 161(416), 163(441), 166(441), 167(418), 175(441, 452), 176(416, 452), 177(451, 452), 178(417), 179(417), 195(448), 197(413, 434), 198(434), *247, 248*

Polak, R. J., 3(143), 54(143), 56(143), 57(143), 58(143), 59(143), 61(143), 63(143), 70(143), 72(143), 73(143), 85(143), 109(143), 140(143), *239*

Ponomarenko, A. A., 56(454, 455), 57(455), 58(454), 60(455), 65(455), 66(455), 67(455), 70(454, 455), 73(455), 75(454, 455), 80(454), 82(455), 85(455), 87(455), 88(455), 89(455), 92(455), 103(455), 109(455), 113(455), *248*

Popp, G., 202(239), 203(239), *242*

Potapova, T. V., 56(209), 198(209, 240), 204(210), *241, 242*

Potenza, J. A., 55(241, 245), 63(247), 64(247), 134(242, 247), 135(242, 243, 244, 247), 136(241, 245, 247), 157(246), 234(20), 235(21), *236, 242*

Poynter, R. L., 3(12), 41(12, 13, 225), 43(12, 13, 225), *236, 242*

Pressley, G. A., 46(324), *244*

Prince, S. R., 35(248), 39(248), 44(248), *242*

Prokai, B., 25(165), 79(270), *240, 243*

Rademaker, W. J., 29(109), 203(109), 204(109, 110), *238*

Reid, J. A., 58(142), 59(142), 60(142), 61(142), 62(142), 65(142), 66(142), 67(142), 69(142), 70(142), 72(142), 73(142), 75(142), 78(142), 81(142), 82(142), 83(142), 85(142), 88(142), 90(142), 91(142), 93(142), 95(142), 98(142), 103(142), 110(142), 113(142), 120(142), 140(142), *239*

Reiner, J. R., 56(260), 100(249), 136(260, 262), 139(262), 141(260, 262), 142(260), 153(249), 154(249), 156(262), 166(249), 168(249), 174(262), *242, 243*

Reintjes, M., 13(137), 46(137), 193(126), 194(137), 197(137), 208(137), 209(137), 210(137), 211(137), 213(137), 214(137), 215(137), 216(137), 217(137), 218(137), 220(137), 221(137), 222(137), *239*

Rigo, P., 147(31), *236*

Ritter, D. M., 24(201), 36(201)

Roberts, M. de V., 3(200), 18(200), *241*

Robinski, J., 233(19), *236*

Robinson, M. A., 131(283), 132(283), 154(283), *243*

Rogovini, Z. A., 95(6), *236*

Roscoe, J. S., 61(250), 62(250), 63(250), 138(250), 153(250), 154(250), 159(250), 167(250), 172(250), 174(250), 179(250), 193(250), 197(250), 198(250), *242*

Rossetto, F., 170(32, 33, 35), 176(34), 187(35), 189(34), *236, 237*

Rotermund, G. W., 35(188, 189), 37(188, 189), 40(188), 42(188), *241*

Rudolph, R. W., 204(343), 205(343), *245*

Ruhle, H. W., 208(133, 251), 209(133), 210(133), 212(251), 213(133), 216(133), 217(133), 218(133), *239, 242*

Rysz, W. R., 56(190), 152(190), 156(190), 177(190), 178(190), *241*

St. Clair, D., 218(251a), *242*

Salinger, R. M., 55(252), 62(252), 120(252), 154(252), 157(252), 169(252), *242*

Sarishvili, I. G., 182(253, 472), 183(4, 5, 285), 189(3), 190(183), 191(284), *236, 240, 242, 243, 249*

Savina, L. A., 59(379), 76(379), 78(379), 83(379), 102(379), 152(379), 163(379), *246*

Scarbrough, F. E., 48(181), *240*

Schaeffer, B. B., 65(236), 121(236), 122(236), 123(236), 183(236), 184(236), 186(236), *242*

Schaeffer, R., 35(248), 39(248), 44(248), 46(50, 51, 52, 53, 324), *237, 242, 244*

Schaffling, O. G., 184(144, 145, 146, 147), 187(263), *239, 243*

Schoenfelder, C. W., 157(256), *242*

Scholer, F. R., 199(178, 257), 200(162), 201(162), 224(160), 225(160), *240, 242*

Schomburg, G., 25(94), 27(94), 28(94), 30(94), 35(94), *238*

Schroeder, H. A., 55(337), 56(7, 205, 260, 264, 281), 61(7), 62(269), 63(205, 247), 64(182, 247), 70(7, 281), 100(249), 115(7, 281), 116(7), 117(9, 281), 127(261), 134(247), 135(205, 247, 281), 136(247, 257a, 260, 262, 281), 137(182), 139(262), 141(182, 260, 262), 142(260, 281), 152(182, 205, 264), 153(249), 154(8, 249, 269), 155(182, 205, 281), 156(182, 262, 264, 281), 157(205, 264, 337), 159(276), 161(281), 166(249), 168(8, 249), 169(261), 170(261), 171(269), 172(281), 173(182), 174(182, 262, 264, 281), 175(182), 177(276, 337), 178(182, 269, 276), 179(182, 261, 276), 180(182, 276, 276a), 184(258, 259), 185(258), 186(258), 187(258, 261, 263), 188(258, 268), 189(7, 258), *236, 240, 241, 242, 243, 245*

Schwartz, N. N., 3(66), 54(66), 56(66, 69, 91), 57(66, 69, 91, 265), 58(66, 69), 60(66, 91), 62(265), 63(91), 65(91), 66(91), 67(91), 69(66, 69, 91), 70(66, 69), 72(91), 73(91), 75(66, 69, 91), 76(66, 69, 91), 77(66), 78(66, 91), 79(91), 80(66, 69), 82(91), 83(91), 85(69, 74, 91), 86(91), 89(91), 92(69, 91), 93(91), 95(91), 98(91), 111(91), 120(265), 124(75, 265), 125(265), 126(265), 193(91), 195(66), *237, 238, 243*

Schwerin, S. G., 193(138), 194(138), 195(138), *239*

Scott, D. R., 235(265a), *243*

Scott, J. E., 28(291), 35(266, 290), 37(266, 290), *243*

Seklemian, H. V., 28(267), 35(267), 37(267), 43(267), *243*

Semenuk, N. A., 62(269), 154(269), 159(276), 171(269), 177(276), 178(269, 276), 179(276), 180(276, 276a), 188(268), *243*

Semin, G. K., 78(446), 134(323), 135(36, 323), 136(446), 172(323), 173(36, 323), *237, 244, 248*

Seyferth, D., 79(270), *243*

Shapatin, A. S., 126(271), 190(272), 191(272), *243*

Shapiro, I., 3(273, 274), 26(275), 33(273, 274), 34(273, 274), 40(273), 41(274), 42(273, 274), 43(274), *243*

Shepilov, I. P., 57(419, 420), 58(419, 420), 60(442), 84(419, 420), 85(419), 90(419, 420), 105(442), 106(442), 108(420), 152(419), 153(442), 162(419, 420), 167(442), *247, 248*

Shiro, M., 193(356), 195(356), *245*

Shkambarnaya, N. I., 120(173), 125(174), *240*

Shkirtil, E. B., 234(27), *236*

Sieckhaus, J. F., 127(261), 143(275a), 159(276), 169(261), 170(261), 177(276), 178(276), 179(261, 276), 180(276, 276a), 187(261), *243*

Silverstein, H. T., 204(333), 205(334), 206(333, 335), 207(333), 226(333), *245*

Simms, B. B., 185(44), 186(44), 192(43), *237*

Siryatskaya, V. N., 126(271), 157(206), 173(338), 191(284), *241, 243, 245*

Smart, J. C., 47(82), 49(82, 83, 277), 50(83), 52(82, 83), 146(277), 147(277), *238, 243*

Smazhok, M. P., 190(272), 191(272), *243*

Smith, G., 4(255), *242*

Smith, H. F., 56(91), 57(91), 59(90), 60(91), 63(91), 65(90, 91), 66(91), 67(91), 69(91), 72(91), 73(91), 75(91), 76(91), 78(91), 79(91), 82(90, 91), 83(91), 85(91), 86(91), 88(90), 89(91), 91(90), 92(90, 91), 93(90, 91), 95(90, 91), 97(90), 98(91), 111(91), 193(91), *238*

Smith, H. D., 56(281), 61(278), 62(282), 70(281), 115(278, 281), 117(281), 118 (279), 119(278, 279), 129(282), 130 (282), 131(282, 283), 132(280, 282, 283), 133(280), 135(281), 136(281), 142(281), 154(282, 283), 155(281), 156(281), 161(281), 167(282), 171(282), 172(281), 174(281), *243*

Smith, I. C., 233(158), *240*

Snyakin, A. P., 90(422a), 144(421, 422), 151(422), 162(422a), 175(421), *247*

Sobolevskii, M. V., 182(253, 472), 183(4, 5, 285), 189(3), 190(183), 191(284), *236, 240, 242, 243, 249*

Solomon, I. J., 24(116, 203), 36(116, 203), *239, 241*

Solov'ev, A. A., 126(271), *243*

Soloway, A. H.,108(286), 110(286), 115(41), 133(120), *237, 239, 243*

Soshka, S. A., 60(442), 105(442), 106(442), 153(442), 167(442), *248*

Sperling, L. H., 186(287, 288), 187(359), *243, 245*

Spielman, F. R., 26(226), 30(226), *242*

Spielman, J. R., 28(291), 35(266, 290), 37(266, 290), 46(48), *237, 243*

Stafford, F. E., 46(324), *244*

Stafford, R. C., 193(136), 194(136), *239*

Stanko, V. I., 23(10), 54(456, 457), 55(10, 314, 315, 321, 457), 56(62, 320, 365, 455, 457), 57(300, 310, 320, 365, 455, 457), 58(293, 298, 300, 313, 372, 373, 463), 59(312, 317, 372, 373, 463), 60(304, 317, 372, 455, 462), 61(312, 317, 365), 62(62, 296, 299a, 307, 318), 63(62, 299a, 307, 318), 64(62, 293, 299a, 307, 318, 461), 65(372, 455), 66(455, 463), 67(292, 455, 458), 68(292, 298), 69(293, 297, 298, 365), 70(320, 365, 455), 72(320, 365), 73(455), 75 (320, 455), 82(455, 458, 463), 83(292, 293, 298), 84(313, 372, 463), 85(304, 317, 372, 455), 86(317, 372), 87(372, 455, 459), 88(365, 372, 455), 89(372, 455), 90(317, 373), 92(365, 455), 93(311), 94(317), 95(372), 97(372), 99(303, 304), 100(317), 103(304, 372, 455, 459), 104(303, 304, 310, 312, 317), 108(300, 365, 462), 109(372, 455), 113(312, 317, 555), 115(303), 134(316, 321, 323, 461), 135(36, 62, 305, 307,

318, 321, 322, 323, 461), 136(294, 295, 318, 322, 460, 461), 137(461), 139(460, 461), 140(294, 295, 296, 299a), 141(306, 460, 461), 143(308), 144(309, 460, 461), 145(299),146(299a), 150(306), 151(306, 314), 152(62, 298, 313), 153(304), 154 (62, 307, 318), 155(62, 299a, 307, 318, 322), 156(62, 307, 314, 315, 318), 157(62, 315), 161(306), 162(302), 163 (298, 313), 166(303, 304), 172(305, 307, 314, 318, 319, 322, 323, 461), 173(36, 62, 305, 307, 322, 323), 174(299a, 461), 175(308), 177(306), 178(307a), 179 (307a), 193(301, 302), 195(25), 196(301, 309), 224(26), *236, 237, 243, 244, 245, 246, 248, 249*

Steck, S. J., 46(324), *244*

Stein, A. A., 185(44), 186(44), 192(43), *237*

Stock, A., 1(325), 233(325), *244*

Streib, W. E., 17(326), 23(22), 50(336), 196(336), 201(336), *236, 244, 245*

Struchkov, Y. T., 62(307), 63(307), 64(307), 135(307), 154(307), 155(307), 156(307), 172(307), 173(307), *244*

Struchkov, Yu. T., 23(10), 54(457), 55(10, 321, 457), 56(457), 57(457), 134(321, 323), 135(321, 322, 323), 136(322), 155(322), 172(322, 323), 173(322, 323), *236, 244, 248*

Syrkin, Ya. K., 56(62), 62(62, 307), 63(62, 307), 64(62, 307), 135(62, 307), 152(62), 154(62, 307), 155(62, 307), 156(62, 307), 157(62), 172(307), 173(62, 307), *237, 244*

Szymanski, J. W., 3(143), 54(143), 56(143), 57(143), 58(143), 59(143), 61(143), 63(143), 70(143), 72(143), 73(143), 85(143), 109(143), 140(143), *239*

Tagliavini, G., 170(32, 33, 35), 176(34), 187(35), 189(34), *236, 237*

Tebbe, F. N., 35(329), 46(84, 327, 328, 329), 47(329, 330), 48(14, 329), 49(329), 50(14, 329), 51(329), 52(327, 329, 330), 135(237), 152(84), 157(329), 193(84, 126, 138), 194(138), 195(84, 138), 196 (14, 84, 327, 329), 200(329), *236, 238, 239, 242, 244, 245*

Templeton, D. H., 209(469, 471), 210(42, 470), 214(469), 215(471), 217(42), 218 (42, 251a), *237, 242, 249*

Thompson, B. C., 74(114), *239*

Titova, N. S., 60(304), 85(304), 99(303, 304), 103(304), 104(303, 304), 115(303), 135(36), 153(304), 166(303, 304), 172 (319), 173(36), *237, 244*

Tobolsky, A. V., 186(287, 288), 187(359), *243, 245*

Todd, L. J., 199(161, 178, 257), 200(161, 162), 201(162), 204(333), 205(199, 334), 206(199, 333, 334, 335), 207(333), 224(160, 179), 225(160, 335), 226(333, 335), 233(332a), *240, 241, 242, 245*

Troscianiec, H. J., 65(236), 121(236), 122(236), 123(236), *242*

Trotz, S. I., 58(142), 59(142), 60(142), 61(42), 62(142), 65(142), 66(142), 67(142), 69(142), 70(142), 72(142), 73(142), 75(142), 78(142), 81(142), 82(142), 83(142), 85(142), 88(142), 90(142), 91(142), 93(142), 95(142), 98(142), 103(142), 110(142), 113(142), 120(142), 140(142), *239*

Tsai, C., 50(336), 196(336), 201(336), *245*

Turco, A., 147(31), *236*

Tzschach, A. Z., 233(163), *240*

Vance, R. L., 24(107), 26(107), 27(107), 30(107), 34(107), 36(107), 40(107), 42(107), 43(107), 45(107), *238*

Vickers, G. D., 55(337), 56(264), 63(247), 64(247), 134(247), 135(247), 136(247), 152(264), 156(264), 157(264, 337), 174(264), 177(337), *242, 243, 245*

Vilkov, L. V., 55(339), 157(206), 173(338), *241, 245*

Vinogradova, L. E., 56(193), 58(380), 59(380), 62(193), 63(193), 64(193), 68(380), 69(380), 71(380), 152(193, 380), 155(193), 156(193), 234(193), *241, 246*

Vishnevskii, F. N., 190(272), 191(272), *243*

Voet, D., 46(342), 55(341), *245*

Vogel, C., 59(90), 65(90), 82(90), 88(90), 91(90), 92(90), 93(90), 95(90), 97(90), *238*

Volkov, A. F., 144(309), 196(309), *244*

Voorhees, R. L., 204(343), 205(343), *245*

Vostrikova, T. N., 135(305), 172(305), 173(305), *244*

Waddington, T. C., 235(344), *245*

Warner, J. L., 200(162), 201(162), *240*

Warren, L. F., 13(137), 46(137), 193(126), 194(137), 197(137, 345), 208(137), 209(137), 210(137, 352), 211(137, 346), 213(137), 214(137), 215(137), 216(137, 345), 217(137, 345), 218(137, 346), 220(137, 345, 346), 221(137, 346, 352), 222(137, 346), 223(346), *239, 245*

Warren, R., 27(289), 28(289, 291), 30(289), 35(290), 37(290), *243*

Watanabe, H., 193(356), 195(356), *245*

Wegner, P. A., 13(137), 18(139), 46(137), 56(135), 57(127, 135), 58(127, 135), 75(127), 82(127), 89(127), 108(127), 152(127), 162(127), 193(126, 136, 138), 194(135, 136, 137, 139), 195(138), 197(137, 139), 198(134, 135), 207(139), 208(137), 209(137), 210(135, 137), 211(137, 347) 213(137), 214(137), 215(137, 139), 216(137), 217(135), 218(137), 220(137), 221(137, 347), 222(137), 235(172), *239, 240, 245*

Weiss, H. G., 17(230), 26(230, 275), 30(230), *242, 243*

Welcker, P. S., 206(335), 225(335), 226(335), *245*

Wiesboeck, R. A., 46(348), 193(348), 194 (348), 196(348), *245*

Wikholm, G. S., 204(333), 206(333), 207 (333), 226(333), *245*

Williams, J., 24(18), 36(18), *236*

Williams, R. E., 3(273, 274), 17(230), 26(226, 230), 28(267), 30(226, 229, 230, 351), 33(47, 273, 274), 34(47, 273, 274), 35(229, 267), 37(267), 40(273), 41(229, 274), 42(229, 273, 274), 43(229, 267, 274), 45(350), 46(48), 48(350), 52(350), 233(349), 234(46), *237, 242, 243, 245*

Williams, R. L., 24(18), 36(18), *236*

Wilson, R. J., 210(352), 221(352), *245*

Wing, R. M., 211(353, 354), 222(353, 354), 223(353, 354), *245*

Wunderlich, J. A., 3(355), *245*

Yoshizaki, T., 193(356), 195(356), *245*
Young, D. C., 13(137), 18(139), 46(137), 47(330), 51(357), 52(330), 193(136, 138), 194(137, 138), 195(138, 357), 196(357), 197(137, 139, 357), 207(139), 208(137), 209(137), 210(137), 211(137), 213(137), 214(137), 215(137), 216(137), 217(137), 218(137), 220(137), 221(137), 222(137), *239, 245*

Zaborowski, R., 61(358), 115(358), 117(358), 119(358), *245*
Zaganiaris, E. J., 186(288), 187(359), *243, 245*
Zakharkin, L. I., 54(456, 457), 55(403, 457), 56(193, 364, 365, 425, 454, 455, 457), 57(364, 365, 382, 388, 410, 419, 420, 425, 455, 457), 58(371, 372, 373, 376, 380, 382, 388, 398, 419, 420, 425, 454, 463), 59(361, 372, 373, 379, 380, 397, 407, 410, 411, 418, 426, 463, 464, 465), 60(202, 362, 372, 381, 398, 411, 418, 426, 430, 435, 436, 437, 438, 439, 442, 455, 462), 61(364, 365, 367, 371, 382, 402, 407, 418, 437, 466), 62(193, 367, 389, 391, 396, 397, 399, 407, 408, 447, 465), 63(193, 389, 390, 391, 397, 407, 447), 64(192, 193, 361, 389, 390, 391, 443, 444, 461), 65(361, 364, 372, 425, 455), 66(376, 377, 455, 463), 67(360, 363, 364, 377, 425, 455, 458), 68(380), 69(217, 365, 375, 380, 423, 453), 70(360, 361, 363, 365, 375, 418, 425, 454, 455), 71(361, 375, 377, 380, 425), 72(364, 365, 377), 73(364, 455), 74(15, 16), 75(377, 425, 443, 454), 76(379, 398, 425), 77(394), 78(379, 394, 446), 80(454), 82(363, 376, 410, 455, 458, 463), 83(375, 378, 379), 84(373, 398, 419, 420, 463), 85(371, 372, 410, 417, 419, 455), 86(364, 370, 371, 372, 436), 87(372, 410, 450, 455, 459), 88(361, 365, 372, 431, 455), 89(372, 398, 427, 455), 90(364, 373, 398, 419, 420, 422a, 92(365, 455), 93(363, 375, 378, 426, 431), 94(202, 362, 430, 432, 436, 439, 450), 95(202, 362, 372, 439), 96(364), 97(202, 363, 364, 372, 427, 439), 99(398,

437, 438), 100(362, 435), 101(378, 435), 102(378, 379, 429, 430, 432), 103(202, 361, 372, 432, 438, 455, 459), 104(202, 362, 432, 435, 439, 440, 449), 105(429, 432, 442, 449), 106(202, 439, 440, 442), 107(411, 418, 440), 108(365, 382, 388, 420, 462), 109(364, 370, 371, 372, 382, 388, 397, 400, 407, 435, 437, 455), 110(402, 410), 111(402, 410), 112(407, 408), 113(364, 370, 371, 375, 432, 437, 449, 455), 114(378, 426, 437), 115(366, 367, 431, 437, 438, 439), 117(367), 118(367, 415), 119(464, 466), 120(366, 367), 127(366, 367), 128(367, 415, 465), 130(465, 465a, 467), 131(465), 134(384, 387, 389, 409, 461), 135(387, 389, 404, 461), 136(385, 394, 406, 424, 444, 446, 460, 461), 137(382, 385, 390, 391, 406, 461), 138(390, 391), 139(444, 447, 460, 461), 140(468), 141(192, 443, 444, 445, 460, 461), 142(192), 143(393, 395), 144(395, 421, 422, 443, 460, 461), 145(423, 424, 447), 146(415, 424, 448), 148(366, 367, 368, 369, 415, 416), 149(368, 369, 415), 150(368, 451), 151(422), 152(193, 202, 376, 379, 380, 382, 388, 398, 419, 428, 433, 441), 153(202, 402, 433, 439, 441, 442), 154(402, 441, 465), 155(193, 383, 389, 423, 441), 156(193, 383, 403, 444), 157(168), 158(383), 159(383, 399a, 405), 160(412, 414, 453), 161(377, 416, 428, 433), 162(202, 382, 388, 401, 419, 420, 422a), 163(376, 379, 441), 164(202, 433, 439), 165(427, 433, 439), 166(202, 433, 439, 441), 167(388, 396, 402, 418, 442), 168(433), 171(465), 172(383, 389, 406, 409, 444, 461), 173(383, 404, 409), 174(385, 403, 423, 461), 175(393, 421, 423, 441, 443, 452), 176(416, 452), 177(451, 452), 178(417), 179(339, 417), 182(472), 193(374, 381, 382, 386, 392, 400, 401), 194(381, 386, 400), 195(374, 448), 196(395), 197(392, 413, 434), 198(434), 209(469, 471), 210(470), 214(469), 215(471), 217(440), 234(193), *236, 240, 241, 245, 246, 247, 248, 249*
Zakharova, G. E., 120(173), 125(174), *240*
Zalkin, A., 209(469, 471), 210(42, 470), 214(469), 215(471), 217(42), 218(42, 251a), *237, 242, 249*

Zamyatina, V. A., 189(3), *236*

Zhigach, A. F., 55(339), 125(174), 126(271), 157(206), 173(338), 182(472), 183(4, 5), 189(3), 190(183), 191(284), 282(253), 183(285), *236, 240, 241, 242, 243, 245, 249*

Zhigareva, G. G., 59(464, 465), 61(466), 62(465), 69(423), 119(464, 466), 128(465), 130(465, 465a, 467), 131(465), 136(424), 140(468), 145(423, 424), 146(424), 154(465), 155(423), 171(465), 174(423), 175(423), *247, 249*

Subject Index

Empirical formulas are included for parent carboranes, carborane anions, and carboranes containing cage heteroatoms (exclusive of transition metals). In all formulas, atoms other than C, B, and H are given first.

Acetylene, *see* Alkynes
Acids, carboranyl, *see* individual carborane cage systems
$AlC_2B_9H_{11}$ (C_2H_5), 202, 204
Alkylboranes, relationship to carboranes, 4
  synthesis of carboranes from, 25, 33, 35, 37–39
Alkynes, synthesis of carboranes from, 24–27, 32–39, 45–46, 54–55, 72, 75, 76, 85, 87–88, 92, 95, 99–100, 103, 109, 113, 120–121, 133, 137–138, 140, 162
Alkynylboranes, synthesis of carboranes from, 37–38
Aluminum, in carborane cage, 202, 204
Antimony, *see also* $1,2\text{-}C_2B_{10}H_{12}$
  in carborane cage, 206–207
Arsenic, *see also* $1,2\text{-}C_2B_{10}H_{12}$
  in carborane cage, 206–207, 226
$AsCB_9H_{10}^{2-}$, π-complexes of, 226
$AsCB_9H_{11}^{-}$, 207
$AsCB_{10}H_{11}$, 206–207

B_2H_6
  alkyl derivatives of, synthesis of carboranes from, 33–39
  bonding in, 15
  reaction with $1,3\text{-}C_2B_7H_{13}$, 47
  reaction with $1,6\text{-}(CH_3)_2\text{-}1,6\text{-}C_2B_8H_8$, 49
B_4H_{10}
  bonding in, 1–2
  carboranes from, 24–25, 27, 33–34, 45
$B_5H_5^{2-}$, 3, 8, 19–21

B_5H_9
  bonding in, 15
  carboranes from, 24, 26–27, 32–36, 39
  structure, 2, 9
B_5H_{11}, carboranes from, 24, 26, 33–34, 45
$B_6H_6^{2-}$, 3, 8, 19–21
B_6H_{10}
  bonding description of, 15
  carboranes from, 45
  molecular orbitals in, 21
  structure, 2, 4, 9
$B_7H_7^{2-}$, 3, 8, 19–21
  bonding description of, 20
$B_8H_8^{2-}$, 3, 8, 19–21
B_9H_{15}, 2, 9
$B_{10}H_{10}^{2-}$, 3, 8, 17–19, 21, 48
  bonding in, 17–19, 21
$B_{10}H_{14}$
  bonding description of, 16
  halo derivatives, halo-*o*-carboranes from, 137–138
  structure, 2
  synthesis of $1,2\text{-}C_2B_{10}H_{12}$ from, 54–55
$B_{11}H_{15}$, 4, 9, 196
$B_{12}H_{12}^{2-}$, 3, 8, 18–19, 21
  molecular orbitals in, 18–19, 21
$BeC_2B_9H_{12}^{-}$, 202–203
Benzocarborane, 71
Beryllium, in carborane cage, 202–203
Bis(*m*-carboranyl), 157, 160
Bis(*o*-carboranyl), 55, 76, 143–144, 157, 160

Boranes
  bonding in, 1–4, 13–21
  carboranes from, 24–27, 32–35, 37–39,
    45–46, 54–55
  structural relationship to carboranes, 1–5,
    8–9, 17
Borazine, o-carboranyl derivatives of, 114

Carbaphosphollide ions, see $PCB_9H_{10}^{2-}$
Carbon vapor, carboranes from, 39
m-Carborane, see $1,7-C_2B_{10}H_{12}$
o-Carborane, see $1,2-C_2B_{10}H_{12}$
p-Carborane, see $1,12-C_2B_{10}H_{12}$
Carboranes, see $also$ specific types and
    individual compounds
  bonding in, 3, 13–21
  $closo$-, see Polyhedral carboranes
  heteroatom, see $also$ Transition metal
    π-complexes, 202–207
  molecular orbitals in, 13, 18–21
  $nido$-, see Open-cage carboranes
  nomenclature, 12
  numbering systems, 6–7, 10, 12–13
  rearrangements in, see Rearrangements,
    polyhedral cage
  structures of unsubstituted, 5–12, 23,
    40–44, 48–53, 55, 156–157, 177
Carboxylic acids, carboranyl, see $1,2-$
    $C_2B_{10}H_{12}$, $1,7-C_2B_{10}H_{12}$, $1,12-$
    $C_2B_{10}H_{12}$, $C_2B_8H_{10}$, and $C_2B_5H_7$
$CB_5H_6^-$, 5, 8, 44
CB_5H_7
  bonding in, 21
  properties, table, 43
  reactions, 44
  structure, 5–6, 8, 44
  synthesis, 35, 39
CB_5H_9
  alkyl derivatives, synthesis, 24–27, 39, 48
  bonding in, 21
  derivatives, table, 30
  reactions, 28, 35, 37
  structure, 9–10, 23
  synthesis, 25, 48
$CB_9H_{10}^-$, 5, 8, 201
$CB_9H_{12}^-$ [$(CH_3)_3NCB_9H_{11}$], 201
$CB_{10}H_{11}^-$, 8, 200–201
$CB_{10}H_{11}^{3-}$, 200–201, 204–206, 224–225
  π-complexes of, 224–225
$CB_{10}H_{13}^-$, 9, 199–201, 206–207

$CB_{11}H_{12}^-$, 5, 8, 201
$1,2-C_2B_3H_5$, derivatives of
  reactions, 40–41
  structure, 6, 8, 40
  synthesis, 33–34, 36
  table, 42
$1,5-C_2B_3H_5$
  derivatives, table, 42
  reactions, 40
  structure, 6, 8, 40
  synthesis, including derivatives, 33–39
$C_2B_3H_7$
  bonding in, 21
  properties, table, 30
  reactions, 31
  structure, 9, 10, 23
  synthesis, 25, 27
$1,2-C_2B_4H_6$
  properties, table, 43
  reactions, 41
  structure, 6, 8, 41
  synthesis, 33–37
$1,6-C_2B_4H_6$
  derivatives, table, 42
  reactions, 41
  structures, 6, 8, 41
  synthesis, including derivatives, 33–37
$C_2B_4H_8$
  bonding in, 17, 21
  derivatives, table, 30
  formation from C_2H_2 and B_5H_9, mech-
    anism studies on, 24
  reactions, 28–29, 36–37
  structure, 9–10, 23
  synthesis, 24, 26–27, 35
$C_2B_5H_7$
  derivatives, 43, 44
  table, 43
  reactions, 41, 44
  structure, 6, 8, 41
  synthesis, including derivatives, 28, 33–39,
    47
$C_2B_6H_8$
  derivatives, table, 52
  reactions, 47–48
  structure, 6, 8, 48
  synthesis, 35, 45, 47
$C_2B_6H_8^{2-}$, π-complexes of, 230–231
$C_2B_7H_9$
  derivatives, table, 52
  reactions, 48

$C_2B_7H_9$—*continued*
  structure, 6, 8, 48
  synthesis, 35, 47
$C_2B_7H_9^{2-}$, π-complexes of, 227–230
$C_2B_7H_{11}^{2-}$, 227–228
$C_2B_7H_{12}^-$, 49
$C_2B_7H_{13}$
  polyhedral carboranes from, 47
  reactions, 35, 47, 227
  structure, 9–10, 46–47
  synthesis, 46
$C_2B_8H_{10}$, 1,6- and 1,10 isomers
  carboxylic acids, acid strength of, 50
  derivatives, table, 52
  metal σ-complex, 49
  reactions, 48–50, 53
  structure, 7, 8, 48–49
  synthesis, including derivatives, 35, 45, 47, 49
$C_2B_8H_{10}^{4-}$, π-complexes of, 218–219
$C_2B_9H_{11}$
  derivatives, table, 52
  Lewis base adducts, 50–51
  reactions, 46, 50–51, 53
  structure, 7, 8, 50
  synthesis, 46, 50, 196
$C_2B_9H_{11}^{2-}$ (dicarbollide ions), (3),1,2- and (3),1,7 isomers
  electronic structure, 207
  insertion reactions, 138, 198–199, 202, 204, 207, 212–224
  structure, 197
  synthesis, 197
  transition metal π-complexes, 207–224
1,2-$C_2B_9H_{12}^-$
  C,C'-dihalo derivatives, 146
  Lewis base adducts, 50–51
  reactions, 46, 195–197, 216–217, 220, 222–224
  structure, 46–47, 194
  synthesis, 46, 51, 193–196
1,7-$C_2B_9H_{12}^-$
  reactions, 46–47, 195–197
  structure, 46–47, 194
  synthesis, 46–47, 162, 193–196
$C_2B_9H_{13}$, 1,2- and 1,7 isomers
  reactions, 196–197, 202
  structure, 9–10, 46, 196
  synthesis, 46, 196–197
1,2-$C_2B_{10}H_{12}$ (o-carborane)
  acid anhydrides, 65, 91, 103

1,2-$C_2B_{10}H_{12}$ (o-carborane)—*continued*
  acridines, 115
  acyl halides
    conversion to aldehydes, 99
    conversion to azides, isocyanates and urethanes, 110
    conversion to ketones, 76
    cyclization, 87
    hydrolysis, 87
    synthesis, 87, 90–91
    table, 59
  acyl pyrazoles, 78
  alcohols
    acylation, 97
    cleavage, 89, 97
    condensation to cyclic ethers, 95–97, 124
    nitration, 108
    oxidation to acids, 87–89, 97
    oxidation to ketones, 97
    polymerization, 98, 182–183
    pyridyl, 114
    synthesis, 73, 79, 92–95, 103–104, 114, 164
    table, 59–60
  aldehydes
    cleavage, 104–107
    reactions, 103–107
    reductions to alcohols, 103–104
    synthesis, 99–100, 114
    table, 60
  alkali metal derivatives, synthesis of, 66–67
  alkenyl derivatives
    alkoxysilanes from, 126
    halogenation, 136
    ozonization, 89, 99
    reactions, 77–81, 89, 99, 126, 136
    synthesis, 54, 70–71, 75–76
    table, 58
  alkoxysilanes, 124–126
  alkyl derivatives, 67, 69–73
    table, 56
  alkynyl derivatives
    synthesis, 55, 76, 81
    table, 58
  amides, 90–91, 109–110, 113
    decarboxylation, 113
    table, 61
  amines, 80, 107–115, 117, 124
    aminoborane derivatives, 110

1,2-C$_2$B$_{10}$H$_{12}$ (o-carborane)—*continued*
  amines—*continued*
    aminophenyl derivatives, 108–110
    aminophosphine derivatives, 116
    B-amino derivatives, 111–112
    properties of, 111
    table, 61
  antimony derivatives, 118
  arsenic derivatives, 116–119
    table, 61
  aryl derivatives, 69–75, 87, 89–90, 108, 111, 135–137
    diazotization, 111
    halogenation, 135–137
    nitration, 89–90, 108
    reaction with alkali metals, 69–70, 73–74
    synthesis, 70–72, 74, 87
    table, 57–58
  azane, diphospha-, 115
  azane, disila-, 121–122
  azides, 110, 116–117
    table, 61
  azo dyes, 111
  benzocarborane, 71
  bonding in, 3–4, 18–21
  borazine derivatives, 114
  bromo derivatives, *see* 1,2-C$_2$B$_{10}$H$_{12}$ halo derivatives
  m-carboranyl-, 160
  carboxylic acids, *see also* 1,2-C$_2$B$_{10}$H$_{12}$ aryl derivatives
    acid strength, 89–90
    table, 84–85
    anilides of, 113
    B-substituted, 89–90
    C,C'-dicarboxylic acid, 91
    cyclic ketones from, 103
    decarboxylation, 90–91
    electronic properties of, 89–90
    reactions, 89–92, 103
    synthesis, 67–68, 81–89, 103
    tables, 58–59, 84–85
  charge distribution, 134–135
  chelate complexes, 118–119, 131–133
  chloro derivatives, *see* 1,2-C$_2$B$_{10}$H$_{12}$ halo derivatives
  comparison with 1,7-C$_2$B$_{10}$H$_{12}$, 151
  cyanides, 80, 99, 106, 109, 113–114
    degradation to C$_2$B$_9$H$_{12}^-$ derivatives, 114

1,2-C$_2$B$_{10}$H$_{12}$ (o-carborane)—*continued*
  cyanides—*continued*
    table, 61
  degradation to 1,2-C$_2$B$_9$H$_{12}^-$, 46, 193–196
  derivatives, table, 56–65
  diazonium salts, 111–112
  dinegative ion, reduction to, 159–160
  epoxides, 78–79, 93, 96
    table, 60
  esters, 75, 85–92, 97, 99
    cleavage, 85–87
    reactions, 75, 85–92, 99, 102
    synthesis, 85–88, 90, 97
    table, 59
  ethers, *see also* 1,2-C$_2$B$_{10}$H$_{12}$ epoxides, 82, 88–89, 91, 95–99
    polymers, 183
    stability, 99
    synthesis, 95–97
    table, 60, 65
  ferrocenyl derivatives, 148–149
  fluoro derivatives, 80–81, 89, 137–138, 175
    cage rearrangement, 159
  germanium derivatives, 120, 127–129
  Grignard reagents, synthesis and rearrangement, 67–69, 72–73, 83, 125
  haloalkyl derivatives, 67–69, 72–73, 77, 83, 105
    table, 56–57
  halo derivatives, *see also* haloalkyl and aryl derivatives, 117, 130, 133–146
    acid strength of, 139, 141–142, 145
      table, 141
    B-decachloro-, 136, 139, 141–142, 175
    B-decafluoro-, 137
    B$_{10}$H$_{14}$ halogen derivatives, synthesis from, 137–138
    cage rearrangements of, 158–159
    C-halogenated, 139–140, 145–146
    degradation of C,C'-dihalo derivatives, 195
    dipole moments of, 135
    halogen migration in, 145–146
    mercuration of, 148
    nucleophilic substitution on, 143
    perchloro-, 139
    polarographic reduction, table, 144
    reactions, 141–146
    table, 62–64
  halogenation
    of alkenyl derivatives, 77–78, 136

1,2-$C_2B_{10}H_{12}$ (*o*-carborane)—*continued*
halogenation—*continued*
at B(3), 112, 138
at carbon, 139–140
electrophilic, 130, 134–136
fluorination, 137
Friedel-Crafts, 134
photochemical, 136–137
sequence of substitution in, 134–135
hydroxy derivatives, B-substituted, 107,
112
table, 59
iodo derivatives, *see* 1,2-$C_2B_{10}H_{12}$ halo
derivatives
isocyanates, table, 61
ketones
cleavage, 94–95, 104–105
cyclic, 87, 90–92, 100
reactions, 78, 94–95, 104–106
synthesis, 76, 87, 90–94, 100–103
table, 60, 65
lactone, 91–92
lithium derivatives, synthesis of, 66–67
magnesium derivatives, synthesis of, 67–69
mercaptans, *see* 1,2-$C_2B_{10}H_{12}$ sulfur
derivatives
mercury derivatives, 141, 147–150, 176
polarographic reduction of, 149–150
metallation, 66–70
nitrates, 107–108
table, 60–61
nitro and nitroso derivatives, *see also*
1,2-$C_2B_{10}H_{12}$ aryl derivatives, 80,
108–109
table, 60–61, 108
nitrophenyl derivatives, 108–109
phosphorus derivatives, 115–119, 132
table, 61
polymers
alcohols, *o*-carboranyl, synthesis from,
98
methacrylates, 190
oxadiazole, 189–190
P–N–P linked, 116–117
polyformals, 183
siloxane, 125, 190–192
properties, table, 56
rearrangement to 1,7-$C_2B_{10}H_{12}$, 55, 156–
159
reduction to anions in liquid ammonia,
69–70

1,2-$C_2B_{10}H_{12}$ (*o*-carborane)—*continued*
Schiff bases, 107
silicon derivatives, 55, 120–126, 183–187,
190–192
cage rearrangement of, 55, 157
polymers, 190–192
table, 62, 65
structure, 7, 8, 55
sulfur derivatives, 106, 129–133
table, 62
synthesis (of parent carborane), 54–55, 66
thermal stability, 55
tin derivatives, 127–129
transition metal derivatives, σ-bonded,
see also 1,2-$C_2B_{10}H_{12}$ chelate complexes, 146–147
tropenylium derivatives, 74–75
zinc derivatives, 69
1,2-$C_2B_{10}H_{12}^{2-}$, 69–70, 74, 111–112, 144,
159–160
1,7-$C_2B_{10}H_{12}$ (*m*-carborane)
acyl halides, 166, 168
table, 153
alcohols, 162, 164–166
table, 153
aldehydes, 164–166
table, 153
alkenyl derivatives, 162
table, 152
alkyl derivatives, 161–162
table, 152
alkynyl derivatives, table, 152
amides, 168
table, 154
amines, 167–168, 171
table, 153–154
aryl derivatives, 161–162, 167
table, 152
azides, 167
table, 154
azo dyes, 167
o-carboranyl-, 160
carboxylic acids, 163
tables, 84–85, 152
charge distribution, 172–173
comparison with 1,2-$C_2B_{10}H_{12}$, 151
cyanide, 171
degradation to 1,7-$C_2B_9H_{12}^-$, 46, 162,
193–195
derivatives, table, 152–156
diazonium salts, 167

1,7-$C_2B_{10}H_{12}$ (*m*-carborane)—*continued*
  dinegative ion, reduction to, 159–160
  epoxides, 164
    table, 153
  esters, 163
    table, 153
  ethers, 164
  germanium derivatives, 169–170, 187–188
  halo derivatives, 165, 167, 172–177
    acid strength, 175
      table, 141
    polarographic reduction, 177
    properties, 175
      table, 154–156
  halogenation, 165, 167, 172–176
    at carbon, 174
    electrophilic, 172–173
    fluorination, 173–174
    photochemical, 174
  ketones, 164–167
    table, 153
  lead derivatives, 170, 187
  mechanism of formation from, 1,2-
    $C_2B_{10}H_{12}$, 158–159
  mechanism of rearrangements, 158–161
  mercury derivatives, 175–177, 189
  metallation of, 161
  nitrates, 167
  nitro derivatives, 167
  phosphorus derivatives, 168
    table, 154
  polymers, 168, 170–171, 176, 183–189
  silicon derivatives, *see also* Polymers,
    carborane, 157, 169–170, 183, 187
    table, 154
  structure, 7, 8, 157
  sulfur derivatives, 171, 188–189
    table, 154
  synthesis, 49, 55, 156–157
  tin derivatives, 170, 187–188
1,7-$C_2B_{10}H_{12}{}^{2-}$, 159–160
1,12-$C_2B_{10}H_{12}$ (*p*-carborane), 177–180
  carboxylic acid, tables, 85, 178
  degradation, resistance to, 193
  derivatives, 178–180
    table, 178
  dinegative ion, reduction to, 179
  halogenation, 179–180
  polymers, 187–189
  structure, 7, 8, 177
  synthesis, 177

1,12-$C_2B_{10}H_{12}{}^{2-}$, 161, 179
$C_3B_3H_6{}^-$, π-complexes of, 231–232
$C_3B_3H_7$
  bonding in, 21
  derivatives of, table, 30
  formation of CH_3 derivatives, mechanism
    studies on, 24–25
  metal π-complexes derived from, 231–232
  reactions of C-alkyl derivatives, 29, 37,
    231–232
  structure, 9, 10, 23
  synthesis of CH_3 derivatives, 24–27
$C_4B_2H_6$
  bonding in, 21
  derivatives of, table, 30
  reactions of alkyl derivatives, 29
  structure, 9, 10, 23
  synthesis of alkyl derivatives, 25, 27, 38
Chelates, *see* 1,2-$C_2B_{10}H_{12}$ chelate com-
    plexes
Chromium
  dicarbollyl complexes, 212–213
    table, 208
  monocarbollyl complexes, 224
Cobalt
  carbaphosphollyl complexes, 225–226
  o-carboranyl chelates, 131
  π-complexes of $C_2B_7H_9{}^{2-}$, 227–230
    table, 227
  dicarbollyl complexes, 217–220
    table, 210
  monocarbollyl complexes, 224–225
Copper
  dicarbollyl complexes, 222–223
    table, 211
  monocarbollyl complexes, 225
Cyclization, of haloalkyl-*o*-carboranes, 71–
    72, 83

Decaborane, *see* $B_{10}H_{14}$
Dexsil polymers, 184–187
Diborane, *see* B_2H_6
Dicarbacanastide ion, 219
Dicarbazapide ion, *see* $C_2B_7H_9{}^{2-}$
Dicarbollide ions, *see* $C_2B_9H_{11}{}^{2-}$
Dicarbollyl-metal π-complexes, *see also*
    individual metals, 207–224
    table, 208–211

Electric discharge reactions, synthesis of
    carboranes in, 33–36, 39

Electron deficiency, 1, 13–15
  definition of, 13
Electron delocalization, in polyhedral
  boranes and carboranes, 18
Electron withdrawal, see inductive effect

Ferrocene, o-carboranyl derivative of, 148–
  149

GaC$_2$B$_4$H$_6$(CH$_3$), 203–204
Gallium, in carborane cage, 203–204
GeCB$_{10}$H$_{11}$⁻, 204
GeCB$_{10}$H$_{11}$(CH$_3$), 204
Germanium, see also 1,2-C$_2$B$_{10}$H$_{12}$ and 1,7-
  C$_2$B$_{10}$H$_{12}$
  in carborane cage, 204
Gold, dicarbollyl complexes of, 222–223
  table, 211

Heteroatoms in carborane cage, see also
  transition metal π-complexes, 4, 202–
  207
Hexaborane, see B$_6$H$_{10}$

Icosahedron, as structural unit in boron
  chemistry, 2
Inductive (–I) effect
  in 1,2-C$_2$B$_{10}$H$_{12}$ derivatives, 65–66, 72–75,
    77–79, 86, 89–90, 105, 108, 112–113,
    130, 135, 192
  in 1,7-C$_2$B$_{10}$H$_{12}$ derivatives, 75, 151, 162–
    163, 166–167, 176
  in 1,12-C$_2$B$_{10}$H$_{12}$ derivatives, 179
Ionization constants, of carborane carboxylic acids, table, 84–85
Iron
  carbaphosphollyl complexes, 225–226
  o-carboranyl chelates, 119
  o-carboranyl σ-complexes, 146–147
  σ-complex of 1,10-C$_2$B$_8$H$_{10}$, 49
  dicarbollyl complexes, 214–217
    table, 209
  monocarbollyl complexes, 224

Lead, see also 1,7-C$_2$B$_{10}$H$_{12}$
  in carborane cage, 204
Linked-cage carboranes, see also bis(o-
  carboranyl) and bis(m-carboranyl)
  C$_2$B$_9$H$_{11}$-C$_2$B$_{10}$H$_{10}$CH$_3$⁻, 51, 53
  C$_2$B$_9$H$_{11}$-C$_2$B$_8$H$_8$C$_6$H$_5$⁻, 51, 53

Manganese
  carbaphosphollyl complexes, 225–226
  complexes of C$_2$B$_6$H$_8$$^{2-}$, 230–231
    table, 227
  complexes of CH$_3$C$_3$B$_3$H$_5$⁻, 231–232
    table, 227
  dicarbollyl complexes, 214
    table, 209
  monocarbollyl complexes, 224–225
Mercury, see 1,2-C$_2$B$_{10}$H$_{12}$ and 1,7-
  C$_2$B$_{10}$H$_{12}$ mercury derivatives
Meta-carborane, see 1,7-C$_2$B$_{10}$H$_{12}$
Metal carbonyls, carborane complexes of,
  212–214, 216–218, 225, 227, 230–232
  o-carboranyl chelates of, 119
Metal chelates, see 1,2-C$_2$B$_{10}$H$_{12}$ chelate
  complexes
Metallocenes, carborane analogs of, 207–232
Molecular orbitals
  in boranes and carboranes, 18–21
  simplified approach, 20–21
  in transition metal π-complexes, 207,
    222–223
Molybdenum
  o-carboranyl chelates, 119
  dicarbollyl complexes, 212–213
    table, 208–209
Monocarbollide ion, see CB$_{10}$H$_{11}$$^{3-}$
Monocarbollyl transition metal π-complexes, 224–225
  table, 224
Monocarbon carborane anions, 5, 8, 199–
  201, 206–207, 224–226
  π-complexes, 224–226

Nickel
  carbaphosphollyl complexes, 225–226
  o-carboranyl chelates, 118–119, 131
  dicarbollyl complexes, 220–222
    table, 210–211
  monocarbollyl complexes, 224–225

Open-cage (nido-) carboranes, see also
  individual compounds
  bonding in, 17–18, 21
  derivatives, table, 30
  reactions, 28–31, 35–37, 47, 196–197, 202,
    227
  structures, 9–10, 23, 46–47, 196
  synthesis, 24–27, 46, 196–197

Organoboron compounds
  relationship to carboranes, 4
  synthesis of small carboranes from, 25,
    27, 33, 35, 37–39
Ortho-carborane, see 1,2-$C_2B_{10}H_{12}$

Palladium
  o-carboranyl chelates, 119
  dicarbollyl complexes, 221
    table, 211
Para-carborane, see 1,12-$C_2B_{10}H_{12}$
$PCB_9H_{10}^{2-}$ (carbaphosphollide ion), 206,
    225–226
  π-complexes, 225–226
    table, 225
$PCB_9H_{11}^-$, 206
$PCB_{10}H_{11}$, 205–206
Pentaborane, see B_5H_9 and B_5H_{11}
Phosphorus, see also 1,2-$C_2B_{10}H_{12}$ and
    1,7-$C_2B_{10}H_{12}$ phosphorus derivatives
  in carborane cage, 205–206, 225–226
Platinum, o-carboranyl σ-complexes of, 147
Polyhedral borane anions, see also individual
    ions
  bonding in, 13–21
  structural relationship to carboranes, 3, 5,
    8
Polyhedral (closo-) carboranes, see also
    individual compounds
  icosahedral, see 1,2-$C_2B_{10}H_{12}$, 1,7-
    $C_2B_{10}H_{12}$, and 1,12-$C_2B_{10}H_{12}$
  intermediate
    bonding in, 18–21
    derivatives, table, 52
    molecular rearrangements in, 49
    reactions, 46–53
    structures, 5–8, 48–50
    synthesis, 45–47
  small
    bonding in, 18–21
    derivatives, table, 42–43
    molecular rearrangements in, 11, 40, 41
    reactions, 40–44
    structures, 5–6, 8, 40–44
    synthesis, 32–39
Polymers, carborane, see also 1,2-
    $C_2B_{10}H_{12}$ and 1,7-$C_2B_{10}H_{12}$, 181–192

Polymers, carborane—continued
  Dexsil type, 184–187
  methacrylates, 190
  polyesters, 182
  polyformals, 183
  siloxanes, 183–187, 190–192
  single atom-linked, 187–189

Rearrangements, polyhedral cage, 11–12, 41,
    49, 55, 156–161, 179, 205, 228–230
  mechanisms of, 11–12, 158–161
Rhenium, dicarbollyl complexes of, 214
  table, 209

$SbCB_9H_{11}^-$, 207
$SbCB_{10}H_{11}$, 206–207
Silicon, see 1,2-$C_2B_{10}H_{12}$, 1,7-$C_2B_{10}H_{12}$,
    $C_2B_5H_7$, $C_2B_4H_8$
Silver, dicarbollyl complex of, 223–224
$SnC_2B_9H_{11}$, 204–205

Tetraborane, see B_4H_{10}
Three-center bond, 14–18
  in carboranes, 17–18
  in polyhedral boranes, 17
Tin, see also 1,2-$C_2B_{10}H_{12}$ and 1,7-
    $C_2B_{10}H_{12}$
  in carborane cage, 204–205
Titanium, o-carboranyl σ-complexes of, 146
Transition metals, see also individual metals
  carbaphosphollyl complexes, 225–226
    table, 225
  o-carboranyl chelates, 118–119, 131
  o-carboranyl σ-complexes, 146–147
  π-complexes of $CH_3C_3B_3H_5^-$, 231–232
  π-complexes of $C_2B_6H_8^{2-}$, 230–231
  π-complexes of $C_2B_7H_9^{2-}$, 227–230
  dicarbollyl complexes, 207–224
    table, 208–211
  monocarbollyl complexes, 224–225
    table, 224
  nonicosahedral complexes, 226–232
    table, 227
Tungsten, dicarbollyl complexes of, 212–213
  table, 208–209

Zinc, o-carboranyl derivatives, 69